Nano-FET Devices: Miniaturization, Simulation, and Applications

(Part 2)

Edited by

Dharmendra Singh Yadav

Department of Electronics and Communication Engineering
National Institute of Technology
Kurukshetra, Haryana 136119
India

&

Prabhat Singh

Department of Electronics and Communication Engineering
Indian Institute of Information Technology Senapati
Manipur-795002
India

Nano-FET Devices:Miniaturization, Simulation, and Applications *(Part 2)*

Editors: Dharmendra Singh Yadav and Prabhat Singh

ISBN (Online): 979-8-89881-030-6

ISBN (Print): 979-8-89881-031-3

ISBN (Paperback): 979-8-89881-032-0

Published by Bentham Science Publishers Pte. Ltd. Singapore,

in collaboration with Eureka Conferences, USA. All Rights Reserved.

First published in 2025.

need for a court order if at any point you breach any terms of this License Agreement. In no event will any delay or failure by Bentham Science Publishers in enforcing your compliance with this License Agreement constitute a waiver of any of its rights.

3. You acknowledge that you have read this License Agreement, and agree to be bound by its terms and conditions. To the extent that any other terms and conditions presented on any website of Bentham Science Publishers conflict with, or are inconsistent with, the terms and conditions set out in this License Agreement, you acknowledge that the terms and conditions set out in this License Agreement shall prevail.

Bentham Science Publishers Pte. Ltd.
No. 9 Raffles Place
Office No. 26-01
Singapore 048619
Singapore
Email: subscriptions@benthamscience.net

CONTENTS

N. Suthanthira Vanitha, K. Radhika, D. Anbuselvi, C. Kathiravan, S. Grace Infantiya
and A. Kalaiyarasan

FOREWORD

Welcome to the world of Nano-FET Devices, where innovation meets the frontier of miniaturization. The evolution from traditional MOSFETs to nanoscale transistors—such as Single Electron Transistors, Carbon Nanotube FETs, Nanowire FETs, and Graphene FETs—has marked a significant leap in semiconductor technology. These advancements have not only enhanced performance and energy efficiency but also expanded applications into areas like biosensing and biomedical engineering. As the demand for faster, smaller, and more efficient devices continues to grow, Nano-FETs have emerged as key enablers of next-generation electronics. Their unique properties offer superior control at the atomic scale, making them ideal for applications ranging from ultra-low-power systems to high-frequency communication.

This book provides a focused exploration of Nano-FET devices—covering their operating principles, simulation techniques, and diverse applications. Emphasis is placed on simulation methodologies, including quantum mechanical and device-level modeling, which are essential for accurate performance prediction and design optimization. With practical insights, real-world examples, and a strong foundation in theory, this book bridges the gap between research and application. It is intended for researchers, engineers, and students eager to explore the transformative potential of Nano-FET technology.

We invite you to begin this journey into the future of electronics—where quantum effects, novel materials, and nanoscale engineering converge to shape tomorrow's devices.

Ravi Ranjan
Tyndall National Institute - Cork
Lee Maltings Complex
Dyke Parade,
Cork, Ireland

PREFACE

Millions of transistors comprise an integrated circuit. Transistors are the essential aspect of all modern electrical components and electronic devices. The size of the transistors has been progressively shrunk as the VLSI industry grows to integrate more functionality onto a silicon wafer and minimize circuit power consumption. Nano-FET devices are being realized using various materials with different structures, with promising results. Novel nano-FET devices should be an excellent candidate to replace the existing technologies for low-power and high-frequency applications with reduced time delay in circuit applications.

The relentless pursuit of miniaturization in semiconductor technology has led to the emergence of nano FETs as pivotal components in modern electronic systems. This book aims to provide a comprehensive overview of nano FET devices, from their theoretical foundations to application implementations. Due to the enormous study of Nano-CMOS and post-CMOS technologies and the lack of a comprehensive guidebook, research articles are now the cornerstone for the knowledge of novel design based on the fundamentals of Nano-FET devices. As a result, this book outlining the essential characteristics of Nano-CMOS and post-CMOS technologies will benefit engineers who must understand the fundamentals of these devices and scholars developing/implementing Nano-CMOS and post-CMOS devices and their applications. This book, Nano-FET Devices: Miniaturization, Simulation, and Applications, is intended to fulfill this requirement of the research community.

In the opening chapters, readers will embark on a journey through the basic concepts of FETs, understanding how these devices operate and their significance in electronic engineering. Building upon this foundation, the book delves into the unique characteristics of nano FETs, including quantum effects, scaling considerations, and material properties that define their behavior at the nanoscale.

This book is a concise benchmark for beginners who are just getting started with Nano-FET Devices and their application with recent advancements, and those who want to design integrated circuits using novel FET devices. We hope that "Nano-FET Devices: Miniaturization, Simulation, and Applications" serves as a valuable resource for researchers, engineers, and students interested in unlocking the potential of nano-FET technology. May this book inspire discoveries, innovations, and advancements at the forefront of electronic engineering.

Dharmendra Singh Yadav
Department of Electronics and Communication Engineering,
National Institute of Technology,
Kurukshetra, Haryana 136119, India

&

Prabhat Singh
Department of Electronics and Communication Engineering,
Dr B R Ambedkar National Institute of Technology,
Jalandhar, Punjab 144008, India

DEDICATION

I would like to dedicate and express my hearty gratitude towards my respected parents, Uncle, aunty, younger brothers, sisters for their affection and persistent efforts in my education. Also dedicated to my wife and our loving son Armaan Singh for their everlasting supports, encouragements and understanding. This work is dedicated to my family and others who have always been as source of my continued efforts for academic excellence. Over to all infinite gratitude flows to the almighty for the countless blessings bestowed upon us.

— Dharmendra Singh Yadav

I dedicate this book to my loving mother, Shakuntala Singh, my father, Dinesh Singh, my wife, Sadhana Singh, and my brother, Prasoon Singh, as a token of deep appreciation for everything you have done and continue to do for me. Your love and support are the foundation of my success, and I am truly blessed to have you in my life. I also wish to dedicate this book to Dr. Nagendra Pratap Singh, Dr. Ashish Raman, and Dr. Navjeet Bagga for their invaluable guidance and unconditional support, which have played a crucial role in helping me achieve this success. This book is a symbol of my heartfelt gratitude for all that you have done.

— Dr. Prabhat Singh

List of Contributors

A. Kalaiyarasan Department of Mechanical, Muthayammal Engineering, Rasipuram, Tamil Nadu, India

Ashish Raman ECE Department, B. R. Ambedkar National Institute of Technology, Jalandhar, Punjab, India

A.K.C Varma Department of ECE, Vishnu Institute of Technology, Bhimavaram, Andhra Pradesh, India

B.V.V Satyanarayana Department of ECE, Vishnu Institute of Technology, Bhimavaram, Andhra Pradesh, India

C. Kathiravan Department of Chemistry, Muthayammal Engineering College, Rasipuram, Namakkal, India

D. Anbuselvi Department of Physics, Muthayammal Engineering College, Rasipuram, Namakkal, India

Gowrishankar J. Department of Computer Science and Engineering (AI), JAIN (Deemed-To-Be University), Bangalore, India

G. Prasanna Kumar Department of ECE, Vishnu Institute of Technology, Bhimavaram, Andhra Pradesh, India

Jyoti Kandpal Deperment of Electronics and Communication Engineering, Graphic Era Hill University, Dehradun, Uttarkhand, India

K. Radhika Department of ECE, Muthayammal Engineering College, Rasipuram, Tamil Nadu, India

M. Parvathi Department of Electronics and Communication Engineering, BVRIT Hyderabad College of Engineering for Women, Hyderabad, Telangana, India

N. Suthanthira Vanitha Department of Electrical and Electronics Engineering, Muthayammal Engineering College, Rasipuram, Namakkal, India

Subarna Mondal ECE Department, Maulana Abul Kalam Azad University and Technology, West Bengal, India

Soumya Sen Computer Science and Engineering Department, University of Engineering and Management, Jaipur, Rajasthan, India

Srividya P. Department of Electronics and Communication Engineering, RV College of Engineering, Bengaluru, Karnataka 560059, India

S. Grace Infantiya Department of Physics, Muthayammal Engineering College, Rasipuram, Namakkal, India

Swagata Devi Faculty of Engineering, Assam downtown University, Panikhaiti, Guwahati, Assam, India

Shonak Bansal Department of Electronics and Communication Engineering, Chandigarh University, Gharuan, Punjab, India

T. Narmadha Department of Computer Science Engineering, JAIN (Deemed-to-be University), Bengaluru, India

T. Divya Department of Computer Science, Velalar College of Engineering Technology, Thindar, India

T.S.S Phani   Department of ECE, Bonam Venkata Chalamayya Engineering College (A),
                          Odalarevu, Andhra Pradesh, India

T. Saran Kumar   Department of ECE, Bonam Venkata Chalamayya Engineering College (A),
                          Odalarevu, Andhra Pradesh, India

CHAPTER 1

Physics and Properties of Single/ Multi-Gate Fets

Subarna Mondal[1,*], Soumya Sen[2] and Ashish Raman[3]

[1] *ECE Department, Maulana Abul Kalam Azad University and Technology, West Bengal, India*

[2] *Computer Science and Engineering Department, University of Engineering and Management, Jaipur, Rajasthan , India*

[3] *ECE Department, B. R. Ambedkar National Institute of Technology, Jalandhar, Punjab, India*

Abstract: In the semiconductor industry, the integration of the Complementary-Metal-Oxide-Semiconductor (CMOS) mechanism into Integrated Circuits (ICs) has resulted in a significant rise in the count of transistors on a single chip. This is made possible by shrinking down the size of Metal-Oxide-Semiconductor-Field-Effect-Transistors (MOSFETs). However, scaling can lead to device performance degradation. To address this, advanced MOSFET designs like multi-gate transistors, junction-less transistors, and Tunnel FETs have been proposed, aiming to sustain Moore's Law and support continued transistor scaling in the coming decade. The key principles of this chapter involved in both single and multi-gate FETs include quantum mechanics, carrier transport, and electrostatics. The scaling of transistors to smaller sizes involves considerations of quantum effects, like tunneling and quantum confinement, which have a significant impact on their behavior. Understanding this chapter is crucial for optimizing their performance, enabling further miniaturization, and enhancing the capabilities of integrated circuits. Additionally, it plays a crucial role in advancing the field of next-generation electronics and computational devices.

Keywords: Electrostatic integrity, Quantum mechanics, Quantum confinement MOSFET, FET, multi-gate FET, Single gate FET.

INTRODUCTION

The advancement of semiconductor technology has been a relentless march to make more capable and efficient electronic devices. A pivotal chapter in this technological journey is the transition from single-gate Field-Effect Transistors (FETs) to multi-gate FETs, a transformation that has redefined the boundaries of modern electronics. This journey reflects not only a remarkable evolution in

*Corresponding author Subarna Mondal: ECE Department Maulana Abul Kalam Azad University and Technology, West Bengal, India; E-mail: subarnamondal.39@gmail.com

Dharmendra Singh Yadav & Prabhat Singh (Eds.)

transistor design but also a testament to human ingenuity and the relentless pursuit of greater computing power. Single-gate FETs, such as the well-known Metal-Oxide-Semiconductor FETs (MOSFETs), have been the linchpin of electronics for decades. They served as the foundational element for integrated circuits and electronic systems by utilizing a lone gate electrode to regulate the current between the source and drain terminals. Yet, as the demands for more compact and energy-efficient devices grew, it became evident that single-gate FETs were approaching their performance limits. The shift towards multi-gate FETs represents a transformative leap in semiconductor technology [1-3]. Multi-gate FETs, also known as multi-gate transistors, feature multiple gate electrodes, enabling enhanced control over the flow of electrons within the channel. The most prominent of these innovations is the FinFET (Fin Field-Effect Transistor), which utilizes a three-dimensional fin-shaped channel and multiple gates to govern the electron flow [4]. This design innovation has set the stage for unprecedented scaling and performance improvements. The progression from single to multi-gate FETs began in response to the semiconductor industry's challenge of maintaining Moore's Law [3]. Renowned for his observation, Intel co-founder Moore noted that the number of transistors on a microchip would undergo a doubling roughly every two years [2], leading to exponential advancements in computing power. However, as transistor dimensions approached the nanoscale, the limitations of single-gate FETs became increasingly apparent. Issues such as leakage current, power consumption, and heat dissipation threatened to stall the progression of Moore's Law [4]. Section 2 focuses on the evolution of FET technology and recalls the history of the single to multi-gate concept. It also highlights the benefits of multi-gate FETs in terms of precise control over short channel effects. Section 3 discusses the basic physics of single-gate FET to multi-gate FET and also analyzes the electrostatics of the multi-gate MOS system. The study also investigates the influence of electron excavating through narrow gate dielectric. Section 4 introduces a different structural view of multi-gate FET in Silvaco atlas TCAD, its working, I_d-V_d characteristics, and Ion-Ioff ratio. Section 5 discusses the profound impact of multi-gate FET in nanotechnology by enabling improved performance, energy efficiency, continued scaling, and sustenance of Moore's law. Section 6 focuses on challenges in simulating Multi-gate FET and structural analysis hurdles and reviews structural modification to address this issue.

PROGRESSION OF FET TECHNOLOGY

In 1965, Gordon Moore's influential paper foresaw a fourfold increase in transistors on a single chip every three years, laying the groundwork for Moore's Law [2]. The semiconductor industry has closely adhered to this law, with collaborative efforts between semiconductor firms and educational institutions working towards refining the accuracy of industry predictions since the early 1990s. This collaboration led to the establishment of the International Technology Roadmap for Semiconductors (ITRS) organization [5]. The ITRS releases an annual outline, serving as a reference for the semiconductor sector, outlining the required technology, tools for design, and equipment to match the swift advancements in semiconductor devices projected by Moore's Law (Fig. **1**).

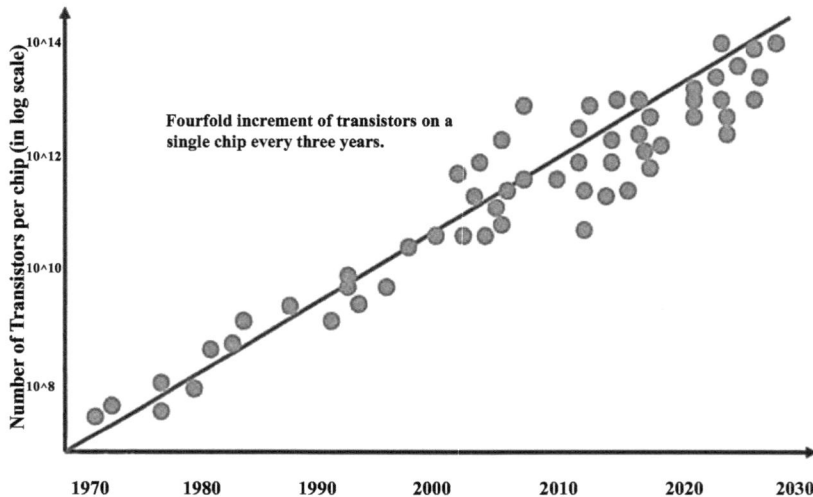

Fig. (1). This picture illustrating the number of transistor vs years according to Moore's theory [5].

Silicon CMOS is the essential technology at the core of the semiconductor industry, and the MOS transistor, also known as the MOSFET (MOS Field-Effect Transistor), stands as its cornerstone. To align with the relentless progression defined by Moore's Law, the physical measurements of transistors have consistently undergone a halving every triennial. The challenge of sub-micron scale dimensions was effectively addressed in the early 1980s. Looking back to 2023, the semiconductor industry considered the introduction of 3nm and 5nm nodes to achieve greater transistor density and performance. These technological strides will empower the development of processors, memory modules, and System-on-Chip (SoC) designs.

The physics of single to multi-gate-field effect transistors (FETs) is a fundamental aspect of modern semiconductor technology. The concept of multi-gate-Field-Effect-Transistors (Mug-FETs), namely double, triple, and quadruple gate SOI MOSFET, has emerged as a favorable choice to align with the scaling trends observed in conventional bulk MOSFET within the nanoscale domain. Multi-gate FET traces its origins back to several researchers and developments in the late 20th century. The first significant innovation that laid the groundwork for multi-gate FETs was the Surrounding Gate Transistor (SGT) proposed by Japanese researcher Toshihiro Sekigawa in the early 1980s. The SGT featured a gate electrode that surrounded the semiconductor channel on all sides, providing improved control over the flow of current. However, the Fin Field-Effect Transistor (FinFET) design, which is a prominent type of multi-gate FET, gained more recognition for its practical implementation. The FinFET was conceived and developed by University of California, Berkeley researchers, led by Dr. Chenming Hu, in the late 1990s. This innovative transistor design used a three-dimensional fin-shaped channel structure with multiple gate electrodes, offering superior electrostatic control and improved performance. It marked a significant step towards realizing the potential of multi-gate transistors. The initial patent for the FinFET concept was filed by the University of California in 1999. Subsequently, various semiconductor manufacturers, including Intel, played a pivotal role in developing and commercializing FinFET technology, making it a standard in the semiconductor industry. Intel, in particular, integrated FinFETs into their microprocessors, which significantly advanced the field of semiconductor manufacturing. So, while the Surrounding Gate Transistor (SGT) was an early concept in the development of multi-gate FETs, the FinFET design, developed at the University of California, Berkeley, is often recognized as a key milestone in the practical realization of multi-gate transistor technology [6-10].

THE EFFECTS ARISING FROM SHORT CHANNEL

Turning down the channel length in a bulk MOSFET leads to the manifestation of a range of phenomena collectively referred to as 'Short Channel Effects.' The primary factor behind these effects is the drop in the influence of the gate over the depletion zone beneath it. In simpler terms, the gate's control weakens over the depletion charge underneath it due to encroachment from the source and drain. As a result, there is a limitation of threshold voltage with declining the channel length, and this phenomenon is known as 'threshold voltage roll-off' [6]. Fig. (**2**) depicts the threshold voltage rattle down in both bulk and SOI devices across distinct silicon thickness values (tsi).

DIBL, which is another manifestation of short channel effects, originates from the mutual sharing of charge connecting the gate to the source/drain junctions. It manifests in both bulk and SOI devices, but similar to the threshold voltage rattling-off in SOI devices [7].

Mitigating short-channel effects involves reducing both the depth of the junction and the thickness of the gate oxide layer. Another approach is minimizing the depletion depth by increasing doping concentration. Designers have long adhered to implicit design rules to make certain production of devices devoid of the effects of short-channel [8].

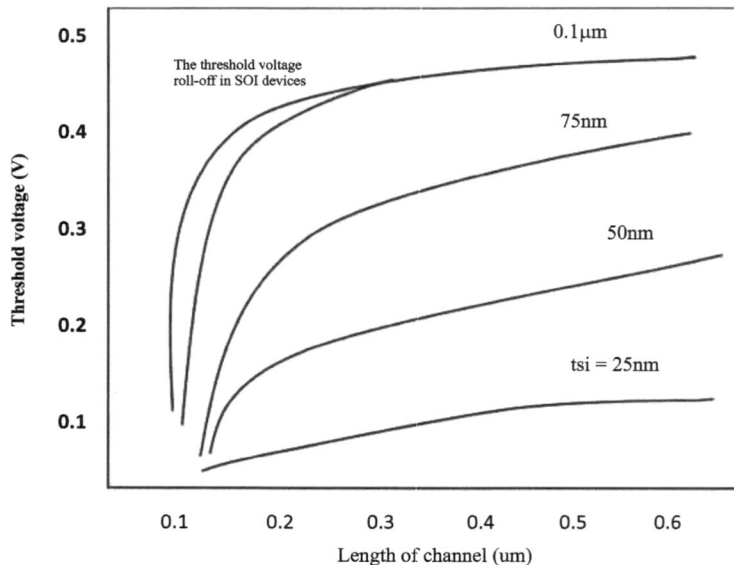

Fig. (2). The discrepancy of threshold voltage with respect to the length of the channel is examined for several silicon thickness values.

PHYSICS OF GATED FET

Many technological solutions to the problems caused by Moore's Law scaling have been put forth recently. These advancements include the use of novel materials, like metal gate electrodes, stressors, and high-k dielectrics, in the CMOS method. By making it easier for mechanical stress to be introduced into the silicon, these materials increase carrier mobility. Apart from these material advances, new transistor architectures have also emerged, one prominent example of which is the silicon-on-insulator (SOI) transistor [11-16].

The extraordinary performance capabilities of SOI have been acknowledged for an extended period. The SOI transistor has changed recently, moving from a conventional flattened single-gate design to a three-dimensional design with multiple gates (double, triple, or quadruple-gate devices). One of the most promising device architectures for extending CMOS scaling into the nanoscale realm is the double-gate MOSFET in particular [9-15].

Classical Physics of Multi-gate FET System

Fig. (**3**) shows the multi-gate FET system in a simple form where the middle of the semiconductor body serves as the transport region, facilitating the movement of electrons starting from the source towards the drain. This area is circumambient by the gate stack [16-21].

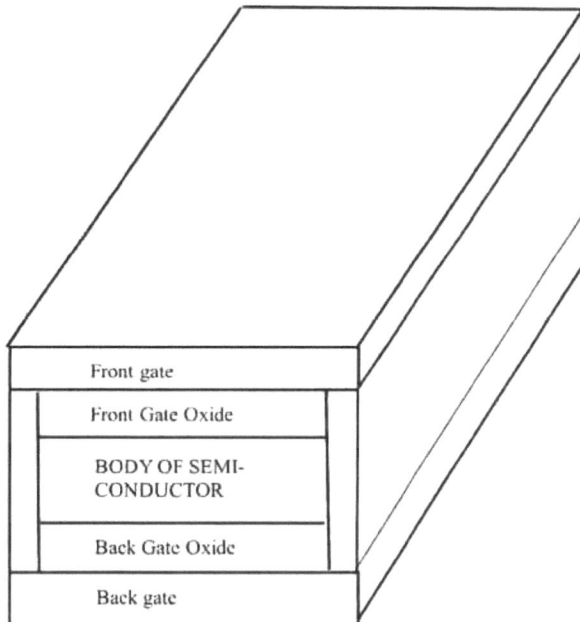

Fig. (3). Geometry of multi-gate FET adopts a quadrilateral channel cross section.

The gates can be either interconnected or electrically isolated. The configuration is regarded as a double-gate device when the side gates are absent. The structure is classified as a single-gate device when the rear oxide layer is noticeably wide. The structure is called a triple-gate device if it has both side gates and a thick back oxide. For clarity's sake, this section uses a quadrilateral channel cross-section to illustrate the physics of the multi-gate system, even though the cross-section can have non-rectangular geometries like cylindrical or triangular [22].

A critical metric in the study of device physics is "Electrostatic Integrity" (EI), which is associated with the 'Short-Channel Effect' (SCE) and 'Drain Induced Barrier Lowering' (DIBL) at threshold voltage roll-off. The efficiency of the gate in electrostatically regulating the channel area can be indicated by the EI. Research has demonstrated that decreasing the thickness of the silicon layer or the gate oxide can both improve Electrostatic Integrity (EI) [10]. We will now expand the investigation to include MuG-FET systems, and for any specific device geometry and boundary conditions, the field configuration inside a device can be determined by solving Poisson's equation. Poisson's equation can be solved using the depletion approximation to obtain the 'Potential Distribution' (φ) within the channel of a fully depleted MuG-FET [4].

$$\frac{d^2\varphi(x,y,z)}{dx^2} + \frac{d^2\varphi(x,y,z)}{dy^2} + \frac{d^2\varphi(x,y,z)}{dz^2} = \frac{qN_a}{\varepsilon} \tag{1}$$

This equation can be modified as

$$\frac{d^2\varphi(x,y,z)}{dx^2} + \frac{d^2\varphi(x,y,z)}{dy^2} + \frac{d^2\varphi(x,y,z)}{dz^2} = C \tag{2}$$

This correlation shows that the total fluctuations of the electric field constituent in the direction of x, y, and z stay constant at any given count (x, y, z) within the channel. As a result, if one element rises, the others must fall. Short-channel effects are caused by the drain force field's incursion into the channel portion, which is represented by the electric field's x-component (Ex). The effects of (Ex) on a small element in the channel area at coordinates (x, y, z) can be lessened by lengthening the channel, L, or by strengthening the dominance provided by top or bottom gates $\frac{d^2(E_y)(x,y,z)}{dy^2}$ or $\frac{d^2(E_z)(x,y,z)}{dy^2}$ the lateral gate. Reducing the fin width (Wsi) and/or thickness (tsi) of silicon can also accomplish these. The step-up of $\frac{d^2(E_y)(x,y,z)}{dy^2} + \frac{d^2(E_z)(x,y,z)}{dy^2}$ turns down the effect of short-channel and gives the gates greater control over the channel. The conflict between the electric fields from the drain and the gates for channel control is shown in Fig. (4) [4,11].

For single and double-gated FET, the term $\frac{d\varphi}{dz} = 0$ & the Poisson's Equation can be changed as [4],

$$\frac{d^2\varphi(x,y,z)}{dx^2} + \frac{d^2\varphi(x,y,z)}{dy^2} = \frac{qN_a}{\varepsilon} \tag{3}$$

A curve shaped like a parabola potential distribution within the silicon sheet along the vertical direction, known as the Y direction, is obtained from a simplified one-dimensional study of a totally depleted device [12]. Poisson's equation is modified as follows [13], assuming an approximated distribution in the Y-direction for a two-dimensional examination:

$$(\varphi)(x, y) = a_0(x) + a_1(x)(y) + a_2(x)(y^2) \tag{4}$$

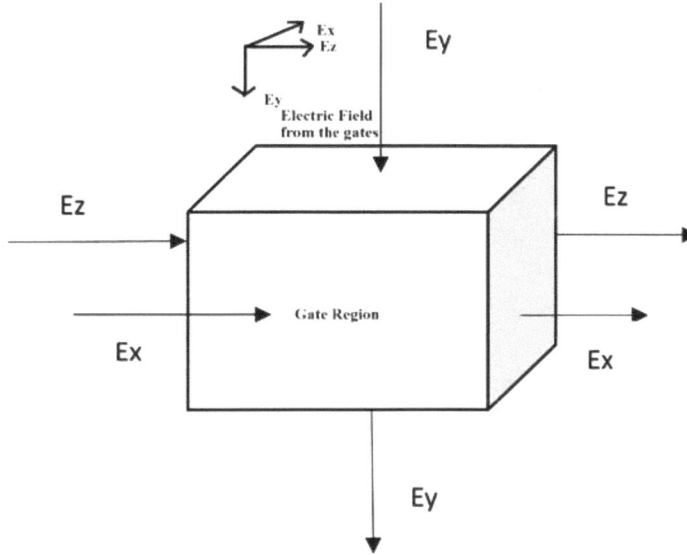

Fig. (4). Co-ordinate and electric field component of multi-gate FET system.

Single Gate SOI FET System

In a single Gate SOI FET system, applying the boundary condition in equation 4 [14]

1. $(\varphi)(x, 0) = (\varphi_f)(x) = a_0(x)$ where $(\emptyset_f)(x)$ is the front gate's surface potential

2. $\dfrac{d\varphi(x,y)}{dy} = \dfrac{\varepsilon_{ox}}{\varepsilon_{si}} \dfrac{\emptyset_f(x) - \emptyset_{gs}}{t_{ox}} = a_1(x)$ where $\emptyset_{gs} = V_{gs} - V_{FB}$ where V_{gs} voltage between the gate and the source and V_{FB} is the voltage of flat band.

3. By incorporating these three boundary conditions into equation 4, we derive:

$$\varphi(x, y) = \varphi_f(x) + \dfrac{\varepsilon_{ox}}{\varepsilon_{si}} \dfrac{\emptyset_f(x) - \emptyset_{gs}}{t_{ox}} y - \dfrac{1}{2t_{si}} \dfrac{\varepsilon_{ox}}{\varepsilon_{si}} \dfrac{\emptyset_f(x) - \emptyset_{gs}}{t_{ox}} y^2 \tag{5}$$

changing equation 5 into equation 3, we get

$$\frac{d^2(\varphi_f)(x)}{dx^2} - \frac{\varepsilon_{ox}}{\varepsilon_{si}}\frac{(\varphi_f)(x)-\emptyset_{gs}}{t_{ox}t_{si}} = \frac{qN_a}{\varepsilon} \tag{6}$$

$\emptyset_f(x)$ can be derived from equation 6 and $\emptyset(x,y)$ determined from equation 5.

Yan *et al.* [15] and Lee *et al.* [11] established an important notion that is called the "natural length," denoted by 'λ', which helps to calculate SCEs in multi-gate transistors. This concept is derived from the Poisson equation. This parameter, which is directly related to the device's electrostatic integrity, describes the depth to which the drain electric field traverses the channel [10, 11]. When a device's channel length surpasses six times its normal length, no SCEs are present [9]. A possible method to calculate the "natural length" is as follows:

$$\lambda_1 = \sqrt{\frac{\varepsilon_{si}}{\varepsilon_{ox}}t_{ox}t_{si}} \tag{7}$$

The thickness of the silicon sheet and the gate oxide define the value λ1. A decrease in the thickness of the silicon film or the oxide of the gate results in a reduction of the natural length. Consequently, this lessens the channel region's exposure to the drain electric field.

Semiconductor devices featuring triple gates are referred to as

The boundary conditions for **equation 4** apply to a double-gate device [14].

1.$\varphi(x,0) = \varphi_f(x,t_{si}) = \varphi_f(x) = a_0(x)$ where $(\varphi_f)(x)$ is the front gate's surface potential;

2. $\frac{d(\varphi)(x,y)}{dy} = \frac{\varepsilon_{ox}}{\varepsilon_{si}}\frac{\emptyset_f(x)-\emptyset_{gs}}{t_{ox}} = a_1(x)$ where $\emptyset_{gs} = V_{gs} - V_{FB}$ where V_{gs} voltage between the gate and the source and V_{FB} is the voltage of flat band.

3.By incorporating these three boundary conditions into **equation 4**, we derive

$$\emptyset(x,y) = \emptyset_f(x) + \frac{\varepsilon_{ox}}{\varepsilon_{si}}\frac{\emptyset_f(x)-\emptyset_{gs}}{t_{ox}}y - \frac{1}{t_{si}}\frac{\varepsilon_{ox}}{\varepsilon_{si}}\frac{\emptyset_f(x)-\emptyset_{gs}}{t_{ox}}y^2 \tag{8}$$

Using the same method as for the single-gate case, the intrinsic length of the double-gated device can be found, leading to

$$\lambda_2 = \sqrt{\frac{\varepsilon_{si}}{2\varepsilon_{ox}} t_{ox} t_{si}} \tag{9}$$

Quadruple-gate FET System

The preliminary research on the 'natural length' idea is limited to single- and double-gate systems. Nevertheless, it is possible to extend this concept to quadruple-gate devices. The centre of the device, where the force field lines from the drain have the greatest effect on the device body, is where this extension is accomplished. The Poisson equation changes to the following in this case [4]:

$$\frac{d^2 \varphi(x,y,z)}{dx^2} + 2 \frac{d^2 \varphi(x,y,z)}{dy^2} = \frac{qN_a}{\varepsilon} \tag{10}$$

Then, the natural length will be

$$\lambda_4 = \sqrt{\frac{\varepsilon_{si}}{4\varepsilon_{ox}} t_{ox} t_{si}} \tag{11}$$

The summary of the natural length of different multi-gate FETs is listed in Table **1**. It has been determined that minimizing the thickness of the silicon layer and gate oxide can lessen the inherent length and the ensuing short-channel effects. Another factor for this decrement is the use of a high-dielectric permittivity gate other than SiO2. Furthermore, the 'natural length' falls as the count of gates rises. Reducing the oxide thickness below 1.5 nm, however, may cause problems with gate tunneling current for very tiny devices. The use of multi-gate devices allows the flexibility of exchanging a thin gate oxide for a thinner fin or silicon film because λ is in proportion to the product of 'tsi' and 'tox'. To expand the definition of the natural length to accommodate any given count of gates, Colinge and colleagues [17] also expressed the idea of the effective gate number, represented as N.

$$\lambda_N = \sqrt{\frac{\varepsilon_{si}}{N\varepsilon_{ox}} \left(1 + \frac{\varepsilon_{ox}}{4\varepsilon_{si}} \frac{t_{si}}{t_{ox}}\right) t_{ox} t_{si}} \tag{12}$$

Accordingly, N in the case of a Pi-gate transistor is 3.14, whereas N in the case of an Omega-gate FET is between 3 and 4, depending on how far the fourth gate delves below the channel portion [18]. Suzuki *et al.* [19] proposed a scaling parameter called "α_N", which makes it easier to assess SC (short-channel) sensitivity for devices with several gate structures:

Table 1. The natural length of different multi-gate FETs.

Single gate SOI FET	$\lambda_1 = \sqrt{\frac{\varepsilon_{si}}{\varepsilon_{ox}} t_{ox} t_{si}}$	[10]
Double Gate FET	$\lambda_2 = \sqrt{\frac{\varepsilon_{si}}{2\varepsilon_{ox}} t_{ox} t_{si}}$	[11]
Triple gate	$\lambda_3 = \sqrt{\frac{\varepsilon_{si}}{3\varepsilon_{ox}} t_{ox} t_{si}}$	[11]
Quadruple-gate FET (square cross-section)	$\lambda_4 = \sqrt{\frac{\varepsilon_{si}}{4\varepsilon_{ox}} t_{ox} t_{si}}$	[12]
Surrounding Gate (circular cross-section)	$\lambda_o = \sqrt{\frac{\varepsilon_{si} t_{Si}^2 \ln\left(1+\frac{2t_{ox}}{t_{si}}\right) + \varepsilon_{ox} t_{Si}^2}{16\varepsilon_{ox}}}$	[16]

$$\alpha_N = \frac{L}{2\lambda_N} \tag{13}$$

A value of 2.2 can be imposed as the minimum gate length to minimize SCE once the silicon thickness (tsi) and oxide thickness (tox) have been determined [18].

EFFECT OF THRESHOLD VOLTAGE

Assuming a fixed doping concentration, the traditional theory predicts that the threshold voltage of a fully depleted Silicon-On-Insulator (SOI) Metal-Oxide-Semiconductor-Field-Effect-Transistor (MOSFET) will fall with a decrease in silicon film thickness [20]. The decline in depletion charge with decreasing film thickness explains this reduction. On the other hand, the depletion charge is negligible and can be disregarded if the film thickness is lower than 10 nm. However, two nonclassical factors that influence the threshold voltage must be considered. The requirement that the inversion carrier concentration be greater than what is predicted by the traditional theory of MOSFET in order to overcome the threshold gives rise to the first component. Consequently, the potential Φ in the thin silicon film is significantly more than the traditional $2\phi f$. The second factor is the increased gate voltage needed to achieve a specific inversion carrier concentration as a result of conduction band splitting. A double gate device's threshold voltage can be written as [4]:

$$V_{TH} = \emptyset_{ms} + \frac{kT}{q}\ln\left(\frac{2kTC_{OX}}{q^2 n_i t_{si}}\right) + \frac{\pi^2\hbar^2}{2qm^*t_{si}^2} \tag{14}$$

Equation 14 is composed of three parts. The difference in work function across the silicon layer and the gate is depicted in the first section. The potential (Φ) in the channel, which is inversely correlated to the silicon film's thickness (tsi), is indicated by the second part. When it comes to very thin films, Φ might be much greater than $2\Phi F$, indicating that the inversion carrier concentration near the threshold is higher than it is for thicker films [21]. The third component is related to the fluctuation of the lowest energy in the band of conduction with respect to the silicon film thickness, which is a prediction that can only be made by quantum-mechanical computations, as Fig. (5) shows.

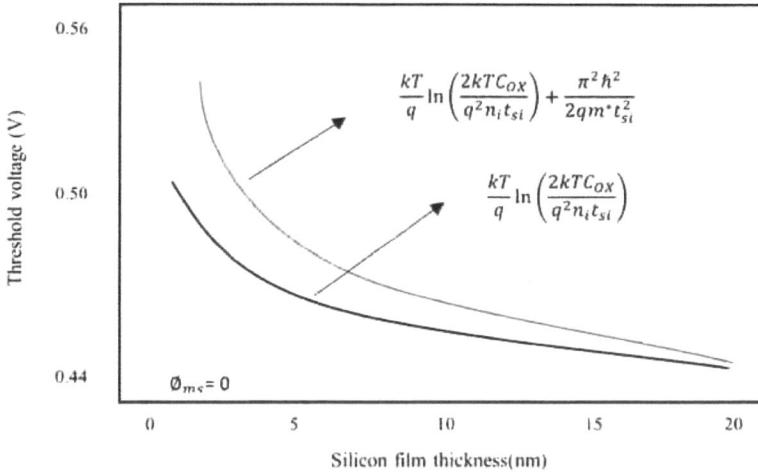

Fig. (5). The threshold voltage is contingent upon the thickness of the silicon layer in a double-gate-transistor. The lower curve corresponds to mechanical factors, while the upper curve encompasses the quantum aspect of the "classical part of Equation (15).

TUNNELING CURRENT

Power consumption increases due to electron tunneling, a quantum mechanical phenomenon that happens between the gate and the channel when gate oxides are around 1.2 nm thick. The Heisenberg uncertainty principle and the wave-particle duality of matter are frequently used to explain quantum tunneling, which occurs when particles pass through a barrier [22, 23]. Gate oxide tunneling current is produced when the oxide is thinned because it creates a stronger electric field across the oxide, which in turn facilitates electron tunneling from the substrate to the gate and vice versa [24]. There are two forms of tunneling that occur between the substrate and gate: direct tunneling and Fowler-Nordheim tunneling. The thickness, height, and construction of a barrier affect the possibility of electron tunneling across it. As a result, different tunneling currents arise from different tunneling probabilities for a single electron in FN tunneling as opposed to direct tunneling. Whereas electrons in direct tunneling move through a trapezoidal well, those in Fowler-Nordheim tunneling move through a triangular potential well. Electrons tunnel straight to the gate through the band gap of the insulator in ultrathin oxide layers, which are less than 3 nm [25].

By solving the Poisson's equation, we get,

$$V_{GS} - V_{POLY} = V_{FB} + \emptyset_S - \frac{q\varepsilon_{si}N_{POLY}t_{ox}^2}{\varepsilon_{ox}^2}\left(1 - \sqrt{1 + \frac{2\varepsilon_{ox}^2(V_G - V_{FB} - \varphi_S)}{q\varepsilon_{si}N_{POLY}t_{ox}^2}}\right) \quad (15)$$

Here, V_{FB} represents flat band voltage, ε_{si} denotes the dielectric constants, and Npoly stands for the doping concentration of the poly-silicon gate; $\emptyset_S$ is the surface potential; t_{ox} indicates the oxide thickness [26, 27].

It is evident that the oxide potential, flat band voltage, and surface potential are all affected by the gate voltage when it is applied. The electric field can be found using

$$E_{OX} = \frac{V_{OX}}{t_{ox}} \quad (16)$$

The equation mentioned above shows that a decrease in semiconductor thickness causes a rise in gate tunnel current while the gate voltage remains constant. When the semiconductor layer is depleted, this impact is especially apparent at lower voltages, especially for semiconductor layers narrower than 20 nm. These results, however, do not fully agree with gate tunnel current measurements made on FinFETs, highlighting the need for more study in this area [25-32].

QUANTUM CONFINEMENT

Both the thickness and width of multi-gate FETs are getting closer to values that are smaller than 10 nanometres. In these situations, electrons in the "channel" of an n-channel device create a 'One-Dimensional Electron Gas' (1DEG) in a triple- or quadruple-gate MOSFET or a 'Two-Dimensional Electron Gas' (2DEG) in a double-gate device. Electrons are limited in the y direction but free to move in the x and z directions in a double-gate device with a thin silicon sheet. On the other hand, electrons are constrained in the y and z directions and can only go in the x direction in a thin and narrow triple- or quadruple-gate device. This causes the "volume inversion" effect [26], which raises the threshold voltage as the devices' width and thickness decrease [27, 28].

IMPORTANT PARAMETERS OF MULTI-GATE FET SYSTEM

Drive-in Current in Multi-gate FET System

The total current drive of a multi-gate FET is a combination of the currents flowing through all of the interfaces that the gate electrode covers. If carriers have the same mobility at each interface, it is equivalent to the current in a single-gate device procreated by the effective count of gates (assuming a square cross-section). Take an example of a single-gate transistor with a matched gate length and width that has a current drive that is half that of a double-gate device. Multi-fin devices are used to achieve greater currents; the current drive of a multi-fin MOSFET can be likened to the product of the current flowing through a single fin and the number of fins present. The schematic view is shown in the top perspective of a FinFET (Fig. **6a**) and a side view of FinFET (Fig. **6b**).

Consider a multi-gate FET, with the top surface mobility denoted as μ_{top}. The mobility at the sidewall interface, denoted as μ_{side}, may differ from μ_{top} mobility and is contingent on the orientation of the crystal sidewall [30]. Then, the current equation in the multi-gate device can be given by [31]

$$I_D = I_{D0} \frac{\theta \mu_{top} W_{si} + 2\mu_{side} t_{si}}{\mu_{top} P} \tag{17}$$

Here, I_{D0} indicates the current in the single-gate flattened device, W_{si} stands for the width of the discrete fin, and t_{si} represents the wideness of the silicon film. In a triple-gate device, θ is set to 1 in cases when conduction occurs across three interfaces, and θ is set to 0 in a FinFET where channels form only at the sidewall interfaces [32].

Corner Effect is a Multi-gate FET System

The three-dimensional architecture of multi-gate devices creates special coupling effects, notwithstanding its benefits in SCE-controlling. Research has shown that silicon body corners have a major impact on the I-V characteristics of SOI-MOSFETs with many gates [32-35]. Corner effects are caused by independent channels emerging at corners that have different threshold voltages than top or sidewall gates. The Ion/Ioff ratio deteriorates when corner elements of the total current reflect a lower threshold voltage, leading to elevated Ioff [36, 37]. The corner radius affects the properties of the device and establishes whether the threshold voltages of planner interfaces and corner sections differ.

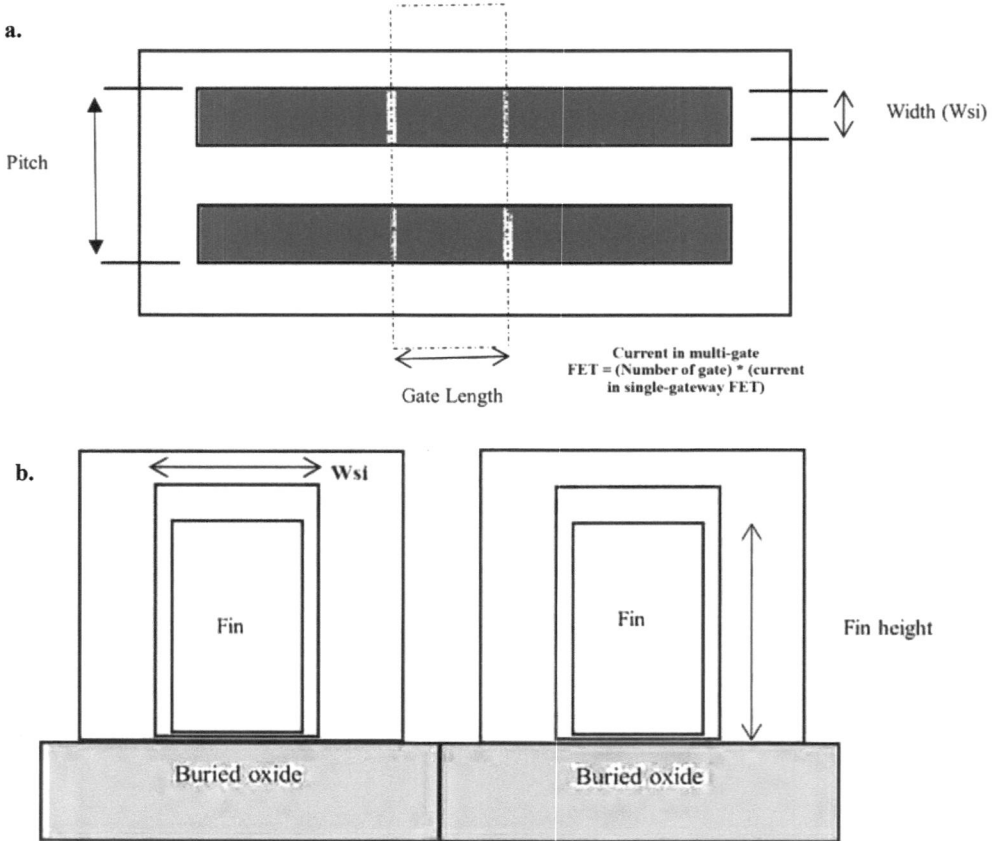

Fig. (6). a. Top perspective of a FinFET. **b.** side view of FinFET.

A study, which examined Pi-gate Silicon-On-Insulator (SOI) transistors, demonstrated that the doping concentration has an inverse relationship with the corner region's extension [35]. It is possible to avoid unwanted double threshold voltages by lowering the doping density. If the device dimensions are small enough, corner effects can be prevented even on highly doped substrates. Gate oxide width reduction and corner rounding were also examined in the study. Reducing electrostatically favorable regions and potential changes along the Si–oxide interface is necessary to eliminate corner effects and stop different device regions from inverting at various gate voltages. As such, it is not expected that corner effects will be important for ultra-small structures. These results cover related architectures, namely tri-gate FETs, W-gate FETs, or Gate-All-Around (GAA) SOI MOSFETs, and go beyond the research on Pi-gate MOSFETs [38-45].

Subthreshold Slope

Limiting the separation between the source and channel can cause the short-channel effect, so it is important to keep it as small as possible. With an increase in gate voltage, there is a corresponding increase in channel current. Therefore, it is necessary to minimize changes in gate voltage as much as possible for each decade. The sub-threshold slope is represented by:

$$SS = \ln(10) \frac{dV_G}{d \ln I_D}$$

$$\text{Since } V_G - V_{FB} = \Psi_S + \sqrt{\frac{2q\epsilon\Psi_S N_A}{C_{OX}}}$$

$$\frac{dV_G}{d\Psi_s} = 1 + \frac{1}{C_{OX}}\sqrt{\frac{q\epsilon N_A}{2\Psi_S}} = 1 + \frac{C_D}{C_{OX}}$$

$$SS = \ln(10)\frac{1}{\beta}\frac{dV_G}{d\Psi_s} = \ln 10 \frac{kT}{q}(1 + \frac{C_D}{C_{OX}}) \tag{18}$$

SS is affected by the increase in channel voltage, leading to an expansion in the sub-threshold slope and, consequently, a minimization in the threshold voltage. Decreasing the threshold voltage, in turn, results in an increase in leakage current [46, 47].

SINGLE AND MULTI-GATE FET – STRUCTURAL ANALYSIS

A. Double-gate FET system:

Semiconductor devices featuring dual gates are referred to as double-gate devices. The characteristics of these devices include [38]:

1. The two gate terminals are interconnected.

2. An established electric field exists between the source and channel beneath the device, terminating at the base gate terminal, preventing its subsequent arrival at the channel region.

3. Field lines extending over the silicon itself can surpass the area of the channel, negatively impacting the small characteristics of the channel. To mitigate this infringement, it is essential to reduce the depth of the silicon film.

4. Initially manufactured double-gate SOI MOSFET devices are entirely depleted with a thin-channel semiconductor (DELTA, 1989). Fig. (**7**) illustrates the internal structure of DELTA MOSFETs.

5. FinFET got its basic structure from DELTA MOSFET. FinFET structure appears as a set of 'Fin'. It comprises a conducting area primarily enclosed by a slender "fin" structure constructed on a silicon insulator, giving rise to the term "FinFET." The effective conduction channel for the device is insistent by the wideness of the fin. Fig. (**8**) shows the structure of basic FinFET.

6. Additional double-gate MOSFET variants encompass the GAA. GAA refers to a planar-MOSFET wherein the electrode of the gate is folded over the channel region.

B. Triple-gate FET System:

Semiconductor devices featuring triple gates are referred to as triple-gate devices. The characteristics of these devices include [45-51]:

1. In triple-gate MOSFETs, there is a thin film with limited separation of silicon between the gate terminals. Triple-gate MOSFETs are a variety of classical MOSFETs with three gate terminals instead of one.

2. Triple-gate FETs include - wire SOI (Silicon-On-Insulator) MOSFET and tri-gate MOSFET.

3. The enhancement of electrostatic effect in triple-gate MOSFETs by modifying the sidewall segments of the gate cathode. It also mentions π-gate devices and Ω-gate MOSFETs, suggesting that these devices have a sustainable number of gates, possibly three or four.

4. The performance of MOSFETs can be improved by using stressed silicon and either metallic gates or dielectric materials with a high dielectric constant (high-k). Stressed silicon is a technique where mechanical stress is applied to the silicon crystal lattice to enhance carrier mobility, and using high-k materials in the gate insulator can improve the transistor's performance [39,40].

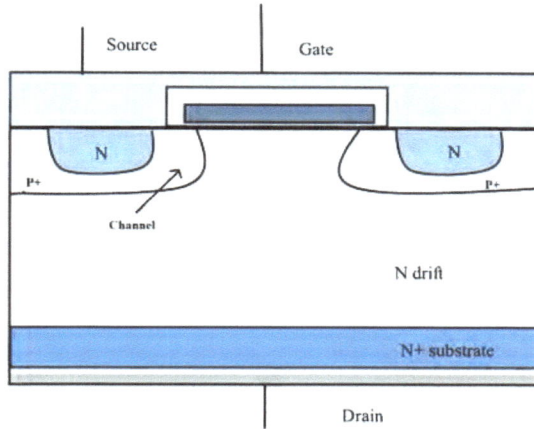

Fig. (7). Structure of DELTA MOSFET.

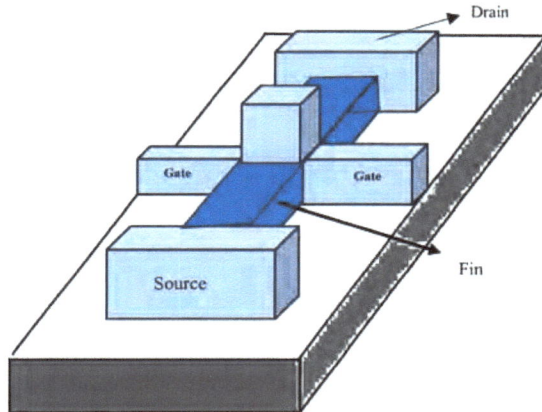

Fig. (8). Basic structure of FinFET.

FINFET AND ITS CHARACTERISTICS

The FinFET, depicted in Fig. (**9b**), is the most promising multi-gate device due to its structural benefit. The crucial feature is its conducting channel confined within a slim silicon fin [41], essentially constituting the device's body. These fins serve as channels connecting the source and drain. The gate terminal surrounds this channel [42], enabling the creation of multiple gate electrodes to mitigate leakage current and enhance drive current [43]. Fig. (**9**) shows the separation between conventional and FinFET structures. The operating principle of FinFET is identical to conventional FinFET. It operates in two modes: enhancement and depletion. The primary distinction is that in enhancement mode, without gate voltage, it remains non-conductive; in depletion mode, with gate voltage, it does not conduct [44]. In enhancement mode, applying voltage to the gate forms a collateral plate capacitor with the oxide layer as the gate. A depletion region is created beneath the oxide layer when a slight positive voltage is put in application to the gate, transforming the region into n-type, forming a region at the Si-SiO2 interface [45]. The applied positive voltage captivates electrons from the source to the drain, establishing an electron-rich channel. Current flow initiates by applying a voltage between the source and drain, and its magnitude depends on the gate voltage [46]. Fig. (**10**) shows the transfer characteristics of FinFET.

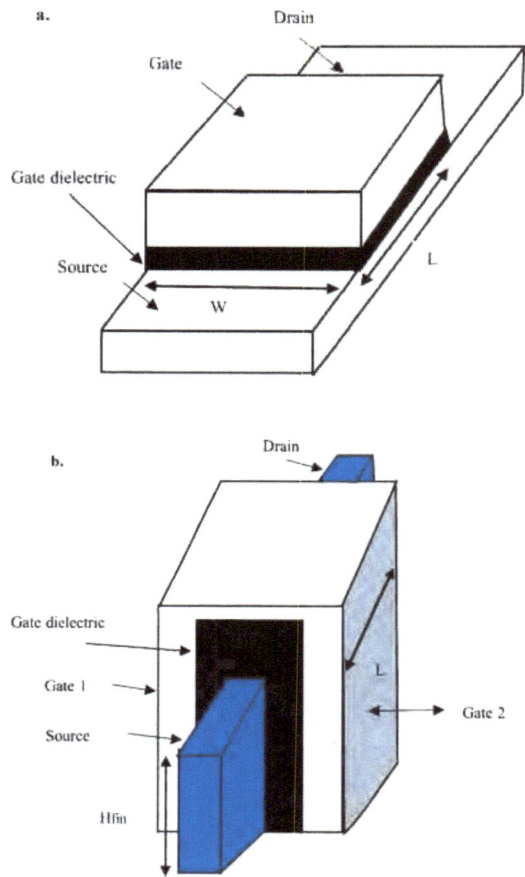

Fig. (9). Structural comparison between   **a.** MOSFET   **b.** FinFET [52,53].

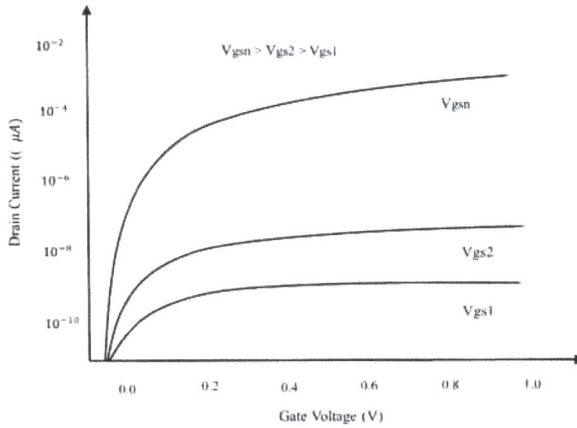

Fig. (10). Transfer characteristics curve of multi-gate FinFET [5].

Performance Analysis of Double-gate and Triple-gate FinFET

The Silvaco TCAD is the tool for simulation employed for examining the performance of both double and triple-gate FinFET, and the device Specification utilized for simulating 3D FinFET structure is mentioned in Table **2**. The DG Fin-FET and Tri-gate Fin-FET construction are nearly alike, except for the contrast that in DG FINFET, the gate oxide layer is broader at the top portion of the fin, resulting in only two gates remaining effective for controlling the channel (Fig. **11**).

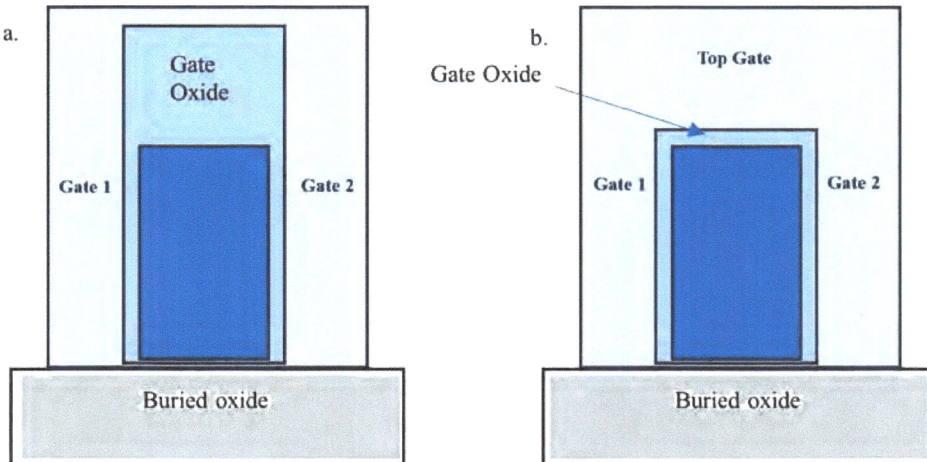

Fig. (11). Structural difference between **a.** Double-gate FinFET **b.** Trigate FinFET [47].

Table 2. Specification utilized for simulating 3D FinFET structure.

Gate Length (Lg)	30nm
Fin Width (Wfin)	10nm
Fin Height (Hfin)	60nm
Thickness of Oxide	1.5nm
Channel Doping	1E18
Drain/Source Doping	1E20
Gate Oxide	SiO2

DG FinFET

In Fig. (**12**), the drain current is illustrated as a varying function of gate voltage across various metal gate work functions [47]. The assessment of DG FinFET performance can be found in Table **3**, revealing that a lower gate work function corresponds to a further down threshold voltage, resulting in higher on and off currents. Conversely, a higher work function value shows minimal alteration in on-current but substantial changes in off-current, exhibiting an almost absolute subthreshold slope.

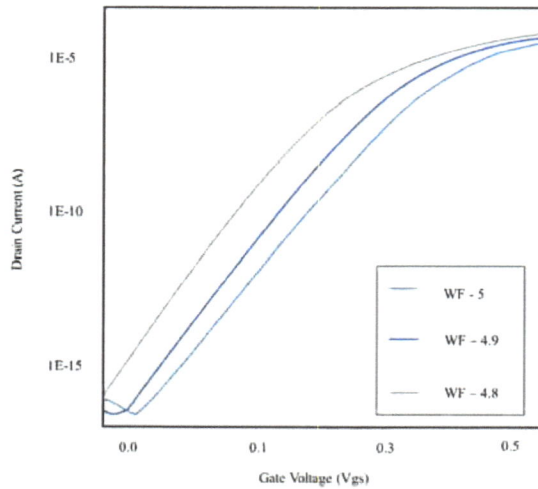

Fig. (12). Subthreshold slope of double-gate FinFET under various gate functions.

Table 3. Performance Analysis of DG FinFET.

Gate Work-function(eV)	Subthreshold Slope	Ion(mA)	Ioff(A)
4.8	65.05	0.3786	$6.20e^{-10}$
4.9	65.04	0.3194	$1.76e^{-11}$
5	64.94	0.2564	$5.03e^{-13}$

Trigate FinFET

As control over the channel improves, better results are seen in TG FinFET. This suggests that the different work functions of the metal gate are used to obtain the maximum threshold voltage and, as a result, the lowest on/off current. TG FinFET's subthreshold properties are demonstrated in Fig. (**13**). It is interesting that the TG FinFET has a greater on/off current than the DG FinFET for an equivalent metal gate work function [47].

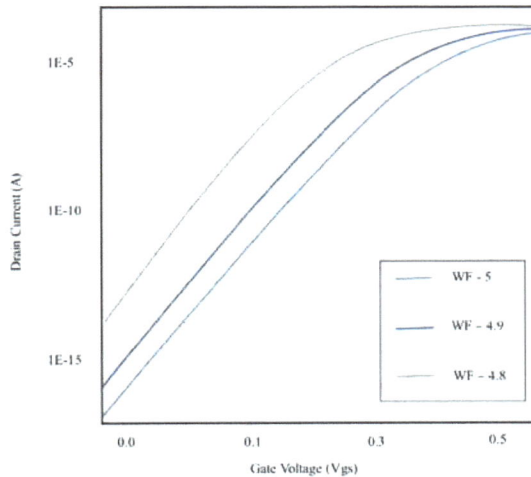

Fig. (13). Subthreshold slope of triple-gate FinFET under various gate functions.

Shortcomings of FINFET Technology

FinFET transistors have a three-dimensional (3D) structure where the gate wraps around the sides of a thin, vertical fin of semiconductor material. This allows for better electrostatic control compared to planar transistors, but the gate controls the channel primarily through the sides of the fin only, and that is not as complete as in GAA transistors, especially as transistors scale down in size. GAA technology is more recent compared to FinFETs and is currently being adopted in advanced semiconductor manufacturing processes due to its better electrostatic control over the channel.

Gate-All-Around FET System

A Gate-All-Around Field-Effect Transistor (GAA FET) is an advanced transistor design that represents a significant evolution beyond traditional transistor architectures. Unlike conventional planar transistors or even FinFETs, where the gate wraps around three sides of a fin-like structure, GAA FETs have a gate that surrounds the channel region completely [48]. This design offers several advantages, making it a promising candidate for the next generation of semiconductor devices. The examination involves simulating GAA FET to explore the potential for further device scaling and providing improved electrostatic control over the flow of electrical current characteristics. Table **4** below presents the device parameters.

Table 4. Device parameters of GAA.

Parameters	Value
Gate Length (Lg)	40nm
Thickness of Oxide	1nm
Channel Doping	$1e^{14}$
Drain/Source Doping	$1e^{20}$
Gate Oxide	HfO2
Metal Work function	4.2 eV

Silvaco simulation tool was used to analyze the electrostatic behavior of a Silicon all-around-gate FET, as depicted in Fig. (**14**). The behavior of the nanowire FET exhibits superior characteristics, especially in the off-state (Fig. **15**). The 'Ion' over 'Ioff' ratio increases as the channel length decreases. Additionally, the subthreshold slope approaches the physical thermal limit of such devices. The summary and overall analysis parameters of the device are shown in Table **5**.

Table 5. Performance analysis of gate-all-around.

Parameter	Result
Ion(mA)	$2.64e^{-5}$
Ioff(A)	$9.55e^{-12}$
Subthreshold Slope	65.93

The presented Table **5** displays key parameters from gate-all-around FET. The outcomes appear highly favorable when compared to theoretical expectations.

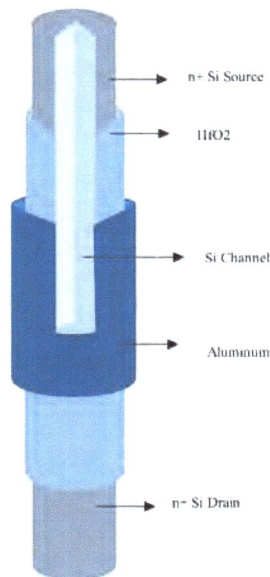

Fig. (14). Structure of Gate-All-Around.

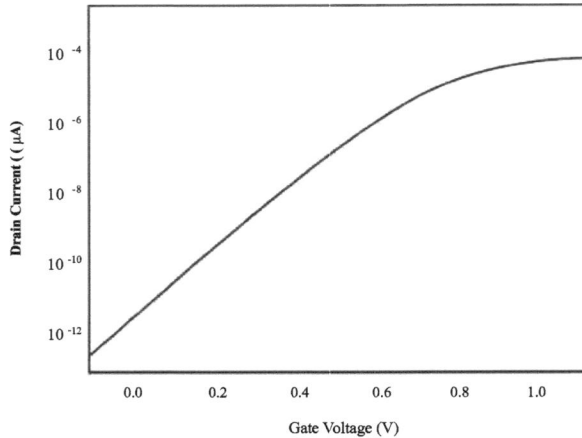

Fig. (15). Subthreshold slope of GAA.

IMPACT OF MULTI-GATE FET ON MODERN TECHNOLOGY

The application of multi-gate field-effect transistors (FETs) in modern technology represents a pivotal chapter in the ongoing evolution of semiconductor devices, ushering in a new era of performance, efficiency, and miniaturization. At the forefront of this revolution are multi-gate structures, exemplified by FinFETs and nanowire FETs, which have become instrumental in shaping the landscape of electronic devices across diverse industries. These transistors introduce a level of precision and control that goes beyond their traditional counterparts, fundamentally altering the capabilities of electronic systems [49].

The heart of the transformative impact of multi-gate FETs lies in their superior electrostatic control over the channel. This attribute translates into remarkable enhancements in processor performance, making them integral to computing devices ranging from personal electronics to high-performance servers. The intricate design of these transistors allows for faster switching speeds, reduced leakage currents, and improved overall efficiency. In an age where computational demands continue to escalate, multi-gate FETs provide a pathway to meet and exceed these requirements, enabling devices to handle complex tasks with unprecedented speed and energy efficiency [50, 51].

The implications of multi-gate FETs extend beyond raw processing power, addressing a critical need for energy-efficient solutions in portable electronic devices. By mitigating leakage currents and refining power control, these transistors contribute to the development of devices that not only perform at high levels but also optimize energy consumption. The result is a significant impact on battery life, a crucial factor in the user experience of smartphones, laptops, and other battery-powered gadgets [52, 53].

As we push the boundaries of transistor scaling, silicon-based transistors continue to face physical and practical limitations. While silicon has been the cornerstone of semiconductor technology for decades, we are approaching a point where further miniaturization may not yield the same performance improvements due to issues such as short-channel effects, leakage currents, and increased power consumption. Emerging materials, particularly two-dimensional (2D) semiconductors like graphene, molybdenum disulfide (MoS_2), and tungsten diselenide (WSe_2), offer promising alternatives. These materials exhibit remarkable electrical properties, including high carrier mobility and a tunable bandgap, which make them suitable for next-generation transistor applications. The atomic-scale thickness of 2D materials allows for the creation of transistors with reduced channel length without compromising on gate control, potentially enabling continued scaling beyond the limits of silicon [54].

Advanced technology nodes, such as 7nm and beyond, leverage the capabilities of these transistors to create smaller yet more powerful integrated circuits. This has profound implications for device design, enabling manufacturers to produce compact, high-performance electronic systems that defy the limitations of yesteryears. Memory devices, both dynamic and non-volatile, benefit immensely from the precision control over charge flow provided by multi-gate FETs, resulting in high-density storage solutions that cater to the escalating data demands of the modern era [55].

The impact of multi-gate FETs extends to the realm of wireless communication, where their high-frequency performance is harnessed to improve signal processing and transmission. In RF circuits, transceivers, and other wireless components, these transistors contribute to the seamless connectivity we experience in our daily interactions with a multitude of wireless devices. Their role is particularly vital in the development of next-generation communication technologies, including 5G and beyond.

In the nascent field of quantum computing, multi-gate FETs represent a foundational element in the control and manipulation of qubits. The precise control over electron flow at the nanoscale is paramount for the creation of stable qubits, the building blocks of quantum information processing. As researchers strive to unlock the immense computational potential promised by quantum computing, multi-gate FETs emerge as key enablers in this groundbreaking endeavor.

Venturing into the realm of nanotechnology, multi-gate FETs play a central role in the development of nanoelectronics, nanosensors, and nanorobotics. Their ability to confine and control electrons at the nanoscale is harnessed to create sensors with unprecedented sensitivity for applications in environmental monitoring, healthcare, and security. In nanorobotics, these transistors serve as nanoscale actuators and controllers, enabling precise manipulation and control of nanoscale devices. This has transformative implications for fields such as targeted drug delivery and nanomanufacturing, where precision at the molecular level is paramount.

The application of multi-gate FETs in modern technology is a multifaceted story of innovation and progress. These transistors have become integral to the very fabric of our electronic devices, from the processors that power our computers to the sensors that enhance our healthcare capabilities. Their impact resonates across industries, driving advancements in computing, communication, energy efficiency, and nanotechnology. As technology continues to advance, the versatile capabilities of multi-gate FETs position them as cornerstones of the digital revolution, propelling us towards a future where electronic devices are not just powerful but also sustainable, efficient, and finely tailored to the dynamic demands of our technological ecosystem.

STRUCTURAL OBSTACLES AND IMPROVEMENT

The shift from single-gate to multi-gate field-effect transistors (FETs) in semiconductor technology is accompanied by a host of structural challenges that demand innovative solutions. In single-gate FETs, a critical obstacle arises from the diminishing thickness of the gate oxide as transistor dimensions approach the atomic scale. This reduction in gate oxide thickness brings about challenges related to quantum tunneling and increased leakage currents, adversely affecting the reliability and power efficiency of the transistors. As the industry pushes the limits of miniaturization in pursuit of higher performance and energy efficiency, overcoming these challenges is imperative for the continued advancement of semiconductor technology.

On the other hand, the structural transition to multi-gate FETs, exemplified by designs like FinFETs and nanowire FETs, introduces its own set of obstacles. One of the primary challenges lies in the intricate three-dimensional architecture of these devices. Fabricating nanoscale fin-like structures or nanowires with precision demands advanced manufacturing techniques and poses challenges in terms of cost-effectiveness. The complexity of creating and aligning these three-dimensional structures adds layers of difficulty to the fabrication process, necessitating a paradigm shift in manufacturing methodologies.

The gate-all-around (GAA) configuration, a prevalent form of multi-gate FET, involves surrounding the channel with the gate material, which significantly complicates the manufacturing process [48]. Achieving uniformity and precision in creating these structures becomes increasingly challenging as the size of the components decreases. Additionally, the adoption of different materials for the gate and channel introduces compatibility concerns and requires novel approaches to ensure seamless integration of these structures into the existing semiconductor processes.

To overcome these structural challenges, researchers and engineers are exploring various strategies. One key focus is the development of advanced materials that can address the limitations imposed by gate oxide thickness. Innovations in materials science aim to introduce compounds with superior electrical properties, reduced leakage currents, and enhanced thermal stability. These materials not only contribute to the overall performance improvement of single-gate FETs but also play a vital role in the success of multi-gate FETs, where material compatibility becomes a critical consideration [50].

In the realm of multi-gate FETs, advancements in manufacturing techniques are pivotal to overcoming structural obstacles. Nanotechnology, with its precision and control at the atomic and molecular levels, is increasingly employed to enhance the fabrication process. Techniques such as extreme ultraviolet lithography (EUVL) enable the creation of smaller features with higher accuracy, addressing the challenges associated with the three-dimensional structures of FinFETs and nanowire FETs. Moreover, innovations in deposition and etching processes contribute to the creation of intricate structures, allowing for the precise alignment and formation of nanoscale components [50-54].

Collaborative efforts within the semiconductor industry also play a crucial role in overcoming structural obstacles. Standardizing processes for the integration of multi-gate FETs into existing technologies is essential to ensure compatibility and ease of adoption. Industry consortia and research partnerships facilitate the sharing of knowledge, best practices, and technological advancements, accelerating the pace at which structural challenges are addressed [55].

Another avenue of exploration involves the design and implementation of novel transistor architectures. Beyond the FinFET and nanowire FET designs, researchers are exploring alternative structures, such as gate-all-around (GAA) FETs, to further optimize performance and mitigate structural challenges. GAA FETs, by virtue of their three-dimensional gate structure, offer improved control over the flow of current and reduced leakage, presenting a promising avenue for overcoming the limitations of traditional transistor designs [51-56].

The structural challenges in transitioning from single-gate to multi-gate FETs are formidable, encompassing issues related to gate oxide thickness, three-dimensional architectures, and material compatibility. However, ongoing research and development initiatives, coupled with advancements in materials science, manufacturing techniques, and collaborative industry efforts, are steadily overcoming these obstacles. As the semiconductor industry continues to push the boundaries of innovation, the assessment from single-gate to multi-gate FETs remains at the forefront, promising enhanced performance, energy efficiency, and the continued miniaturization of electronic devices.

CONCLUSION

The physics of multi-gate FETs, such as FinFETs and nanowire FETs, offer significant advantages over traditional MOSFETs. By leveraging multiple gates to control the channel, these devices enhance electrostatic control, reduce leakage currents, improve gate control over short channel effects, and enable better scaling beyond the limitations of planar transistors. As a result, multi-gate FETs hold immense promise for advancing the performance and efficiency of future electronic devices and integrated circuits.

Learning Objectives/Key Points

(1) **Enhanced Control:** Multi-gate FETs provide enhanced control over the flow of current compared to traditional single-gate FETs. This is achieved by having multiple gates surrounding the semiconductor channel, allowing for better modulation of the channel conductivity.

(2) **Improved Electrostatic Integrity:** The presence of multiple gates helps in improving the electrostatic integrity of the device. This reduces leakage currents and improves device reliability.

(3) **Various Configurations:** Multi-gate FETs can be configured in different ways, such as double-gate (DG), triple-gate (TG), and FinFETs (also known as tri-gate FETs). Each configuration offers specific advantages in terms of performance and scalability.

(4) Better Subthreshold Behavior: Multi-gate FETs exhibit improved subthreshold behavior compared to single-gate FETs. This leads to lower off-state leakage currents and better energy efficiency, making them suitable for low-power applications.

(5) Improved Gate Capacitance Control: With multiple gates, it is possible to control the gate capacitance more precisely. This allows for better tuning of device performance, including speed and power consumption.

(6) Reduced Short-Channel Effects: Multi-gate FETs help in mitigating short-channel effects such as drain-induced barrier lowering (DIBL) and subthreshold slope degradation. This is particularly important for scaling down device dimensions and achieving higher packing densities.

(7) Increased Drive Current: The multiple gates in multi-gate FETs enable higher drive currents compared to single-gate FETs. This is beneficial for high-performance applications where increased current-carrying capacity is required.

(8) Improved Gate-Channel Control: Multi-gate FETs offer better gate-channel control, leading to improved electrostatics and reduced variability in device characteristics. This contributes to better device matching and overall circuit performance.

Multiple Choice Questions (MCQs)

1. What is the primary advantage of multi-gate FETs over traditional FETs?

a) Lower power consumption

b) Higher integration density

c) Faster switching speeds

d) Greater linearity

Answer: b) Higher integration density

2. In a FinFET structure, the channel is wrapped around by gates on how many sides?

a) One

b) Two

c) Three

d) Four

Answer: c) Three

3. What is the purpose of having multiple gates in a multi-gate FET?

a) To increase channel doping

b) To reduce channel length

c) To increase gate capacitance

d) To decrease drain-source resistance

Answer: b) To reduce channel length

4. Which multi-gate FET structure provides better control over the channel and, hence, improved electrostatic integrity?

a) FinFET

b) SOI FET

c) Tri-gate FET

d) Trench FET

Answer: c) Tri-gate FET

5. In a multi-gate FET, which gate is typically used to control the channel current?

a) Source gate

b) Drain gate

c) Top gate

d) Bottom gate

Answer: c) Top gate

6. Which multi-gate FET structure is known for its excellent electrostatic integrity and superior control over short-channel effects?

a) Trench FET

b) FinFET

c) SOI FET

d) Tri-gate **FET**

Answer: d) Tri-gate FET

7. The term "fin" in FinFET refers to:

a) The number of gates surrounding the channel

b) The thin vertical silicon structure resembling a fin

c) The fineness of the channel doping

d) The final stage of fabrication

Answer: b) The thin vertical silicon structure resembling a fin

8. What is the primary reason for the improved electrostatic integrity of multi-gate FETs?

a) Reduced drain-source capacitance

b) Enhanced gate-source capacitance

c) Increased gate-source voltage

d) Reduced gate-source leakage

Answer: b) Enhanced gate-source capacitance

9. Which of the following statements about multi-gate FETs is true?

a) They are immune to short-channel effects.

b) They have a larger footprint compared to traditional FETs.

c) They have higher parasitic capacitance.

d) They are only suitable for low-power applications.

Answer: a) They are immune to short-channel effects.

10. Which of the following is NOT a key challenge in the fabrication of multi-gate FETs?

a) Achieving uniform gate control

b) Minimizing parasitic capacitance

c) Controlling channel depth

d) Ensuring low leakage current

Answer: d) Ensuring low leakage current

11. In a multi-gate FET, how does scaling down the gate length affect the threshold voltage?

a) Decreases

b) Increases

c) Unchanged

d) Depends on the gate material

Answer: a) Decreases

12. What is the natural length of a multi-gate FET?

a) Equal to the gate width

b) Greater than the gate width

c) Smaller than the gate width

d) Independent of the gate width

Answer: c) Smaller than the gate width

13. What factor determines the natural length of a multi-gate FET?

a) Gate material

b) Gate-source voltage

c) Channel material

d) Gate oxide thickness

Answer: d) Gate oxide thickness

14. What is the primary mechanism responsible for tunneling current in multi-gate FETs?

a) Fowler-Nordheim tunneling

b) Smith-Robinson tunneling

c) Shockley-Read-Hall tunneling

d) Ohmic tunneling

Correct Answer: a) Fowler-Nordheim tunneling

15. How does temperature affect tunneling current in multi-gate FETs?

a) Tunneling current decreases with increasing temperature

b) Tunneling current increases with increasing temperature

c) Tunneling current remains constant regardless of temperature

d) Temperature has an unpredictable effect on tunneling current

Correct Answer: b) Tunneling current increases with increasing temperature

REFERENCES

[1] K. Sivasankaran, and P.S. Mallick, *Multigate Transistors for High Frequency Applications.* Springer Nature Singapore: Germany, 2023.

http://dx.doi.org/10.1007/978-981-99-0157-9

[2] G. Moore, "Cramming more components onto integrated circuits", *Electronics (Basel),* vol. 38, p. 114, 1965.

[3] M.S. Equbal and S. Sahay, "Scaling the MOSFET: detrimental short channel effects and mitigation techniques," *Micro and Nano Technologies,* pp. 11–37, 2023.

[4] B. Majkusiak, "Physics of the Multigate MOS System", In: J.P. Colinge, Ed., *FinFETs and Other Multi-Gate Transistors.* Springer: Boston, MA, 2008.

http://dx.doi.org/10.1007/978-0-387-71752-4_4

[5] S. Sen, A. Raman, and M. Khosla, "Design and Challenges in TFET," *Advanced Field-Effect Transistors*, 1st ed. Boca Raton, FL: CRC Press, pp. 1–23, 2023.

http://dx.doi.org/10.1201/9781003393542-2

[6] Y. Taur, and T.H. Ning, *Fundamentals of modern VLSI devices.* Cambridge University Press, 2009.

http://dx.doi.org/10.1017/CBO9781139195065

[7] G.K. Celler, and S. Cristoloveanu, "Frontiers of silicon-on-insulator", *J. Appl. Phys.,* vol. 93, no. 9, pp. 4955-4978, 2003.

http://dx.doi.org/10.1063/1.1558223

[8] S. Cristoloveanu, "From SOI basics to nano-size MOSFETs," in *Nanotechnology for Electronic Materials and Devices.* Springer, pp. 67–104, 2007.

http://dx.doi.org/10.1007/978-0-387-49965-9_2

[9] P. Vimala and N.B. Balamurugan, "Quantum mechanical compact modeling of symmetric double-gate MOSFETs using variational approach," *Journal of Semiconductors*, vol. 33, no. 3, pp. 1–6, 2012.

http://dx.doi.org/10.1088/1674-4926/33/3/034001

[10] J.P. Colinge, "Multiple-gate SOI MOSFETs", *Solid-State Electron.,* vol. 48, no. 6, pp. 897-905, 2004.

http://dx.doi.org/10.1016/j.sse.2003.12.020

[11] C.W. Lee, S-R-N. Yun, C-G. Yu, J-T. Park, and J-P. Colinge, "Device design guidelines for nano-scale MuGFETs", *Solid-State Electron.,* vol. 51, no. 3, pp. 505-510, 2007.

http://dx.doi.org/10.1016/j.sse.2006.11.013

[12] H.K. Lim, and J.G. Fossum, "Threshold voltage of thin-film silicon-on insulator (SOI) MOSFETs", *IEEE Trans. Electron. Devices,* vol. 30, no. 10, p. 1244-1251, 1983.

[13] K.K. Young, "Analysis of conduction in fully depleted SOI MOSFETs," *IEEE Trans. Electron. Devices*, vol. 36, no. 3, pp. 504–506, 1989.

http://dx.doi.org/10.1109/16.19960

[14] R.H. Yan, A. Ourmazd, and K.F. Lee, "Scaling the Si MOSFET: from bulk to SOI to bulk," *IEEE Trans. Electron. Devices*, vol. 39, no. 7, pp. 1704–1710, 1992.

http://dx.doi.org/10.1109/16.141237

[15] R.H. Yan, A. Ourmazd, and K.F. Lee, "Scaling the Si MOSFET: from bulk to SOI to bulk", *IEEE Trans Electron. Dev.,* vol. 39, no. 7, pp. 1704-1710, 1992.

http://dx.doi.org/10.1109/16.141237

[16] C.P. Auth, and J.D. Plummer, "Scaling theory for cylindrical, fully-depleted, surrounding-gate MOSFET's", *IEEE Electron. Device Lett,* vol. 18, no. 2, pp. 74-76, 1997.

http://dx.doi.org/10.1109/55.553049

[17] I. Ferain, C.A. Colinge, and J.P. Colinge, "Multigate transistors as the future of classical metal–oxide–semiconductor field-effect transistors", *Nature,* vol. 479, no. 7373, pp. 310-316, 2011.

http://dx.doi.org/10.1038/nature10676 PMID: 22094690

[18] P. Kollamudi and S.R. Karumuri, "Technology behind junctionless semiconductor devices," in *Field Effect Transistors*, P. Suveetha Dhanaselvam, K. Srinivasa Rao, S.B. Rahi, and D.S. Yadav, Eds., pp. 105–124, 2025.

[19] K. Suzuki, T. Tanaka, Y. Tosaka, H. Horie, and Y. Arimoto, "Scaling theory for double-gate SOI MOSFET's", *IEEE Trans Electron. Devices.,* vol. 40, no. 12, pp. 2326-2329, 1993.

http://dx.doi.org/10.1109/16.249482

[20] H.K. Lim, and J.G. Fossum, "Threshold voltage of thin-film silicon-on insulator (SOI) MOSFETs", *IEEE Transactions on Electron. Devices,* vol. 30, no. 10, p. 1244, 1983.

[21] W. Xiong, C.R. Cleavelin, T. Schulz, K. Schrufer, P. Patruno, and J.P. Colinge, "MuGFET CMOS process with midgap gate material", *Abstracts of NATO International Advanced Research Workshop Nanoscaled Semiconductor-on-Insulator Structures and Devices,* 2006, p. 96

[22] J.P. Colinge, "From gate-all-around to nanowire MOSFETs," in *2007 International Semiconductor Conference*, Sinaia: Romania, 2007, pp. 11–17.

http://dx.doi.org/10.1109/SMICND.2007.4519637

[23] D. A. Neamen, *Semiconductor Physics and Devices.* New York: The McGraw-Hill companies, 1992.

[24] K. Roy, S. Mukhopadhyay, and H. Mahmoodi-Meimand, "Leakage current mechanisms and leakage reduction techniques in deep-submicrometer CMOS circuits", *Proc IEEE,* vol. 91, no. 2, pp. 305-327, 2003.

http://dx.doi.org/10.1109/JPROC.2002.808156

[25] A.A. Khan, A. Audhikary, M.F. Al-Fattah, M.A. Amin, and R. Nandi, "A comparative analytical approach for gate leakage current optimization in silicon MOSFET: A step to more reliable electronic device," in *Proc. 3rd Int. Conf. Electrical Engineering and Information Communication Technology (ICEEICT)*, Dhaka, Bangladesh, 2016.

[26] F. Balestra, S. Cristoloveanu, M. Benachir, J. Brini, and T. Elewa, "Double-gate silicon-on-insulator transistor with volume inversion: A new device with greatly enhanced performance", *IEEE Electron. Device Lett,* vol. 8, no. 9, pp. 410-412, 1987.

http://dx.doi.org/10.1109/EDL.1987.26677

[27] Y. Omura, S. Horiguchi, M. Tabe, and K. Kishi, "Quantum-mechanical effects on the threshold voltage of ultrathin SOI nMOSFETs", *IEEE Electron. Device Lett,* vol. 14, no. 12, p. 569, 1993.

[28] T. Poiroux, M. Vinet, O. Faynot, J. Widiez, J. Lolivier, T. Ernst, B. Previtali, and S. Deleonibus, "Multiple gate devices: advantages and challenges", *Microelectron Eng.,* vol. 80, pp. 378-385, 2005.

http://dx.doi.org/10.1016/j.mee.2005.04.095

[29] P. Kiran Kumar, B. Balaji, M. Suman, P. Syam Sundar, E. Padmaja, and K. Girija Sravani, "A detailed roadmap from conventional-MOSFET to nanowire-MOSFET," in *Machine Learning for VLSI Chip Design*, A. Kumar, S.L. Tripathi, and K. Srinivasa Rao, Eds., 2023

http://dx.doi.org/10.1002/9781119910497.ch5

[30] E. Landgraf, W. Rösner, M. Städele, L. Dreeskornfeld, J. Hartwich, F. Hofmann, J. Kretz, T. Lutz, R.J. Luyken, T. Schulz, M. Specht, and L. Risch, "Influence of crystal orientation and body doping on trigate transistor performance", *Solid-State Electron.,* vol. 50, no. 1, pp. 38-43, 2006.

http://dx.doi.org/10.1016/j.sse.2005.10.041

[31] J.P. Colinge, "Novel Gate Concepts for MOS Devices", *Proceedings of ESSDERC,* vol. 45, 2004

[32] T.S. Park, S. Choi, and D.H. Lee, "Fabrication of body-tied FinFETs (Omega MOSFETs) using bulk Si wafers", *IEEE Symposium on VLSI Technology,* 2003, pp. 135-136

http://dx.doi.org/10.1109/VLSIT.2003.1221122

[33] R. Ritzehnthaler, O. Faynot, C. Jahan, A. Kuriyama, S. Deleonibus, and S. Cristoloveanu, "Coupling effects in FinFETs and triple-gate FETs," in *Proc. 7th Eur. Workshop on Ultimate Integration of Silicon (ULIS)*, 2006, pp. 25–28.

[34] M. Stadele, R.J. Luyken, M. Roosz, M. Specht, W. Rosner, and L. Dreeskornfeld, "A comprehensive study of corner effects in tri-gate transistors," in *Proc. 34th Eur. Solid-State Circuits Conf. (ESSCIRC)*, 2004, pp. 165–168.

http://dx.doi.org/10.1109/ESSDER.2004.1356515

[35] F.J. Garcia Ruiz, A. Godoy, F. Gamiz, C. Sampedro, and L. Donetti, "A comprehensive study of the corner effects in Pi-gate MOSFETs including quantum effects", *IEEE Trans Electron. Dev.,* vol. 54, no. 12, pp. 3369-3377, 2007.

http://dx.doi.org/10.1109/TED.2007.909206

[36] P. Singh and D.S. Yadav, "Assessment of temperature and ITCs on single gate L-shaped tunnel FET for low power high frequency application," *Eng. Res. Express*, vol. 6, no. 1, 2024.

http://dx.doi.org/10.1088/2631-8695/ad32b0

[37] J.G. Fossum, J.W. Yang, and V.P. Trivedi, "Suppression of corner effects in triple-gate MOSFETs", *IEEE Electron. Device Lett,* vol. 24, no. 12, pp. 745-747, 2003.

http://dx.doi.org/10.1109/LED.2003.820624

[38] J.P. Colinge, J.W. Park, and W. Xiong, "Threshold voltage and subthreshold slope of multiple-gate SOI MOSFETs", *IEEE Electron. Device Lett,* vol. 24, no. 8, pp. 515-517, 2003.

http://dx.doi.org/10.1109/LED.2003.815153

[39] D. Hisamoto, *Electron. Devices Meeting. Technical Digest,* 2001, p. 429.https://semiengineering.com/knowledge_centers/integrated-circuit/transistors/3d/gate-all-around-fet/

[40] J. Kedzierski, M. Ieong, T. Kanarsky, Y. Zhang, and H-S.P. Wong, "Fabrication of metal gated FinFETs through complete gate silicidation with Ni", *IEEE Trans Electron. Dev.,* vol. 51, no. 12, pp. 2115-2120, 2004.

http://dx.doi.org/10.1109/TED.2004.838448

[41] P. Singh, and D.S. Yadav, "Impact of work function variation for enhanced electrostatic control with suppressed ambipolar behavior for dual gate L-TFET", *Curr. Appl. Phys.,* vol. 44, pp. 90-101, 2022.

http://dx.doi.org/10.1016/j.cap.2022.09.014

[42] S.S. Richter, S. Trellenkamp, M. Schmidt, A. Schäfer, K.K. Bourdelle, Q.T. Zhao, and S. Mantl, "Strained silicon nanowire array MOSFETs with high-k/metal gate," in *Proc. 13th Int. Conf. Ultimate Integration on Silicon (ULIS)*, 2012, pp. 75–78.

http://dx.doi.org/10.1109/ULIS.2012.6193355

[43] L. Knoll, A. Schäfer, S. Trellenkamp, K.K. Bourdelle, Q.T. Zhao, and S. Mantl, "Nanowire and planar UTB SOI Schottky barrier MOSFETs with dopant segregation," in *Proc. 13th Int. Conf. Ultimate Integration on Silicon (ULIS)*, 2012, pp. 67–70.

http://dx.doi.org/10.1109/ULIS.2012.6193353

[44] P. Singh and D.S. Yadav, "Design and investigation of F-shaped tunnel FET with enhanced analog/RF parameters," *Silicon*, vol. 14, pp. 6245–6260, 2022.

[45] Tsu-Jae King Liu, "Introduction to Multi-Gate MOSFETs", *Department of Electrical Engineering and Computer Sciences, University of California, Berkeley, CA.*

[46] M.A. Guillorn, "A 0.021 µm2 trigate SRAM cell with aggressively scaled gate and contact pitch", *Symposium on VLSI Technology,* 2011, pp. 64-65.

[47] V. Basker, "A 0.063 µm2 FinFET SRAM cell demonstration with conventional lithography using a novel integration scheme with aggressively scaled fin and gate pitch", *Symposium on VLSI Technology,* , 2010pp. 19-20

[48] P. Singh and D.S. Yadav, "Assessment of temperature and ITCs on single gate L-shaped tunnel FET for low power high frequency application," *Eng. Res. Express*, vol. 6, no. 1, 2024.

http://dx.doi.org/10.1088/2631-8695/ad32b0

[49] S. Cao, J.H. Chun, A.A. Salman, S.G. Beebe, and R.W. Dutton, "Gate-controlled field-effect diodes and silicon-controlled rectifier for charged-device model ESD protection in advanced SOI technology", *Microelectron Reliab,* vol. 51, no. 4, pp. 756-764, 2011.

http://dx.doi.org/10.1016/j.microrel.2010.11.013

[50] S. L. Tripathi, Ramanuj Mishra, and R. A. Mishra, "Characteristic comparison of connected DG FINFET, TG FINFET and Independent Gate FINFET on 32 nm technology",

[51] Nagalakshmi Yarlagadda, Yogesh Kumar Verma, Manoj Singh Adhikari, and Varun Mishra, "A Review of Multi-material, Multi-gate MOSFET Structures", *Opto-VLSI Devices and Circuits for Biomedical and Healthcare Applications,* pp. 29-40, .

http://dx.doi.org/10.1201/9781003431138-3

[52] M.S. Narula and A. Pandey, "A comprehensive review on FinFET, gate all around, tunnel FET: Concept, performance and challenges," in *Proc. 8th Int. Conf. Signal Process. Commun. (ICSC)*, Noida, India, 2022, pp. 554–559.

http://dx.doi.org/10.1109/ICSC56524.2022.10009504

[53] Y. Sun, X. Yu, R. Zhang, B. Chen, and R. Cheng, "The past and future of multi-gate field-effect transistors: Process challenges and reliability issues", *J. Semicond,* vol. 42, no. 2, p. 023102, 2021.

http://dx.doi.org/10.1088/1674-4926/42/2/023102

[54] J. Lin, "Advancement and Challenges of Field Effect Transistors based on Multi-gate Transistor", In: *Journal of Physics: Conference Series,* vol. 2370. 2022, no. 1, p. 012004.

http://dx.doi.org/10.1088/1742-6596/2370/1/012004

[55] R.K. Maurya, and B. Bhowmick, "Review of FinFET Devices and Perspective on Circuit Design Challenges", *Silicon,* vol. 14, no. 11, pp. 5783-5791, 2022.

http://dx.doi.org/10.1007/s12633-021-01366-z

[56] R.S. Pal, S. Sharma, and S. Dasgupta, "Recent trend of FinFET devices and its challenges: A review," in *Proc. Conf. Emerging Devices and Smart Systems (ICEDSS)*, Mallasamudram, India, 2017, pp. 150–154.

http://dx.doi.org/10.1109/ICEDSS.2017.8073675

CHAPTER 2

Emerging and Future Prospective of Carbon Nanotube FETs

Srividya P.[1]

[1]Department of Electronics and Communication Engineering, RV College of Engineering, Bengaluru, Karnataka 560059, India

Abstract: In the last few years, the semiconductor industry has brought about a drastic revolution in the existing technology in order to realize a larger on-chip integration, enhance performance, increase operating speed, and decrease energy consumption. Delay and power consumption have become the most vital performance parameters of any digital circuit. One of the methods devised to achieve this is by scaling the feature size of a transistor. However, when the channel length is reduced beyond 45nm in metal-oxide-semiconductor field-effect transistors (MOSFETs) technology, it gives rise to perilous complications and challenges such as decreased gate control, short channel effect, increased power density, higher sensitivity to process deviation, higher manufacturing cost, and increased leakage current. This draws a limit on the transistor size and demands for new transistor structures and technologies to overcome the drawbacks. Technologies like benzene rings, single electron transistors (SET), Quantum-dot cellular automata (QCA), and carbon nanotubes are slowly rising as alternatives to reduce the problems associated with CMOS. New technologies demand faster processors, smaller sizes, and less power consumption. Advances in 5G networks have increased the pressure to improve the battery life of smartphones, their performance, spectral efficiency, and many more. The potential to achieve these is the use of Carbon Nanotube Field Effect Transistors (CNTFETs). They have higher carrier mobilities and direct band gaps that enhance the band-to-band tunneling and optical properties. These features make CNTFETs suitable to be used in future novel electronic devices. Hence, this chapter focuses on the emerging and future trends of CNTFETs. The constructional aspects, features, types, designs, and applications of CNTFETs are dealt with in detail in the forthcoming sections of the chapter.

Keywords: CNTFET, CNTs, Chiral vectors, Dielectric materials, High K dielectrics, Short channel effects.

* **Corresponding author Srividya P.:** Department of Electronics and Communication Engineering, RV College of Engineering, Bengaluru, Karnataka 560059, India; E-mail: subarnamondal.39@gmail.com

Dharmendra Singh Yadav & Prabhat Singh (Eds.)

INTRODUCTION

The consistent scaling of both active and passive electronic components that are unified on ICs has paved the way for an exponential growth of silicon-based microelectronics in the last few decades. Scaling of the feature size has reduced the size of the transistor, decreased the burden on testing, reduced the cost, and has made the switching faster, in addition to other advantages like low power and reliable designs. The growth is in accordance with Moore's law, which says that the integration of transistors on a chip doubles every 18 months [1].

At present, the technology node has reached a scale of 14nm, and further scaling down the channel length will lead to critical short-channel effects and reliability issues. Increment in the metallic on-chip interconnects, and power dissipation also makes the MOSFETs unsuitable for advanced applications. To overcome the short-channel effects of traditional MOSFETs, double-gate MOSFETs and FINFETs have emerged. In such devices, the gate is placed on two or three sides of the channel in order to establish higher control over the channel. The arrangement also brings about a considerable decrease in the leakage current that exists between the drain and the source [2, 3].

Propagation delay, area, cost, and reliability are the main concerns at present for VLSI design engineers. With the increased usage of portable devices like mobile and laptops, lower energy-consuming devices are in huge demand. As compared to microelectronic technology, developments in battery technology are slower. The situation has been further worsened by the fact that the clock speed of the microprocessor has reached the GHz limit, which paves the way to increased energy consumption. Hence, there is a requirement to develop VLSI circuits to reduce power dissipation, and in the near future, the technology, however, is expected to drift from conventional silicon-based MOSFETs and FINFETs to Nanowire Field Effect Transistor (NWFET) [1], Graphene Field Effect Transistor (GFET), and Carbon Nanotube Field Effect Transistor (CNTFET), owing to the critical issues faced by FINFETs like the difficulty in manufacturing thin fins and controllability of the channel in advanced nodes [4].

NWFET is a device in which the channel is surrounded by a gate all around and a semiconducting nanowire of diameter about 0.5nm is used as channel material. The nanowire may be made with Si, SiC, germanium, and ZnO semiconductors. The tiny diameter of the nanowire helps in reducing the short channel effects, as well as in 1-D conduction. NWFET has the potential to replace MOSFET owing to its improved performance and low power when the technology nodes are to be scaled beyond 7nm. NWFETs can be designed both horizontally (H-NWFET) and vertically (V-NWFET), as depicted in Fig. (1). The circuit design of H-NWFET is similar to that of FINFET. This feature makes it a potential candidate to replace FINFET. In contrast, V-NWFET has a radical design style and occupies a small area.

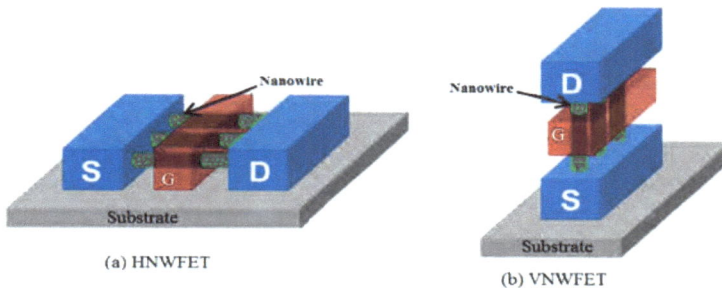

(a) HNWFET

(b) VNWFET

Fig. (1). H-NWFET and V-NWFET structures [6].

However, the challenge faced by the device is in the diffused PN junction fabrication. Although a metal drain-source junction is used, it still faces problems with a large off-state current.

2D materials have the advantages of being small in size and having low cost and mass fabrication capacity. They also have excellent mechanical strength and are highly flexible; hence, they have become promising blocks for many devices. Graphene is a classic example of a 2D material.

The graphene field effect transistor consists of a source, drain, and top or back gate as a MOSFET, as depicted in Fig. (2). But unlike the MOSFET structure, it has a thin graphene channel between the source and drain, which is usually tens of microns thick. Graphene is usually found as a single carbon atom layer material that is tightly packed in a 2-D honeycomb hexagonal lattice. Graphene offers superior electrical properties, higher thermal conductivity, good optical properties, and excellent chemical properties. It has the capacity to deliver 100 to 1000 times better performance than Si-based MOSFET. Despite advantages, graphene-based FETs have setbacks like low on-off current ratio, no bandgap, which makes it work as a metal, and insufficient saturating current that stops the device from attaining maximum voltage gain and oscillation frequency in Radio Frequency (RF) applications [7, 8].

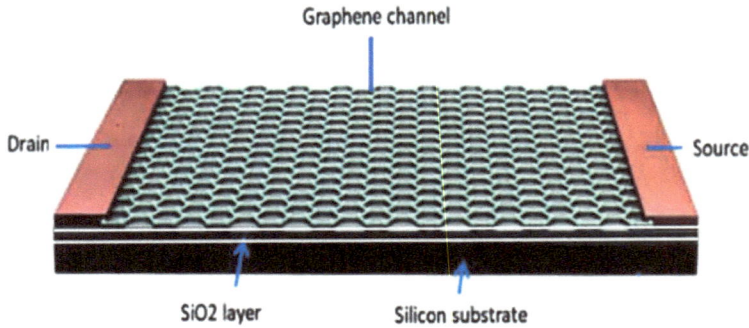

Fig. (2). Graphene field effect transistor.

CNTFETs use a cylindrical structure made of carbon atoms called carbon nanotubes as channels. The carbon atoms are arranged in a hexagonal lattice. CNTFETs use single or multiple semiconducting single-walled carbon nanotubes as channel material. On top of the CNT channel, a gate electrode is placed by separating it using a gate dielectric material. The diameter of CNT varies between 1nm and 10nm. The schematic edifice of CNTFET is depicted in Fig. (**3**). CNTFETs are apt for numerous applications as they exhibit matchless electrical, thermal, and mechanical properties [9-14].

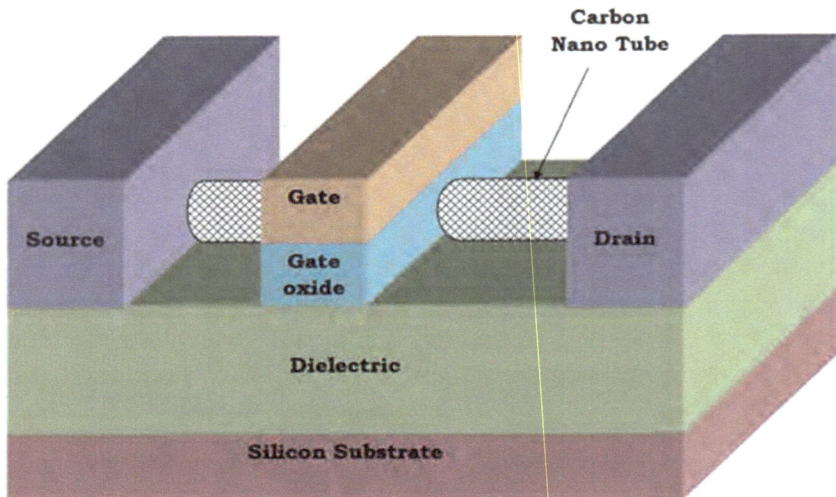

Fig. (3). Carbon Nanotube FET structure.

CNTFET has proven to be a potential candidate to replace silicon devices owing to its outstanding properties like superior performance, low power design, high current density, higher trans-conductance, larger sub-threshold slope, low off current, high carrier mobility, high-K dielectric material integration, and ballistic transport operation in the channel [15]. These properties lead to achieving strong chemical bonding, higher thermal conduction, higher chemical stability, good matching of P-type and N-type CNTFETs, and better gate electrostatics, finally leading to smaller transistor sizing while designing complex circuits [16]. Comparison between MOSFET and CNTFET is listed in Table **1**.

Table 1. MOSFET and CNTFET comparison.

Characteristics	MOSFET	CNTFET
Switching	By altering channel resistivity	By modulation of contact resistance
Drive current	Lower	Typically three times higher than MOSFET
Trans conductance	Lower	Four times higher than MOSFET
Carrier velocity	Lower	Twice compared to MOSFET

Traditional MOSFETs use bulk silicon, whereas CNTFETs use Carbon nanotubes (CNTs) sandwiched between the source and the drain of the MOSFET structure. This ensures CNTFETs are able to provide higher current carrier mobility and higher drive current density. CNTFETs that consist of CNTs exhibit great promise for the forthcoming generation of integrated transistors in addition to the traditional CMOS [17, 18].

CNTs are continuous nanotubes that are made with single or manifold layers of graphene sheets rolled about a central axis. The predominant advantage of the structure is that it is lightweight and has a perfect hexagonal assembly structure. Its distinguished electronic transport property makes it a potential candidate for nanodevices. Another distinguishing feature of CNTs is they are atomically thin. This provides higher electrostatic control over the channel and makes it suitable when the device is scaled down. Compared to other technologies, CNTFET-based devices have higher sensitivity and selectivity, lower operating temperature, quicker response, smaller recovery time, and better stability. Owing to its detection capability and extraordinary performance, CNTFETs are anticipated to play a vital role in many sensitive fields. Added advantages of CNTFETs include a low probability of pulverization as they are one-dimensional. This ensures that the device operates in a ballistic regime [19-21].

The major advantages of CNTFET are as follows:

1. Higher electron mobility

2. Lower power consumption

3. Channel formation has better control

4. Threshold voltages are lower

5. Reduction in gate leakage current

6. No direct tunneling

The disadvantages of CNTFET are as follows:

1. It cannot operate at high electric current and higher temperatures

2. Its lifespan is less

3. Difficult to fabricate

4. Production costs are higher

CARBON NANOTUBE FIELD EFFECT TRANSISTORS (CNTFETs)

Carbon nanotube was founded by S.Ijima at NEC Fundamental Research Laboratory, Japan, in 1991. It is a three-terminal device with a semiconducting nanotube, source, and drain. The tube acts as a channel that can be turned on or off by the gate voltage. When voltage is applied to the gate electrode, it modulates the conductivity of the carbon nanotube channel, allowing the control of current flow [22-24].

The nanotube consists of carbon atoms or carbon allotropes (graphene) linked in a hexagonal shape and rolled as a tube. Although the tubes are small, they are very strong and flexible. Commonly used methods to make CNTs are Arc discharge, laser ablation of graphite, and Chemical Vapor Deposition (CVD) [25].

CNTFETs are categorized into several types based on various parameters. One such classification is based on its geometry. CNTFETs are classified as planar and coaxial based on their geometry, as depicted in Fig. (4).

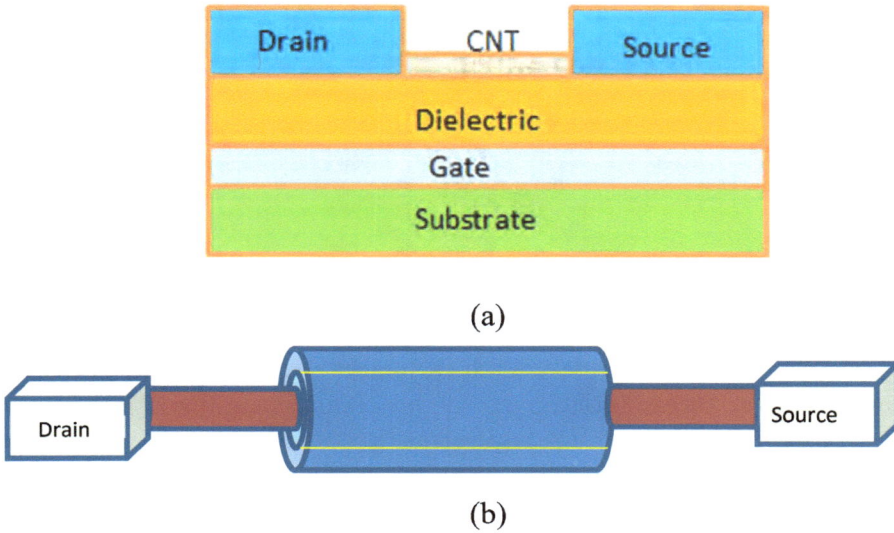

(a)

(b)

Fig. (4). CNTFET classification based on geometry [2].

Planar geometry is most commonly used owing to its simple construction and compatibility with the existing technology. However, the coaxial geometry of CNTFET helps increase capacitive coupling between the gate electrode and the surface of the nanotube. This provides additional channel charge and helps to alleviate short-channel effects [26, 27].

Planar geometry can further be categorized into two types based on the gate position: top gate CNTFET and bottom gate CNTFET, as depicted in Fig. (5).

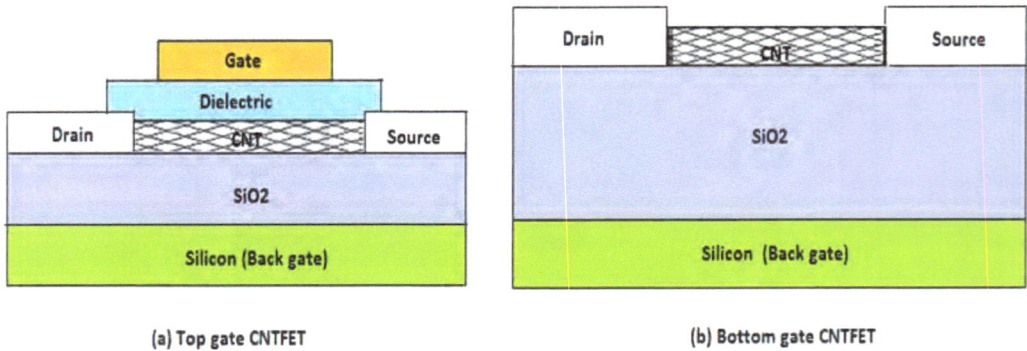

(a) Top gate CNTFET   (b) Bottom gate CNTFET

Fig. (5). Planar geometry of CNTFET.

In the bottom gate CNTFET, SiO_2, which is the gate oxide, is fabricated over the silicon layer, and over SiO_2, single-walled carbon nanotube is placed between the source and drain by the photolithography process. Since the CNT is placed randomly over the SiO_2, there exists high parasitic contact resistance, lower trans-conductance, and low drive currents. In the top gate CNTFET, a layer of dielectric is laid on CNT by atomic layer deposition or by evaporation, and on the top, the gate is fabricated. Electron beam lithography is used to pattern the source and drain. This type of geometry increases the drain current and the trans-conductance. Hence, it is preferred over bottom-gate CNTFET [28-30].

Depending on the number of layers of graphene used in construction, CNTFETS are categorized into two types: Single-walled carbon nanotube (SWNT) and Multi-walled Carbon nanotube (MWNT), as depicted in Fig. (6).

(a)   (b)

Fig. (6). Types of CNTFET: **(a)** SWNT **(b)** MWNT [2].

Chiral vector is a term associated with CNTFET that is used to signify the manner in which a graphene sheet is rolled up to form a nanotube [31]. Chirality introduces added quantization in the circumferential direction that introduces an energy gap that fluctuates between 0 and 1eV based on the structure of the CNT. CNT with zero energy gap becomes metallic, and the others become semiconducting nanotubes. Thus, the electric transport properties can be adjusted, deprived of any need for doping as opposed to conventional semiconductors [32]. This feature has made CNTFET very attractive for future electronics. Added to this, CNTs also exhibit high modulus of elasticity and thermal conductance. They have less delay and power consumption compared to silicon-based MOSFTs [33].

SWNT- It consists of only one layer of continuous graphene cylinder [34]. A unique electronic property of this is that it changes considerably with the chiral vector, $n_1 a + n_2 b$. Here n_1, n_2 are integers that indicate the chirality of the tube, and a and b are the lattice unit vectors, as depicted in Fig. (7).

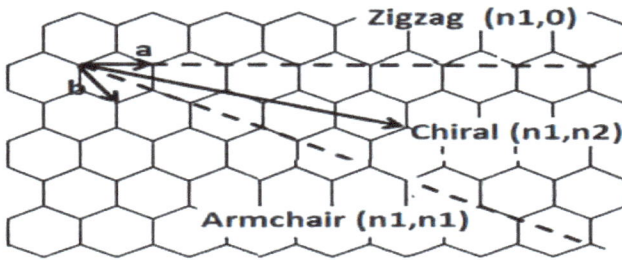

Fig. (7). Planar sheet of graphene.

Depending on the values of n_1, n_2, SWNT behavior can further be categorized into three types – Armchair ($n_1 = n_2 = n$), Zigzag ($n_1 = 0$ *or* $n_2 = 0$), and chiral structure ($n_1 - n_2 \neq 0$) [35, 36]. If the difference between chiral vector ($n_1 - n_2$) is multiple of 3, the SWNT behaves like a metal else it behaves like a semiconductor [37]. The behavior is summarized in Table **2**.

Table 2. CNT types with Chiral vector (n_1, n_2) values.

Type of CNT	n_1, n_2 Values	Properties
Armchair	$n_1 = n_2$	Metallic
Zigzag	$n_2 = 0$	Semiconducting
Chiral	$n_1 - n_2 \neq 3 * integer$	Large band gap
	$n_1 - n_2 = 3 * integer$	Small band gap

The structure of CNT varies with the changes in the chiral vector. The variation of the structure is depicted in Fig. (**8**) and the characteristics exhibited for various variations are tabulated in Table **3**.

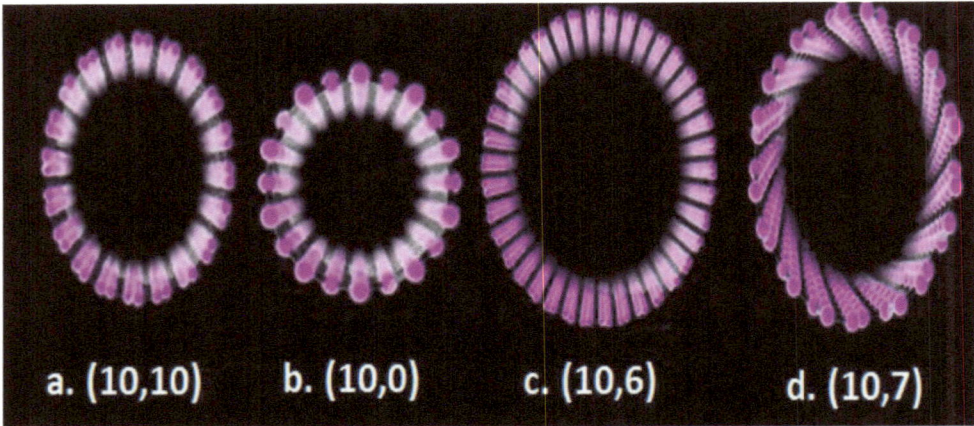

Fig. (8). CNT structure for various values of chiral vector [3].

Table 3. CNT structure variations for various chiral vector values.

	Chiral Vector (n_1, n_2) *values*	Characteristics exhibited	Reason
a.	(10,10)	Metallic	Overlapping of energy band
b.	(10,0)	Semiconducting	Non-overlapping energy band
c.	(10,6)	Large band gap	-
d.	(10,7)	Small band gap	-

The energy band diagrams and the density of state for the above-mentioned chiral vectors are depicted in Figs. (**9** and **10**), respectively.

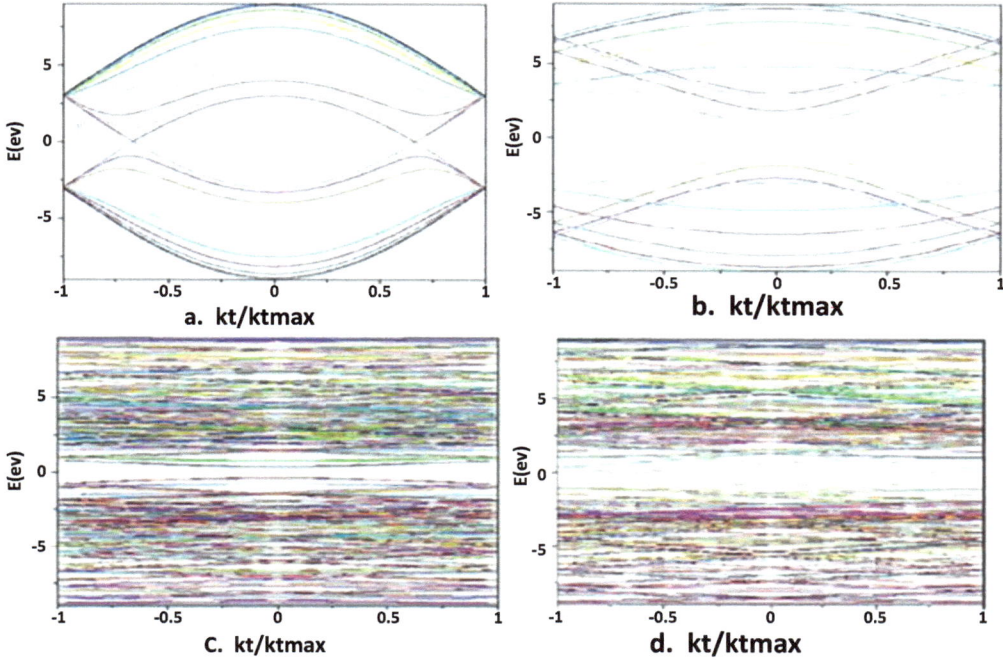

Fig. (9). Energy band diagrams for various values of the chiral vector [5].

The overlapping of energy bands in Fig. **(9a)** shows the CNT's metallic characteristics, while the semiconducting behavior is exhibited in (b) by non-overlapping energy band diagrams. Fig. **(9c)** illustrates the properties of the large band gap, and Fig. **(9d)** illustrates the properties of the small band gap.

The diameter of the nanotube is expressed as given in Equation 1 [5, 38].

$$d_1 = \sqrt{3}a_{c-c} * \sqrt{\frac{n_1^2 + n_1 n_2 + n_2^2}{\pi}} = \frac{C_h}{\pi} \tag{1}$$

Where $a_{c-c} = $ C − C bond length (1.42 Å) *and* $C_h = $ *length of carbon nanotube*

The gate width of CNTFET is expressed as given in Equation 2 [5, 39].

$$W_{gate} = (W_{min}, N * S) \tag{2}$$

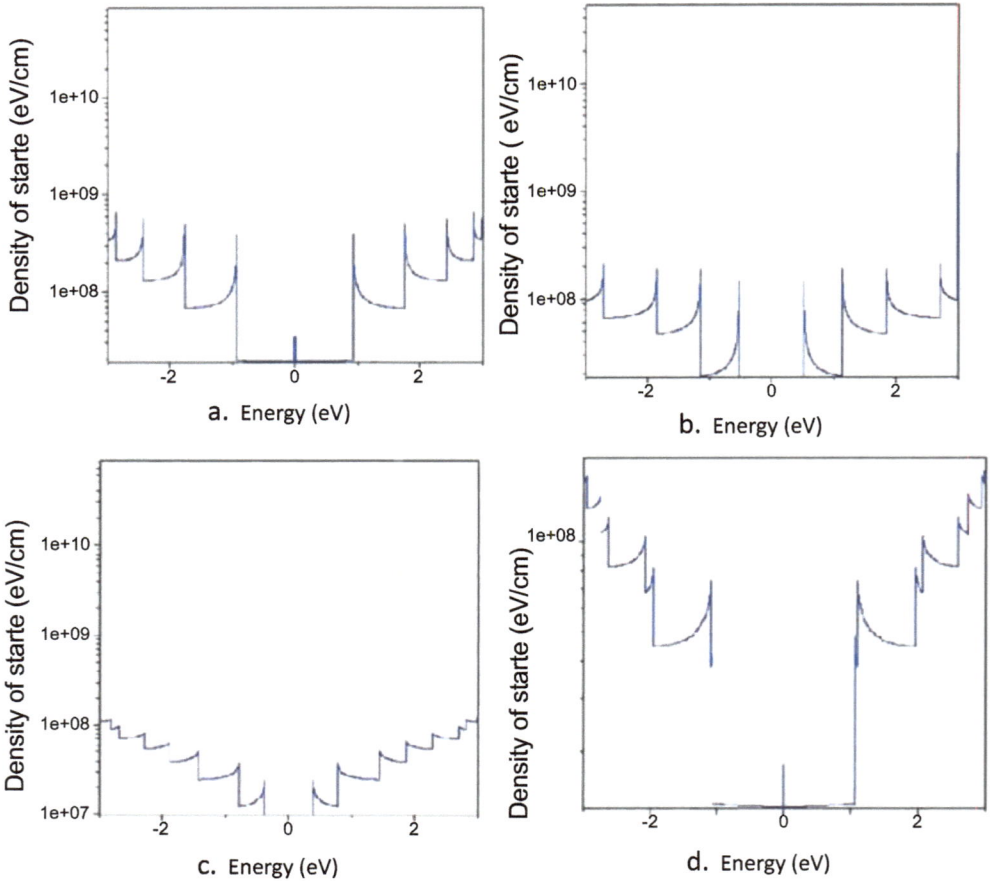

Fig. (10). Density of state for various values of the chiral vector [5].

Where W_{min} represents the minimum gate width, N represents the number of CNTs, and S is the distance between the centers of two adjacent CNTs present under the same gate (Pitch).

The threshold voltage is the minimum voltage essential to turn on the transistor through the gate and is given by Equation 3.

$$V_{th} = \frac{a\,(v_\pi)}{\sqrt{3}(qD_{CNT})} \qquad (3)$$

Where $v_\pi = 3.033eV$ represents the carbon π-π bond energy, a represents carbon to carbon atom distance (0.249nm), q is the unit electron volt charge, and D_{CNT} is the CNT diameter that depends on the chirality (n_1, n_2). D_{CNT} is given by Equation 4.

$$D_{CNT} = \frac{\sqrt{3}a\sqrt{n_1{}^2+n_2{}^2+n_1 n_2}}{\pi} \qquad (4)$$

From the equations, it is vibrant that the threshold voltage is inversely proportional to D_{CNT}, which in turn is directly proportional to chirality of the CNT. As chirality increases, D_{CNT} increases and threshold voltage decreases. The threshold voltage of P-CNTFET and N-CNTFET remains the same but with opposite signs.

MWNT- It consists of graphene rolled as multiple layers. MWNT is not well defined because of its complex structure compared to SWNT. The benefits of MWNT over SWNT are increased thermal and chemical stability and lower production price per unit [40].

Another category of CNTFET is double-walled carbon nanotubes (DWNT), which is a combination of SWNT and MWNT. It exhibits intermediate properties of two types. It consists of exactly two concentric nanotubes that are separated by a spacing of 0.35nm – 0.40nm. It also has a sufficient band gap to be used in FETs.

Short Channel Effects in CNTFET

Factors that assist CNTFET to prove itself better in comparison with the silicon-based devices in mitigating the short channel effects are-

• Higher carrier mobility of the CNTs – The device performance improves since the electrons can travel more quickly through the channel and offers better control over the channel charge. Higher mobility of the charge carriers helps in the reduction of the short channel effect.

• Reduction in doping levels – This decreases the abrupt junctions and contributes to the reduction of the short-channel effect.

• Higher electrostatic control – CNTs offer good electrostatic control on the channel owing to its size. This assists in having better control of the gate and overcoming the short-channel effect.

• Quantum confinement – CNTs exhibit matchless quantum properties that can be used in controlling the electron flow very precisely. This aids in controlling the short-channel effect.

• Device scalability – CNTFETs are better scalable than traditional silicon-based MOSFETs owing to their small size and better mechanical strength. This aspect addresses the short-channel effect.

Although CNTFETs help in reducing short-channel effects, they still face the challenges associated with short-channel effects. The effect becomes more dominant as the channel length decreases. It leads to issues like drain-induced barrier lowering (DIBL) and velocity saturation. Some of the methods to overcome the effect of CNTFETs include:

• **Usage of high-K Dielectrics** – High-k materials have higher permittivity as compared to traditional SiO2. This enables it to have better control over the gate and helps in addressing short-channel effects.

• **Gate structure** - The gate structure helps to mitigate the short-channel effects. Better gate structures like dual gates or wrap-around gates provide better electrostatic control over the channel and improve the gate control.

• **Channel modification** – It involves the usage of multiple CNTs or different channel materials to obtain improved electrical properties.

• **Application of strain to the channel** - Strain application to the channel can improve carrier mobility and device performance as it alters the electronic properties of the CNT structure.

• **Usage of advanced techniques of fabrication** – Possible defects during manufacturing can be reduced. CNTs can be placed precisely, and a controlled annealing process can be undertaken in order to achieve better control over the device and minimize the short-channel effects.

• **Variations in the materials used** – Alternative materials can be explored for CNTFETs in order to diminish the short-channel effects.

Investigating and developing new techniques and materials to overcome short-channel effects in CNTFETs are efforts to improve nanoelectronics.

Types of CNTFET Devices

There are six types of CNTFET devices – Schottky barrier CNTFET (SB-CNTFET), MOSFET like CNTFET (M-CNTFET), Band to Band Tunneling

CNTFET (T-CNTFET), Double gated CNTFET (DG-CNTFET), Gate-All-Around CNTFET (GAA-CNTFET), and Suspended CNTFET (S-CNTFET).

SB-CNTFET – It works according to the principle of a direct tunnel through a Schottky barrier that is present near the source/drain (made of metal) and the channel junction, as depicted in Fig. (**9**). The device is made by providing direct contact between semiconductor CNT and metal. The Schottky barrier between CNT and metal junction confines the current delivery capacity that, in turn, makes the device unsuitable for high-speed applications. The device also exhibits strong ambipolar behavior that limits its usage in complementary circuits that use transistors.

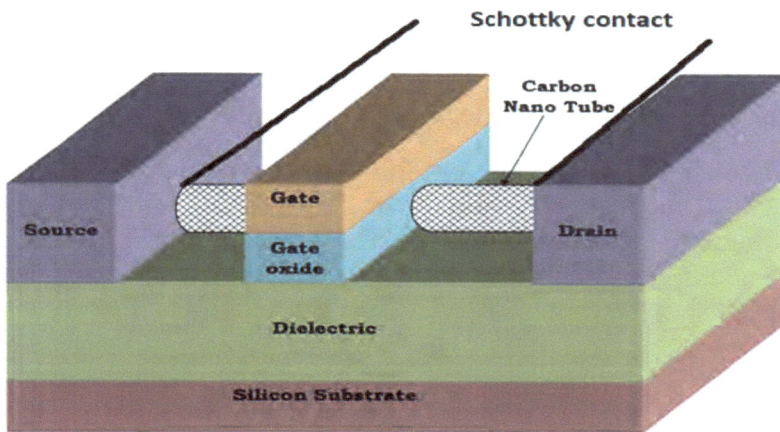

Fig. (9). SB-CNTFET structure.

M-CNTFET – This device overcomes the drawbacks of SB-CNTFET and operates like a normal MOSFET with lower power consumption and high speed. The device structure of M-CNTFET is depicted in Fig. (**10**). It is made with highly doped source and drain CNT regions. The device has a higher on-current owing to the lack of Schottky barrier at the source/drain channel junction. This makes it suitable for designs demanding ultra-high performance.

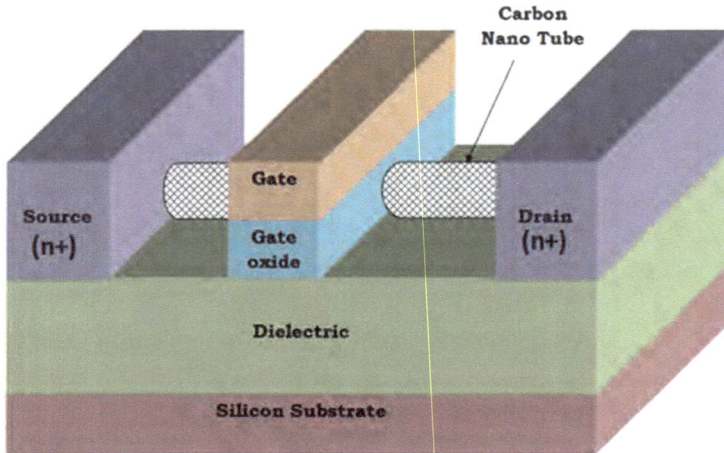

Fig. (10). M-CNTFET structure.

T-CNTFET –This device has low on-current and also exhibits good off characteristics. This makes it suitable for ultra-low power designs and sub-threshold designs. The structure of T-CNTFET is depicted in Fig. **(11)**. A highly doped n+ region inside the channel at the junction of the source and channel helps to boost the ON state performance.

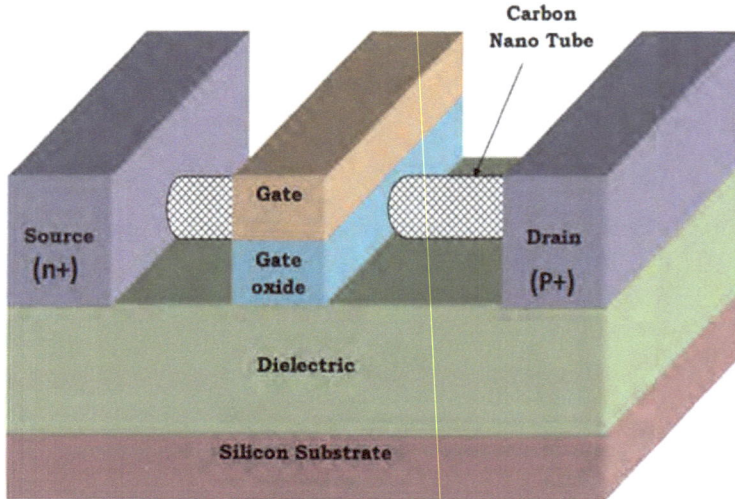

Fig. (11). T-CNTFET structure.

DG-CNTFET – In this structure, a gate is present on the top and at the bottom, as depicted in Fig. (**12**). The device is found to have better switching and retention time. This property makes it suitable for memory-based circuits.

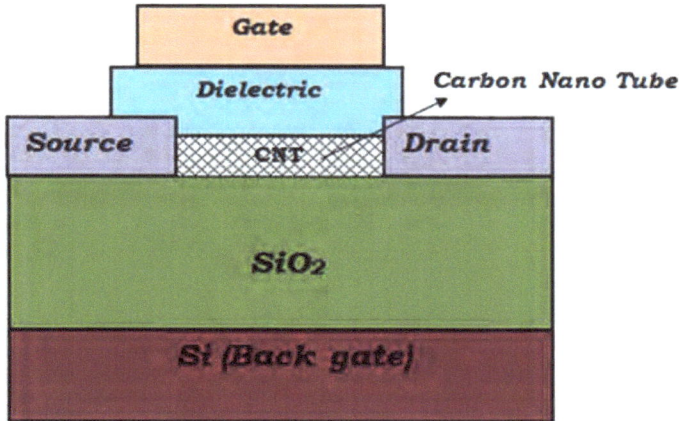

Fig. (12). DG-CNTFET structure.

GAA-CNTFET – In all of the types indicated above, only part of the nanotube is gated. In GAA-CNTFET, the gate covers the entire nanotube, as depicted in Fig. (**13**). By doing so, the leakage current reduces and the electrical performance improves. The fabrication starts by first wrapping the CNT with a dielectric layer by an atomic layer deposition process and then partially etching the dielectric layer to expose the nanotube ends and finally depositing the source, drain, and gate contacts.

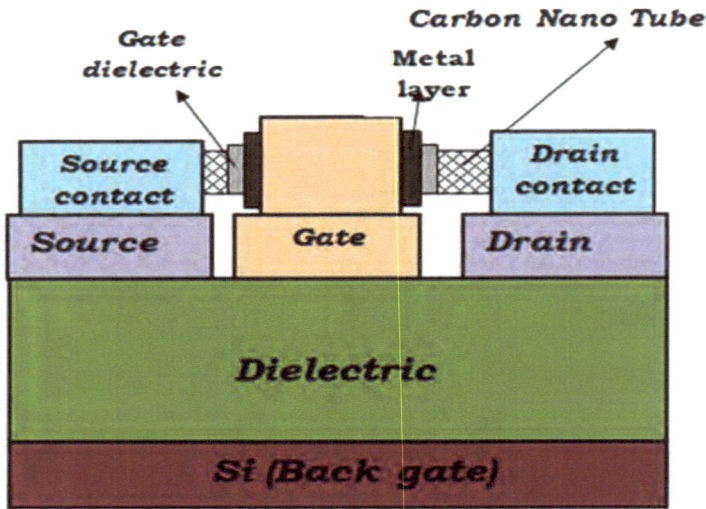

Fig. (13). GAA-CNTFET structure.

S-CNTFET- Here, the gate is suspended on a trench so that the contact with the substrate and gate oxide reduces as shown in Fig. (**14**). This improves the device's performance. But the downfall of this type is that gate dielectric is a vacuum or air. Only short-length tubes can be used. Longer ones will touch the metal contact and might lead to short in the device. Hence the device is still under research and not available commercially.

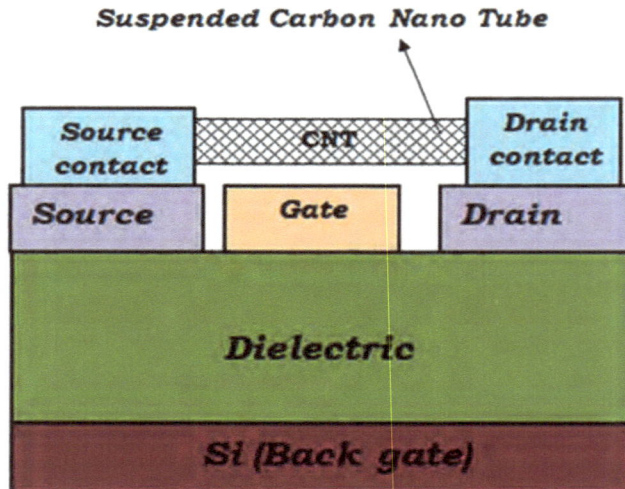

Fig. (14). S-CNTFET structure.

Properties of CNTFET

Excellent properties of CNTFET that make it a prospective candidate for building electronic systems with low power and high performance are:

1. The high carrier mobility of $10^3 - 10^4 \text{cm}^2/\text{V} - \text{S}$. This offers a high ON current of greater than $1\text{mA}/\mu\text{m}$.

2. The absence of dangling bonds provides easier integration with high-K dielectric material.

3. Long dispersion path that leads to lower delay and lower heating.

4. Higher current density $(10^{10}\text{A}/\text{cm}^2)$ due to strong chemical bonds and chemical stability.

5. Complementary CNTFETS are better adaptable, i.e., P-CNTFET and N-CNTFET of the same dimensions offer the same carrier mobility and the same conduction current.

6. Outstanding electrical properties due to ballistic transport in the metallic tubes.

7. High thermal conductivity.

8. Highest mechanical strength due to the highest Young's modulus.

Although CNTFETs offer numerous advantages, they also possess certain below listed limitations, owing to which they are not practical for commercial purposes-

1. Misalignment of CNT

2. Packing density of CNT

3. Metallic CNT growth

4. Diameter and density variation of CNT

Dielectric material for CNTFET

The dielectric material is placed between the gate electrode and the CNT channel. It serves as an insulator between the gate and the channel and provides good electrostatic control. In order to achieve optimal performance of CNTFET, proper choice of dielectric material becomes significant.

Since the discovery of MOSFET in 1960, SiO_2 has been used as a gate dielectric material due to its good electrical and thermal stability. To improve the performance of MOSFET, scaling of the SiO_2 dielectrics was considered the most effective approach. In the past few decades, the reduction of SiO_2 dielectric thickness has increased the integration of transistors and has brought about substantial enhancement in circuit functionality. However, as the devices approach the scale of around 45nm, they demand an oxide thickness of 1nm. However, this results in a large leakage current as a result of quantum tunneling and also increases the power dissipation. As an alternative to the above problem, gate oxide with a higher dielectric constant can be used. This will keep the performance of the transistor intact while retaining the same dielectric thickness. The potential candidates for replacing the SiO_2 are nitride SiO_2, Hf, and Zr oxides. The basic problem with a gate based on high K was low crystalline temperature. This made it hard to integrate it into traditional CMOS process. This problem can be solved by adding elements like N, Si, Al, Ti, Ta, and La into high-K dielectrics.

The selection criteria for dielectric materials are listed below:

1. It must establish a good electrical interface with Si.

2. It must possess good kinetic stability.

3. It must possess a high-K dielectric. This ensures good electrostatic control over the channel, provides effective coupling between the channel and the gate, and also reduces the voltage required for transistor modulation. The overall device performance thus increases.

4. It must have low solid charge density.

5. It must have long life with good reliability.

6. It must have thermodynamically good stability.

7. It must have a higher bandgap with a larger barrier height than the metal gate and silicon substrate so that the leakage current is reduced.

8. It must have a low leakage current so as to prevent undesired charge leakages between the channel and the gate. This improves the energy efficiency and the switching characteristics of the transistor.

9. It must have good thermal stability so as to exhibit stable operation over the device operating range and ensure reliability of operation.

10. It must offer good adhesion between the CNT channel and the dielectric material to ensure higher device performance.

The most suitable dielectric materials for CNTFETs are

1. Materials with high-K dielectrics like hafnium oxide (HfO_2), aluminum oxide ($Al2O_3$), and zirconium oxide (ZrO_2).

2. Organic dielectric materials like poly methyl methacrylate owing to their flexible substrate.

3. Stacks of various dielectric materials like SiO_2 and high-K materials can be used to grab the advantages of both.

The selection of dielectric materials will have a better impact on the device's performance and the CNTFET characteristics. Research on exploring new dielectric materials and their fabrication techniques is still in progress.

High K Dielectric for Gate Oxide

The dielectric constant value (K) for the gate oxide material can range between 20 to 30. The K value of the oxide material is inversely proportional to the band gap, and hence, a relatively low K value has to be accepted. Marginal fields from drain to gate or from drain to source are generated with high K dielectrics, and this degrades the performance of a nano-scale transistor. For dielectric gate applications using Si substrate, simple dielectric oxide materials with conduction band offset of less than 1eV will be unsuitable. The dielectric gate materials exhibit a general tendency to have an inverse association between dielectric constant and band gap size. This makes it difficult to meet the leakage current requirements. The dielectric materials must also possess the capacity to avoid dopant (boron and phosphorous) spread and should also have good thermal stability and high recrystallization temperature. Table **4** lists some of the materials that can be used as gate oxides, along with their K value.

Table 4. Gate oxide materials with their K value.

S. No	Gate Oxide Material	K value
1	Silicon dioxide (SiO2)	3.9
2	Zirconium Silicate (ZrSiO4)	5
3	Zirconium dioxide (ZrO2)	24
4	Hafnium Silicate (HfSiO4)	11
5	Hafnium dioxide (HfO2)	24

Variation of threshold voltage for various dielectric values and various oxide layer thicknesses is depicted in Figs. (**15** and **16**), respectively. From the plots, it can be observed that threshold voltage rises with higher values of dielectric material and with lower oxide layer thickness.

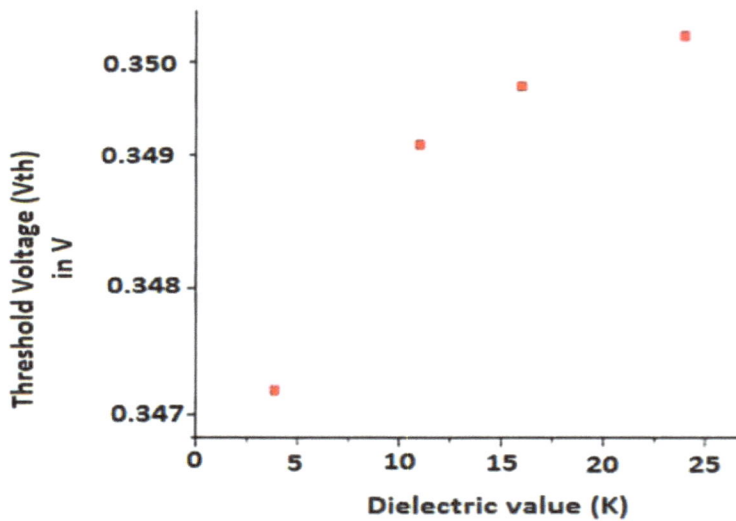

Fig. (15). Vth *vs.* Dielectric value.

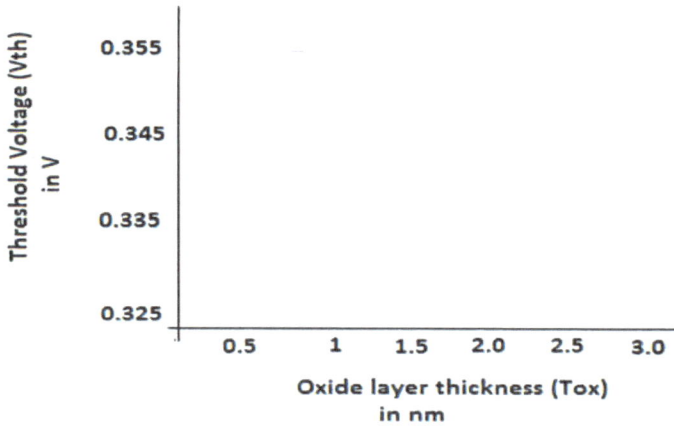

Fig. (16). Vth *vs.* Oxide layer thickness.

Electron mobility also rises with the rise in dielectric value; this, in turn, raises the drain current. Thus, it can be concluded that high K dielectric material is better suitable for CNTFET-based devices.

CNTFETs I-V Characteristics

When the voltage applied is below the threshold voltage (minimum gate voltage required to turn on the transistor), the drain current almost remains zero [3]. But when the applied voltage increases and crosses the threshold voltage, the drain current starts increasing, as depicted in Fig. (**16**).

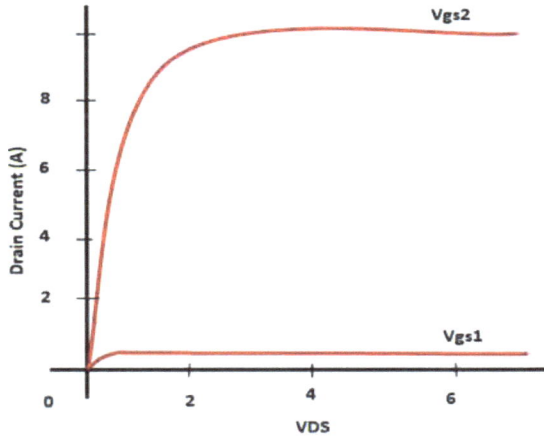

Fig. (16). CNTFETs I-V Characteristics.

Applications of CNTFETs

Being a high-performance and low-power device, CNTFETS finds applications in the following:

1. In digital ICs, it can be used in memory devices and building logic gates [4]. They are most appropriate for high performance computing and in faster computation chips owing to their high electron mobility and lower power consumption.

2. In analog electronics, they can be used to build amplifiers, oscillators, and converter circuits, as they have a high potential to perform analog signal processing.

3. As CNTFETs can be integrated into flexible substrates, they are suitable for wearable devices.

4. The sensitivity of CNTFETs and their higher surface areas makes them suitable for detecting environmental changes and making sensors like gas sensors.

5. CNTFET can be used in bioengineering applications like storage of energy, energy conversion devices, sources of radiation, and hydrogen storage media. They can also be used to capture and transform the surrounding energy into electrical energy by integrating them into energy harvesting devices.

6. CNTFETs are capable of interacting with light and also possess unique optical properties. This makes them suitable in photonics as optoelectronic devices. They can be used as optical modulators and detectors.

7. CNTFETs are suitable for wireless systems and radar systems owing to their capabilities to operate at high frequencies.

8. It is observed that CNTs are capable of acting as quantum bits (QBits). This has drawn attention to checking the feasibility of CNTFETs for quantum computing.

Major Challenges Faced by CNTFETs

The major challenges faced in the fabrication of CNTFET-based circuits are listed below.

1. Difficulty in varying the diameter of the tubes.

2. The existence of metallic tubes.

3. Misalignment of the tubes during the fabrication process.

CONCLUSION

The chapter discussed the features of CNTFET, suitable materials to build CNTFET, types of CNTFET, their characteristics, and applications. Although CNTFETs have greater advantages over silicon-based devices, they also possess scalability issues and face challenges during the manufacturing processes and in providing consistent electrical properties. These issues are to be addressed in order to commercialize the CNTFET-based devices. Research towards overcoming the shortfalls of CNTFETs is to be carried out in order to discover the full capability of the technology.

Learning Objectives/Key Points

(1) MOSFET sets a limit on transistor sizing. Scaling further will give rise to challenges such as decreased gate control, short channel effect, increased power density, higher sensitivity to process variation, higher manufacturing cost, and increased leakage current. To overcome the shortfalls of MOSFETs, alternative transistors are looked upon and this chapter suggested one such option – CNTFET.

(2) CNTFET has higher carrier mobility and a direct band gap that enhances the band-to-band tunneling and optical properties. These features make CNTFET suitable to be used in future novel electronic devices.

(3) CNTFET uses carbon nanotubes between the source and the drain of the MOSFET structure. This ensures CNTFET provides higher current carrier mobility and higher drive current density.

(4) Compared to other technologies, CNTFET-based devices will have higher sensitivity and selectivity, lower operating temperature, quicker response, smaller recovery time, and better stability.

Although CNTFETs help in reducing short-channel effects, they still face the challenges associated with them. The effect becomes more dominant as the channel length decreases. It leads to issues like drain-induced barrier lowering and velocity saturation. Efforts are being made in order to reduce these effects.

Multiple Choice Questions (MCQs)

1. The critical problems and challenges associated with MOSFET when the channel length is reduced beyond 45nm are caused due to _____

(a) Short channel effects   **(b)** Increased manufacturing cost

(c) Requirement for skilled designer   **(d)** All of the above

Answer: a) Short-channel effects

2. Current conduction in CNTFET is through _____

(a) Bulk silicon   **(b)** Carbon nanotube   **(c)** Source   **(d)** Drain

Answer: b) Carbon nanotube

3. Identify the technology in which the channel is surrounded by gates all around

(a) NWFET   **(b)** MOSFET   **(c)** GFET   **(d)** SET

Answer: a) NWFET

4. Identify the two types of NWFET.

(a) Horizontal and lateral NWFET

(b) Vertical and lateral NWFET

(c) Horizontal and vertical NWFET

(d) None of the above

Answer: c) Horizontal and vertical NWFET

5. In CNTFETs to construct nanotubes, _____ sheets are used.

(a) Copper   **(b)** Silicon   **(c)** Graphene   **(d)** Germanium

Answer: c) Graphene

6. _____ Geometry of CNTFET helps to increase the capacitive coupling between the gate electrode and the surface of the nanotube.

(a) Coaxial   **(b)** Planar   **(c)** Top gate   **(d)** Bottom gate

Answer: a) Coaxial

7. In the top gate CNTFET, _____ is used to pattern the source and drain.

(a) Electron beam lithography

(b) Tunneling

(c) Electron microscopy

(d) Etching

Answer: a) Electron beam lithography

8. _____ is a term that signifies the manner the graphene sheet is rolled up to form a nanotube in CNTFET.

(a) Chiral table   **(b)** Chiral vector   **(c)** Chiral loop   **(d)** Chiral roll

Answer: b) Chiral vector

9. CNT with zero energy gap becomes _____

(a) Conducting   **(b)** insulating   **(c)** semiconducting   **(d)** metallic

Answer: d) metallic

10. In Armchair type SWNT, n_1, n_2 that indicates the chirality of the tube should be _____

(a) $n_1 = n_2 = n$

(b) $n_1 = 0 \; or \; n_2 = 0$

(c) $n_1 - n_2 \neq 3 * integer$

(d) $n_1 + n_2 \neq 0$

Answer: a) $n_1 = n_2 = n$

11. In Zigzag type SWNT, n_1, n_2 that indicates the chirality of the tube should be _____

(*a*) $n_1 = n_2 = n$

(b) $n_1 = 0$ or $n_2 = 0$

(c) $n_1 - n_2 \neq 3 * integer$

(d) $n_1 + n_2 \neq 0$

Answer: b) $n_1 = 0$ or $n_2 = 0$

12. In Chiral structure type SWNT, n_1, n_2 that indicates the chirality of the tube should be _____

(a) $n_1 = n_2 = n$

(b) $n_1 = 0$ or $n_2 = 0$

(c) $n_1 - n_2 \neq 0$

(d) $n_1 + n_2 \neq 0$

Answer: c) $n_1 - n_2 \neq 0$

13. _____ between CNT and metal junction makes SB-CNTFET unsuitable for high-speed applications

(a) Schottky barrier   **(b)** Silicon layer   **(c)** Source   **(d)** Drain

Answer: a) Schottky barrier

14. The type of MOSFET with low current and exhibiting good off characteristics is ____

(a) SB- CNTFET   **(b)** M-CNTFET   **(c)** DG-CNTFET   **(d)** T-CNTFET

Answer: d) T-CNTFET

15. In _____, the gate covers the entire nanotube.

(a) GAA- CNTFET   **(b)** M-CNTFET   **(c)** DG-CNTFET   **(d)** T-CNTFET

Answer: a) GAA- CNTFET

REFERENCES

[1] T. Song, "Opportunities and Challenges in Designing and Utilizing Vertical Nanowire FET (V-NWFET) Standard Cells for Beyond 5 nm", *IEEE Trans Nanotechnol.,* vol. 18, pp. 240-251, 2019.

http://dx.doi.org/10.1109/TNANO.2019.2896362

[2] N. Anitha and P. Srividya, "Comparative study of different technologies to replace CMOS technology," *Int. J. Curr. Res.*, vol. 9, no. 9, pp. 58019–58035, 2017.

[3] N. Anitha and P. Srividya, "Parameter analysis of CNTFET," *Int. J. Recent Technol. Eng. (IJRTE)*, vol. 8, no. 2, pp. 1–5, 2019.

[4] N. Anitha and P. Srividya, "Balanced XOR/XNOR circuits using CNTFET," I*nt. J. Innov. Technol. Explor. Eng. (IJITEE)*, vol. 8, no. 10, pp. 746–751, 2019.

[5] R. Kumar and S. Singh, *Overcoming the Limitations of Conventional FETs Using Carbon Nanotube Field Effect Transistors, M.Tech. thesis, Vellore Institute of Technology*, India, 2018.

[6] K. Mukhopadhyaya and P. Srividya, "Trends in performance characteristics and modelling of oxide based TFT," *Mater. Today: Proc.*, vol. 55, pt. 2, pp. 414–418, 2022.

http://dx.doi.org/10.1016/j.matpr.2021.12.596

[7] P. Srividya, "SOI technology in designing low-power VLSI circuits," In: *Energy Systems Design for Low-Power Computing*, R. R. Gatti, C. Singh, P. Srividya, and S. Bhat, Eds. IGI Global, Mar. 2023, pp. 17–28.

[8] D. Lee, C.T. Ho, I. Kang, S. Gao, B. Lin, and C.K. Cheng, "Many-Tier Vertical Gate-All-Around Nanowire FET Standard Cell Synthesis for Advanced Technology Nodes", *IEEE J. Explor. Solid-State Comput. Devices Circuits,* vol. 7, no. 1, pp. 52-60, 2021.

http://dx.doi.org/10.1109/JXCDC.2021.3089095

[9] Anil Kumar, and Dharmender Dubey, "Single Electron Transistor: Applications and Limitations", *Advance in Electronic and Electric Engineering,* vol. 3, no. 1, 2013.

[10] A. Raychowdhury, and K. Roy, "Carbon Nanotube Electronics: Design of High-Performance and Low-Power Digital Circuits", *IEEE Trans Circ. Syst.,* 2007.

[11] A. Benfdila, M. Berd, and A. Lakhlef, "Carbon Nanotube Field Effect Transistors Development and Perspectives", *Fourth IEEE Conference on Nanotechnology, Munich,* 2004 Germany

[12] C-H. Chang, J. Gu, and M. Zhang, "A review of 0.18- m full adder performances for tree structured arithmetic circuits. IEEE Trans. Very Large Scale Integr. (VLSI)", *Syst.,* vol. 13, p. 686, 2005.

[13] S. Goel, M.A. Elgamel, M.A. Bayoumi, and Y. Hanafy, "Design methodologies for high–performance noise-tolerant XOR– XNOR circuits. IEEE Trans. Circuits yst. I", *Reg Papers,* vol. 53, p. 867, 2006.

[14] T. Nikoubin, F. Eslami, A. Baniasadi, and K. Navi, "A new cell design methodology for balanced XOR/XNOR circuits for hybrid CMOS logic", *J. Low Power Electron.,* vol. 5, no. 4, pp. 474-483, 2009.

http://dx.doi.org/10.1166/jolpe.2009.1046

[15] F. Obitea, "b, Geoffrey Ijeomahc and Joseph Stephen Bassia, Carbon nanotube field effect transistors: toward future nanoscale electronics", *Int. J. Comput. Appl.,* 2018.

[16] A. Al-Shaggah, M. Khasawneh, and A. Rjoub, "Carbon nanotube field effect inverter: Delay time and power consumption analysis," in *Proc. 9th Jordanian Int. Elect. Electron. Eng. Conf. (JIEEEC),* Amman, Jordan, 2015.

http://dx.doi.org/10.1109/JIEEEC.2015.7470750

[17] Jie Zhang, Albert Lin, Nishant Patil, Hai Wei, Lan Wei, H.-S. Philip Wong, and Subhasish Mitra, "Carbon Nanotube Robust Digital VLSI", *IEEE Transactions on Computer-Aided Design of Integrated Circuits and Systems,* vol. 31, no. 4, 2012.

[18] L. Jacques, "A CNFET-based Characterization Framework for Digital Circuits", *IEEE International Conference,* 2011

[19] M.J. Kumar, "Molecular diodes and applications", *Recent Pat. Nanotechnol.,* vol. 1, no. 1, pp. 51-57, 2007.

http://dx.doi.org/10.2174/187221007779814790 PMID: 19076020

[20] "A Compact SPICE Model for Carbon-Nanotube Field- Effect Transistors Including Nonidealities and its Application—Part I: Model of the Intrinsic Channel Region", *IEEE Trans Electron. Dev.,* vol. 54, no. 12, p. •••, 2007.

[21] Jie Zhang, Albert Lin, Nishant Patil, Hai Wei, Lan Wei, H.-S. Philip Wong, and Subhasish Mitra, "Carbon Nanotube Robust Digital VLSI", *IEEE Transactions on Computer-Aided Design of Integrated Circuits and Systems,* vol. 31, no. 4, pp. 453-471, 2012.

[22] Kunal K Sharma, and P. Reena Monica, "Implementation of Mod-16 Counter using Verilog-A Model of CNTFET", , vol. 1, no. 2, 2013.

[23] O.M. Khakifirooz, "Nayfeh and D. Antoniadis, A simple semi-empirical short-channel MOSFET current-voltage Model Continuous Acrss All Regions of Operation and Employing only Physical Parameters", *IEEE Trans Electron. Dev.,* vol. 56, no. 8, pp. 1674-1680, 2009.

http://dx.doi.org/10.1109/TED.2009.2024022

[24] Jacques L. Athow, Come Rozon, Dhamin Al-Khalili, and J. M. Pierre Langlois, "A CNTFET-based Characterization Framework for Digital Circuits", *IEEE International Conference,* 2011

[25] S. Biswas, K. M. Jameel, R. Haque, and M. A. Hayat, "A novel design and simulation of a compact and ultra fast CNTFET multivalued inverter using HSPICE," in *Proc. 14th IEEE Int. Conf.*, 2012, pp. 28–30.

[26] Vinay Pratap Singh, Arun Agrawal, and Shyam Babu Singh, "Analytical Discussion of Single Electron Transistor (SET)", *International Journal of Soft Computing and Engineering,* vol. 2, no. 3, 2012.

[27] P. Yeole, "Design of Basic logic gates using Carbon Nano Tube Field Effect Transistor and Calculation of Figure of Merit", , 2015

http://dx.doi.org/10.1109/ICETET.2015.15

[28] Pranay Kumar Rahi, Shashi Dewangan, and Nishant Yadav, "Design & Simulation of 2-bit Full Adder Using Different CMOS Technology", , vol. 2, no. 3, 2015.

[29] Krishnan, R. (2014). Single Electron Transistors. *Int. J. Sci. Eng. Res.*, 5(9). ISSN 2229-5518.

[30] M. Masoudi, M. Mazaheri, A. Rezaei, and K. Navi, "Designing high-speed, low-power full adder cells based on carbon nanotube technology," *Int. J. VLSI Des. Commun. Syst. (VLSICS)*, vol. 5, no. 5, pp. 31–41, 2014

[31] M. Schröter, M. Claus, P. Sakalas, M. Haferlach, and D. Wang, "Carbon nanotube FET technology for radio-frequency electronics: State-of-the-art overview," *IEEE J. Electron Devices Soc.*, vol. 1, no. 1, pp. 9–20, 2013.

[32] M. Sereda, "Electroplating of carbon nanotube yarns and tapes with five metals," *SF Surface Technol. White Papers*, vol. 80, no. 8, pp. 1–11, May 2016.

[33] A. Benfdila, M. Berd, and A. Lakhlef, "Carbon nanotube field effect transistors development and perspectives," *Proc. 4th IEEE Conf. Nanotechnology*, 2004.

[34] Diwakar Agrawal, and Bahniman Ghosh, "Quantum Dot Cellular Automata Memories", *International Journal of Computer Applications,* vol. 46, no. 5, 2012.

[35] D. Chaudhary, R. Yadav, and N. Kr, "A Simulation Based Analysis of Lowering Dynamic Power in a CMOS Inverter", *Proc of Int. Conf on Advances in Signal. Processing and Communication,* 2013

[36] G.Y. Tseng, and J.C. Ellenbogen, "Architectures for molecular electronic computers: 3. Design for a memory cell built from molecula electronic devices", In: *MITRE McLean.* Virginia MP, 1999.

[37] Rasmit Sahoo, and R.R. Mishra, "Simulation of Carbon Nanotube Field Effect Transistors", *International Journal of Electronic Engineering Research,* pp. 117-125, 2009.

[38] Reena P. Monica, and V.T. Sreedevi, "A Low Power And Area Efficient CNTFET Based GDI Cell For Logic Circuits",

[39] SnehLata Murotiya, and Anu Gupta, "Design of High Speed Ternary Full Adder and Three-Input XOR Circuits Using CNTFETs", *IEEE 14th International Conference,* 2012p. 28–30

[40] G.L. Snider, A.O. Orlov, I. Amlani, X. Zuo, G.H. Bernstein, C.S. Lent, J.L. Merz, and W. Porod, "Quantum Dot Cellular Automata", In: *University of Notre Dame.* 1999.

[41] Available from: https://www.learnelectronicswithme.com/2020/09/cntfet-types-i-v-characteristics.html

CHAPTER 3

The Future Outlook for Field Effect Transistors Using Carbon Nanotubes

C. Kathiravan[1,*], **Gowrishankar J.**[2], **S. Grace Infantiya**[3], **D. Anbuselvi**[3] and **N. Suthanthira Vanitha**[4]

[1]*Department of Chemistry, Muthayammal Engineering College, Rasipuram, Namakkal, India*

[2]*Department of Computer Science and Engineering (AI) JAIN (Deemed-To-Be University), Bangalore, India*

[3]*Department of Physics, Muthayammal Engineering College, Rasipuram, Namakkal, India*

[4]*Department of Electrical and Electronics Engineering, Muthayammal Engineering College, Rasipuram, Namakkal, India*

Abstract: Carbon Nanotube Field Effect Transistors (CNTFETs) are potential nano-scaled devices for realising high-performance, very dense, and low-power circuits. A Carbon Nanotube Field Effect Transistor is a FET that uses a single CNT or an array of CNTs as the channel material rather than bulk silicon as in a standard MOSFET configuration. A carbon nanotube is at the heart of a CNTFET. This paper provides an overview of CNTFETs-pH sensor based on carbon nanotubes (CNTs)-FETs-pH measurement range of 1.34 to 12.68, reliability, and low hysteresis, indicating a promising application prospect in harsher testing environments. The determination of carbamate pesticides-adjusting the V_{TH} revealed that carbaryl and carbofuran additions had a favorable effect on the CFO/s-SWCNT-FET and structure. In this chapter, modeling, fabrication, and applications have been discusseddevices.

Keywords: Carbon nanotubes or Buckytube, FET-Uni-polar transistors, nanoelectronics, pH sensors, environmental stability, s-SWCNTs, CFO/s-SWCNT.

INTRODUCTION

Integrated circuits are widely used in our daily lives, especially personal computers and mobile gadgets. As economic conditions and electronic technology advance, people demand more powerful and lighter gadgets [1]. Carbon Nanotube Field Effect Transistors (CNTFETs) or Carbon Nanotube Uni-polar transistors hold

---

* **Corresponding author C. Kathiravan:** Department of Chemistry, Muthayammal Engineering College, Rasipuram, Namakkal, India; E-mail: kathiravan.c.chem@mec.edu.in.

Dharmendra Singh Yadav & Prabhat Singh (Eds.)

significant promise as a nanoscale technology for developing densely packed, low-power, high-performance circuits. A CNT-Uni-polar transistor is a semiconductor device that, unlike MOSFETs, typically uses bulk silicon as the channel material that is used in carbon nanotubes [2-10].

A carbon nanotube forms the core of a CNTFET. Uni-polar transistor biosensors have sparked considerable attention due to their unique high sensitivity. Traditional Uni-polar transistor (FET) biosensors often lack stability due to the intricate connection between the sample solution and the transistor. The harsh acidic or basic conditions in which they operate can lead to connection failures in Uni-polar transistor-based pH sensors, making it difficult to monitor pH consistently, correctly, and with low hysteresis, we present a pH sensor based on carbon nanotubes (CNTs)-FETs that have better environmental stability due to the addition of a HfO_2 layer to the gate insulator. The proposed technologies exhibit the necessary resistance to external influences, including a wide range of chemical and sensible constituents. Mostly, we demonstrated that this CNT-FET device could be utilized to continuously detect pH. It has 68 mV / pH sensitivity, a pH value of 1.0 -13, a minimal hysteresis of 600 pA, and pH altering loops of 4-10 [11-15].

Because of their distinctive geometries with high surface-to-volume and aspect ratios, Buckytube Uni-polar transistors, in particular, with single-wall carbon nanotube connecting channels, present exciting possibilities for very sensitive and label-free biosensors and chemical sensors [16-25]. SWNTs have all of their atoms on the surface and are accessible to the environment. As a result, even small shifts in the charge environment may significantly change their electrical properties [2]. Unipolar transistors based on semiconducting single-walled carbon nanotubes have demonstrated several advantages, including high transporter ability to adapt, either high on or high off ratio, electron transport that is semi-ballistic, name-free discovery, and steady reaction. Thus, cobalt ferrite oxide (CFO) nanoparticle-adorned s-SWCNTs were arranged and employed to connect the source and channel cathodes. For non-enzymatic carbaryl and carbofuran recognition, cobalt ferrite oxide /s-SWCNT/FET was used as planned. When used as a detection stage, the cobalt ferrite oxide /s-SWCNT half-breed film demonstrated great responsiveness and selectivity with the least restriction of discoveries for the compound like carbaryl and carbofuran (0.10fM) and (0.067fM) respectively, and a broad direct range of location from 10 to 100fM respectively. Moreover, this sensor was utilized to distinguish between carbaryl in experiments on tomatoes and cabbage, validating its accurate identification. The charge move response on the s-SWBTs/FET conduction channels progressed as a result of the execution, which may be attributable to the forceful synergist movement of CFO oxidizing carbamate. The sophisticated CFO/s-SWCNT-based detection methodology described in this paper may be applied to assess pesticide residues in food testing [26-37].

Traditional silicon-based complementary metal oxide semiconductor devices are approaching their basic scalability restrictions. This has prompted a lot of interest among researchers in looking into cutting-edge device technologies that use diverse materials to keep the scaling boundaries of modern integrated circuits. Carbon nanotubes have been extensively researched as a prominent material alternative across various fields due to their advantageous properties such as reduced short channel effects, high mobility, and high normalized driving currents. CNTs make up the majority of carbon nanotube Uni-polar transistors, which are regarded to be the most practicable alternatives to silicon transistors [38-40].

CARBON NANOTUBE FIELD-EFFECT TRANSISTOR

MOSFETs of today are comparable to carbon nanotube Uni-polar transistors. The source, drain, and gate are the three terminals they need. The current flow at the drain terminal and source is regulated by the gate. Current can pass via a channel and across the source when the gate is activated. The use of carbon nanotubes as the channel in CNFETs versus heavily doped silicon in MOSFETs is the primary difference between the two types of transistors. Complementary devices, such as semiconductors, are essential to both technologies. They are reliable, boost gain, use less power, and integrate readily into logic circuits. Holes are conducted via a p-type transistor and an n-type transistor carries electrons. Both sorts, as well as another type of device, will be investigated, with an emphasis on their production and swapping.

Current CNFET Designs' Blueprint

Development

Because the goals of these devices were essentially proof of concept and rudimentary comprehension, the first CNFETs were constructed in the most basic manner imaginable. Fig. (**1**) [3] shows first-generation CNFETs. The source and drain are represented by the two gold electrodes over which the nanotube is wrapped. It is possible to place the gate underneath or to the side of the transistor. Moreover, an insulator, like silicon dioxide, separates the gate from the nanotube. The electric field can be used to switch on and off the transistor because the gate is isolated. The transistor would become unusable if it was not protected from short circuits.

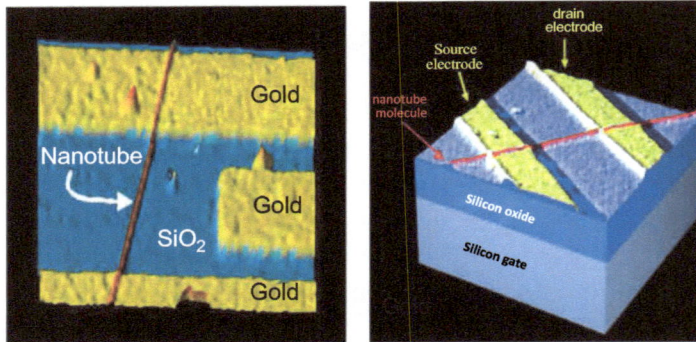

Fig. (1). Uni-polar transistors using carbon nanotubes, first generation [3].

These designs have certain issues. The first is that the transistor can only operate as a p-type as we shall see, because the nanotube is exposed to the air. Additionally, there are issues with the gate's positioning. The gate oxide must be at least 100 nm thick, regardless of whether it is above or below the nanotube. This means that significant voltages are required to break through the oxide and switch the transistor. Because all transistors share the same gate, placing the gate underneath complicates things further by causing all transistors to switch at the same time, making large circuits or chips impossible to create.

Top Current Style

The second-generation transistors are vastly superior to the first. Fig. (**2**) depicts the redesigned design.

The placement of the gate electrode is the key distinction between the initial and subsequent generations. The gate is now secured to the nanotube. This design changes the shuttles on the nanotube, preventing air from pushing the semiconductor to behave as a p-type semiconductor. An additional advantage is that the gate oxide is significantly thinner, at 14.9×10^{-9}m, allowing much lower currents to be employed. Fig. (**3**) [4] displays a plot of drain current vs drain-to-source voltage.

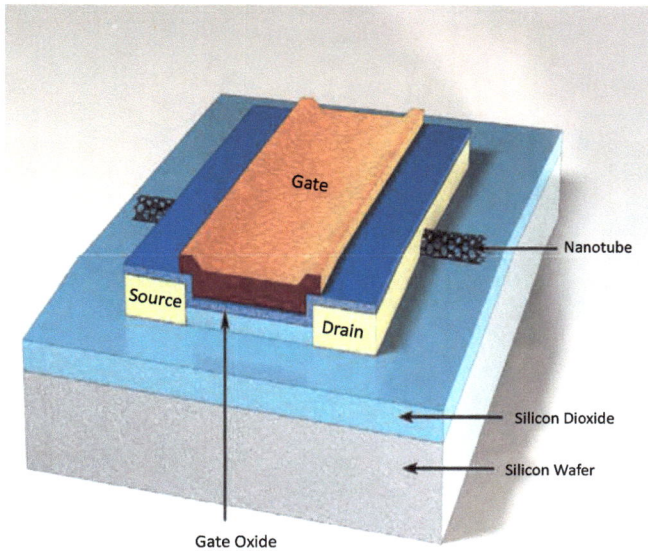

Fig. (2). Second Generation CNTFET [3].

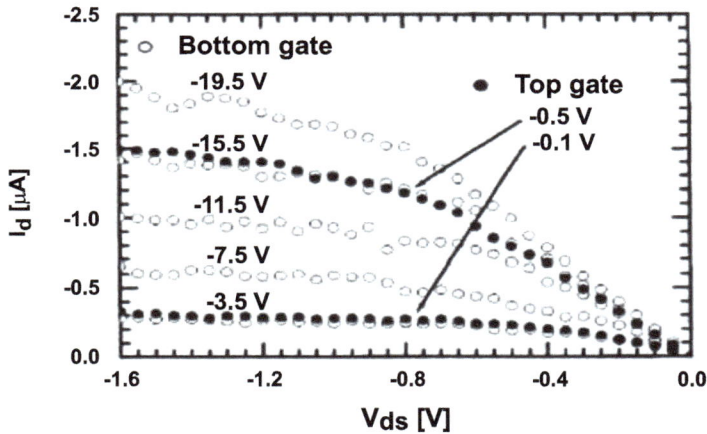

Fig. (3). Shows drain-source I-V curves for both first and second generation CNFETs [4].

The top gate of the second generation and the bottom gate of the first generation are contrasted in the graph. The top gate layout, as expected, employs substantially lower gate voltage levels. The individual transistors can be used in the same manner as MOSFETs. This is a result of the striking similarities in MOSFET architecture. A sample is shown in Fig. **(4)** [5].

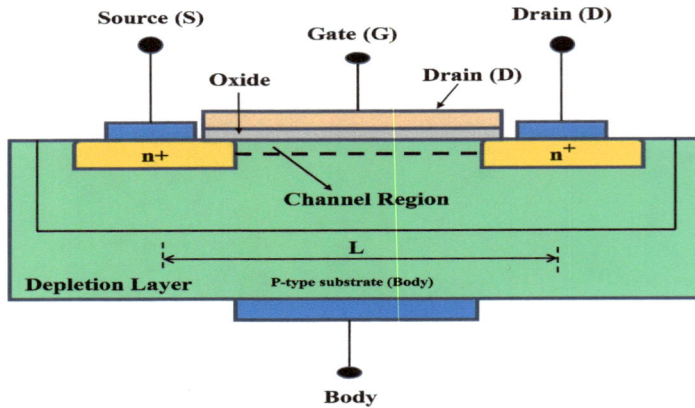

Fig. (4). Diagram of a Field-Emitting Transistor Device [5].

A top-gated framework is incorporated into both designs. Actually, the channel is the only thing that differs. The second-generation CNFET's ability to be a good fit for high-frequency operations with few modifications is another benefit. The first generation is unable to accomplish this due to the significant capacitance overlap resulting from the three electrodes. Another method for creating the 2nd generation design is to manufacture the gate out of an electrolyte. Fig. (**5**) [6, 41] shows a schematic diagram.

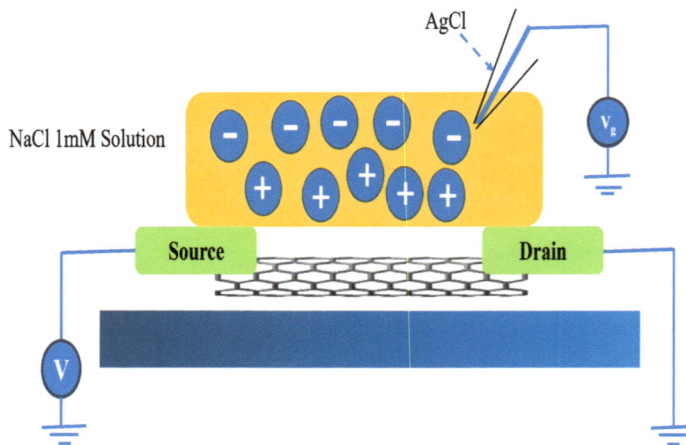

Fig. (5). Schematic of a CNFET with an electrolyte solution as the gate [6].

When an electrolyte solution is used as the gate, the capacitance between the nanotube and the gate is improved in comparison to first-generation CNFETs.

Performance improves as capacitance increases. Fig. (**6**) depicts the current-voltage curves for an electrolyte-gated CNFET.

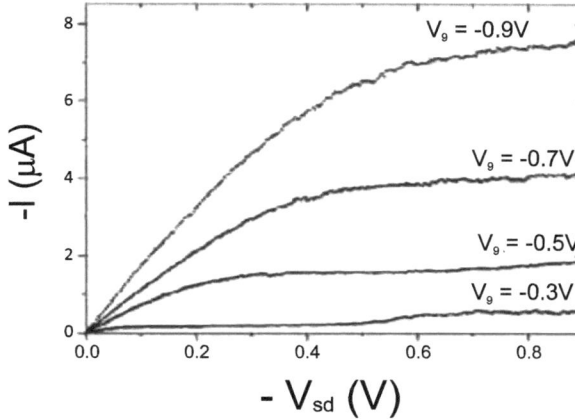

Fig. (6). I-V curves for the electrolyte-gated CNFET [6].

The shapes are very similar to those of MOSFET and metal top-gate designs. In the metal top gate design, the saturation threshold is reached earlier than in the MOSFET architecture. This indicates the possibility that this design choice would use less power than the other two varieties of transistors. However, there are disadvantages to employing a liquid electrolyte. To avoid leakage and air from leaking through and turning an n-type into a p-type, the electrolyte needs to be sealed, requiring extra manufacturing steps.

pH Sensors

Introduction

Low-dimension nanomaterial-based Uni-polar transistor (LD-FET) biosensors have gained a lot of attention in recent years due to their potential applications in various fields such as medicine, chemistry, and industry [7-11]. Among the various materials available, semiconducting carbon nanotubes (semi-CNTs) offer several advantages including extremely thin dimensions, excellent electrical properties and stability, and strong compatibility with biological systems. These inherent advantages make CNT-FETs particularly well-suited for biosensing applications [12-16]. One commonly used type of biosensor is the pH sensor, which utilizes changes in channel conductivity to indicate the pH level of the sample solution. FET pH Sensors face a more challenging operational environment compared to standard Uni-polar transistor devices because they necessitate exposing the semiconductor channels to the analyzing solution [36]. Complex interactions at the interface of the sensor, like the permanent harm caused by strong bases or acids, and electrical disruptions generated by ion loading, often lead to significant stability issues. These issues manifest as inconsistent signal responses, limited testing ranges, high hysteresis, and lack of repeatability [17-21]. Therefore, establishing stability in the environment is crucial when developing pH sensors based on CNT-FETs. Some groups have tackled this issue by utilizing Uni-polar transistors with extended gates to detach the channel from the solution phase, thereby enhancing equilibrium within the ecology or ecosystem [22–26]. However, this method often compromises high performance in favor of stability [27,28]. Another strategy involves employing transistors with floating gate field-effect (FG-FETs) [29,30]. By applying a nanometer-thick high-κ dielectric layer to coat carbon nanotubes, the biosensing interface transitions from a CNT solution to an insulator, leading to enhanced stability with a minimal impact on sensing performance. Consequently, establishing a stable biosensing interface utilizing floating gate CNT-FETs is crucial for developing environmentally stable FET biosensors [36]. In current CMOS technology, HfO2 serves as the substance used as a gate dielectric to shield the semiconductor channel [31,32] owing to its high dielectric constant and chemical and physical stability. Applying a thin layer of HfO_2 through the gate dielectric enhances the sensitivity of contact.

Our research introduces a floating gate- CNT-FET biosensor that utilizes chemical vapor deposition (CVD) for the sensing surface and Y_2O_3 as a seed layer for the growth of HfO_2 on buckytubes [36]. After undergoing rigorous physical and chemical treatments, the proposed device exhibited outstanding environmental stability. To further validate its sensing capabilities, a biosensor for pH sensing was developed using this floating gate CNT-FET technology. It has accomplished an extensive detection range, notable sensitivity, the ability for continuous monitoring, and minimal hysteresis [36]. Overall, the data demonstrate that this floating gate CNT-FET biosensor has extensive applications and is particularly effective in challenging conditions and when combined with other advanced approaches.

The pH biosensor FG CNT-FET

Dynamic testing of the device's channel was conducted by plating and introducing solutions (10.0 L) with varying pH values (ranging from 1.5 to 13) into the system. Microfluidic channel devices were utilized for this purpose. Subsequently, the apparatus underwent nitrogen drying and cleaning with deionized water (DI water). Following this, a repeatability test was performed using solutions with pH values of 13.5, 7.4, and 2.3, respectively. Moreover, a hysteresis test was repeated using pH 4.23 and pH 10.23 solutions, each stabilized for 90 seconds in solution At a consistent source and the gate and drain power settings. All the trials were conducted with a bias voltage of Vds = -0.30 V, with (Hg/Hg$_2$Cl$_2$) taken as an electrode for reference. The electrode applied a voltage in the range of -1.8 V to -0.40 V at 21°C to the solution. The measurements were documented using the Keithley instrument 4400A-SCS throughout the experiments [42-45].

RESULTS AND DISCUSSION

The Supplementary Information (Fig. **7**) outlines the construction methods of FG buckytube -FETs using the widely adopted complementary- Metal-Oxide-Semiconductor called the semiconductor technique. Fig. (**7a**), illustrates the structural schematic diagram of the floating gate buckytube-FET device, while Fig. (**7b**) showcases a four-inch wafer containing a buckytube-FET array. Fig. (**7c**) presents an optical microscope view of an individual unit (0.90 mm x 0.90 mm) among an array of 17 FET devices. Fig. (**7d**) presents a scanning electron microscope image of a single FET device featuring a channel with dimensions of 21 μm in length and 51 μm in width. The region with the yellow dots has carbon nanotube coverage or Hafnium Oxide. Prior to this, the buckytube surface was seeded with Y$_2$O$_3$, facilitating tight binding of the Hafnium Oxide high-layer on the buckytube, as depicted in Fig. (**7e**). Fig. (**7f-h**) demonstrate the characteristics of carbon nanotube-FET devices electrically.

The typical curve of transfer in Fig. (**7f**) illustrates a typical p-type FET device with a 104 on or off ratio. Carrier transportation in the hole domain is facilitated by the interaction of inert metals like palladium, and the high on or off ratio is attributed to the high efficiency of semiconductors of the carbon nanotube material. Sub-threshold swing was investigated for 17 devices post gate oxide deposition, with an average SS of 120 mV/dec, slightly higher than in previous investigations[7]. The Y$_2$O$_3$/HfO$_2$ thin film in FG carbon nanotube -FETs enhances SS by creating an electric double layer coupled with a sequence of capacitance, acting as a connecting capacitance between the solution and sensing layers. It is worth noting that improvements in other devices' performance outweigh this slight SS preference [46].

The output characteristic curve at room temperature (Fig. **7g**) displays excellent source and drain electrodes making an ohmic contact and carbon nanotube channel evident from the linear curve. Fig. (**7**) also demonstrates minimal hysteresis, indicating a strong carbon nanotube-dielectric interaction [36]. The durability assessment of floating gate carbon nanotube-FET biosensors in different chemical environments involved immersing 17 floating gate carbon nanotube-FET devices on a single chip unit in various chemical solutions for 30 minutes. These solutions encompass pH solutions, diverse organic phases, and ionic species and strengths. To mimic typical bioassay samples, an organic solution comprising $(CH_3)_2SO$, $(CH_3)_2N-CH$, N-Methyl-2-pyrrolidone, and CH_3OH was chosen due to its inclusion of both liquid and non-liquid phases. Solutions with elevated concentrations of ions utilizelike calcium, magnesium, sodium, potassium and chloride to simulate saline environments [36]. Multiple pH values were employed to assess the impact of pH variations on the devices. By using transfer curves for testing, the essential electrical properties were assessed and contrasted pre-and post-treatment. The results are depicted in Figs. (**8a–d**). A linked chart illustrates the differences in on-state current among the 17 devices, while Figs. (**8e–g**) display the SS as determined from Figs. (**8a–d**). The treatments showed a minimal impact on all devices, maintaining the gating effectiveness of the device under consideration and highlighting the impressive practical stability of the HfO_2 thin film. A slight shift in A threshold voltage (5th) was detected, attributed to the existence of hydroxyl groups on the HfO_2 surface [36]. This adjustment does not indicate a deterioration in device performance but underscores the importance of certain procedures at the FET sensing interface configuration to generate an appropriate density of hydroxyl groups, crucial for effective bridging between FET devices and bio probes during assembly. Additionally, SEM imaging in Fig. (**7**) confirmed that the gate electrical layer remained intact even after exposure to various chemical solutions [46-50].

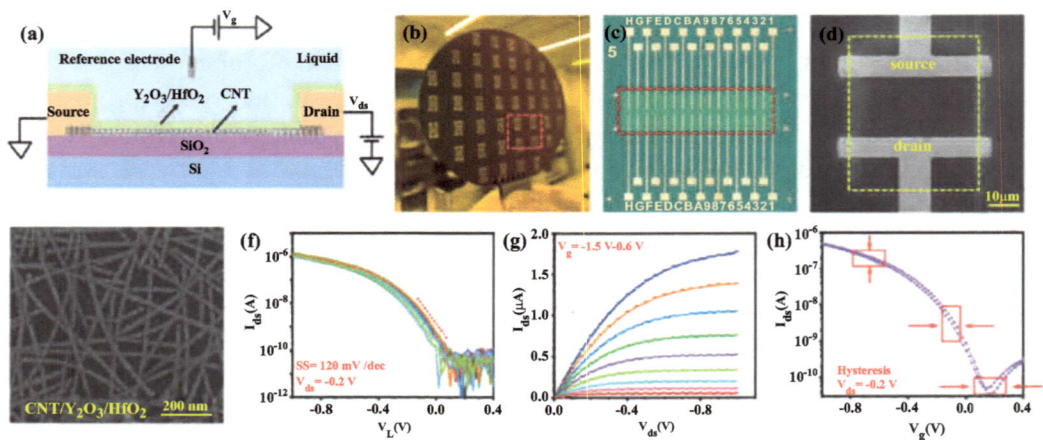

Fig. (7). Wafer-scale fabrication of FG CNT-FET biosensors [36].

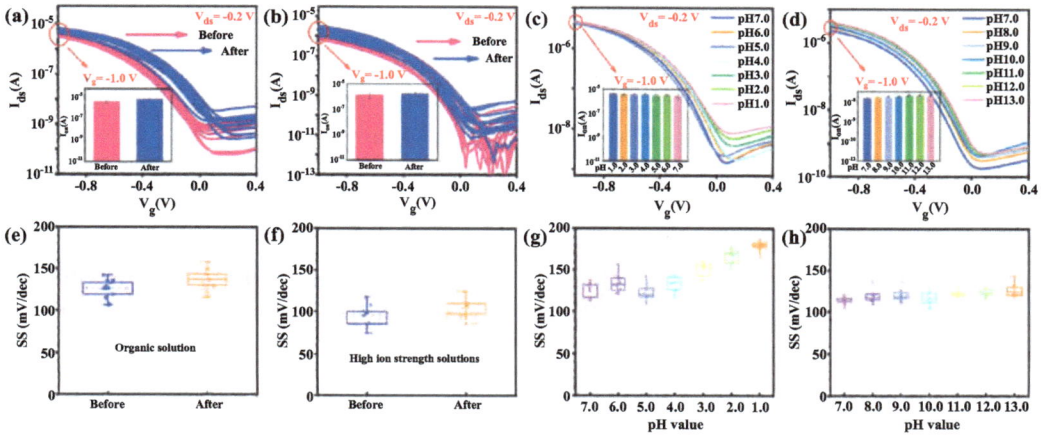

Fig. (8). (a) A structural diagram. **(b)** Photograph shot of first-generation CNT-FET biosensor arrays on a 4-inch wafer. **(c)** One of the biosensor arrays, as seen via an optical microscope. **(d)** SEM image of FG CNT-FET biosensor (size bar = 10 μm). **(e)** Carbon nanotubes in the channel of an FG CNT-FET biosensor coated in Y2O3/HfO2, as seen through a partial SEM. The scale bar is 200 nm. **(f)** Transfer characteristics, **(g)** output characteristics, and **(h)** Hysteresis curves of the FG CNT-FET device.

(b) Transfer curves of characteristic for FG CNT-FET biological sensors treated with **(a)** biological solvent, **(b)** strongly ionic solution, and **(c, d)** varying pH solutions. Insert the current value of FG CNT-FET biological sensors before and after various condition treatments at Vg = 1.0 V (n = 17). The SS values of the FG CNT-FET device after treatment with **(e)** organic solvent, **(f)** highly ionic solution, and **(g, h)** various pH solutions [36].

Exceptional chemical stability was observed in the floating gate-carbon nanotube. Uni-polar transistor biosensors can be attributed to the intrinsic characteristics of the HfO2 material [33]. This stability, combined with its reliable biological sensing interface, makes the biosensor suitable for a wide range of application scenarios. In addition to chemical stability, physical stability is also crucial for device performance. It is important to take into account how the device will be affected by oxygen plasma therapy, extended storage, and temperature variations in the surrounding area.

The oxygen plasma therapy of the carbon nanotube-FET surface is a critical step in integrating microfluidic devices, making it essential for the practical utilization of uni-polar transistor biosensors [36]. This treatment allows the precise modification of liquids at the micro-scale, significantly enhancing the functionality of FET biosensors. Before deployment in real-world applications, investigations into the device's shelf life and storage temperature are necessary to ensure long-term reliability. Due to its superior acid-base stability, the proposed floating gate CNT-FET device is well-suited for the measurement of potential of hydrogen applications. The HfO_2 interface, which contains a significant amount of hydroxyl compounds, experiences protonation and deprotonation of hydroxyl groups with variations in pH levels [34]. Fig. (**9**) depicts this mechanism, where the addition of hydrogen of hydroxyl groups into $-OH^{2+}$ in acidic solutions results in increased positive charges on the HfO_2 surface, leading to a higher potential barrier for holes and a reduced on-state current. Conversely, in alkaline environments, the removal of hydrogen of -OH groups into -O- results in an increased on-state current due to the lowered Schottky barrier to holes caused by the buildup of negative charges present on the HfO_2 surface. This pH-dependent charge modulation mechanism is crucial for accurate pH sensing with the buckytube -FET biosensor.

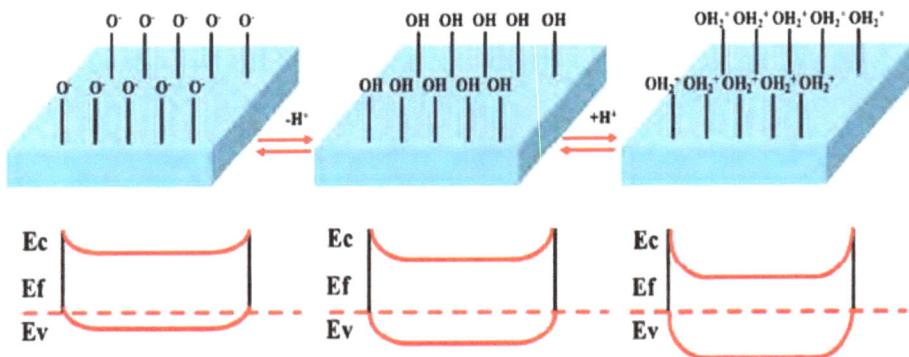

Fig. (9). Schematic representation of the sensing mechanism for pH measuring in an FG-FET biosensor [36].

The ion current (Ion) consistently increases with rising pH values, as shown in Fig. (**10a**), aligning with the pH detection theory utilized in the suggested carbon nanotube Uni-polar transistor (CNT-FET) biosensor. Proton electrostatic doping on carbon nanotubes causes a gradual shift in threshold voltage (Vth) to the right. Overall, the floating gate buckytube -FET biosensor maintains consistent electrical characteristics with changes in ion current and Vth shifts across different pH solutions. Fig. (**10a**) presents the Vth values at various pH levels, which are then fitted in Fig. (**10b**). The responsiveness was calculated and displayed in Fig. (**10b**) as 68 mV/pH (1 ~ 4.5), 33 mV/pH (5.45 ~ 8.67), and 46.67 mV/pH (8.8 ~ 12.87), respectively. These values closely approach the Nernst limit of 60 mV/pH, indicating distinct responses and high sensitivity. The proposed CNT-FET biosensor demonstrates good linearity, as depicted in Fig. S4, with correlation coefficients (R2) of hydrogen ion concentration 1.0 (1.33 ~ 5.55), 1.0 (5.55 ~ 8.88), and 0.977 (8.77 ~ 12.66) [36].

The carbon nanotube-FET biosensor uses the protonation and deprotonation process of hydroxyl molecules on the HfO2 surface as its pH-detecting mechanism [36]. The acidity or alkalinity of a solution influences the rate of surface protonation and deprotonation phenomena, thereby affecting the device's sensitivity. The biosensor exhibits higher sensitivity in acidic environments due to faster protonation rates than deprotonation rates, while in alkaline environments, deprotonation rates dominate. Near-neutral pH conditions result in lower sensitivity due to balanced protonation and deprotonation rates. Furthermore, an evaluation was conducted on the pH sensing capabilities of CNT-FETs fabricated with either Yttrium oxide or Alumina as the gate dielectric. Fig. (**10**) presents the electrical testing results, with microscopic and characterization findings also depicted. After testing with multiple pH sample solutions, irregular responses and device failure were observed. Microscopic characterization revealed dielectric layer erosion as the cause of failure in both Yttrium oxide and Alumina devices. These results underscore the importance of using exceptional stability in gate dielectric components (such as Hafnium Oxide) for Uni-polar transistor-type pH sensors. Real-time monitoring is achievable with the floating gate CNT-FET biosensor at a low operating voltage of 0.8 V. Dynamic test results with various pH solutions are illustrated in Figs. (**10c and d**), where pH variations influence real-time source and drain currents (Ids). The device's IDs show rapid variations with changes in pH, indicating sensitivity to proton disruption. Additionally, repeatability and hysteresis testing demonstrate consistent and reliable performance of the biosensor across multiple cycles and different pH levels, as evidenced in Fig. (**10e**).

The goal of the investigation is to use numerous dynamic testing cycles, repeatability between pH 4.22 and pH 9.92 was used to understand the interplay between ions at the solid-liquid interface and potential sensing interface defects. [36]. Hysteresis was quantified by computing the difference among the currents at the pH value that is both the highest and minimum. As depicted in Fig. (**10f**), the device demonstrated high resistance to hysteresis, with a hysteresis window of 600 pA observed at pH 4.29 and pH 10.3 during pH switching loops (4.3-10.3-4.3-10.3-4.3). This hysteresis window is smaller than that reported for most previous pH sensors [35]. The floating gate CNT-FET biosensor exhibited exceptional performance and a stable HfO2 sensing interface with minimal flaws. This study utilized a CMOS-compatible approach to fabricate FG carbon nanotube Uni-polar transistor based pH sensors, incorporating a 10 nm thick layer of robust HfO_2 on CNTs using Y_2O_3 as the seed layer. The device demonstrated excellent functionality under challenging conditions, including various chemical and physical operations. Additionally, it shows potential compatibility with a PDMS module, enabling the creation of a microfluidic sensing device. The biosensor's capability for pH detection was validated, showcasing a wide pH measurement range of 1.0 to 13, along with reliability and low hysteresis.

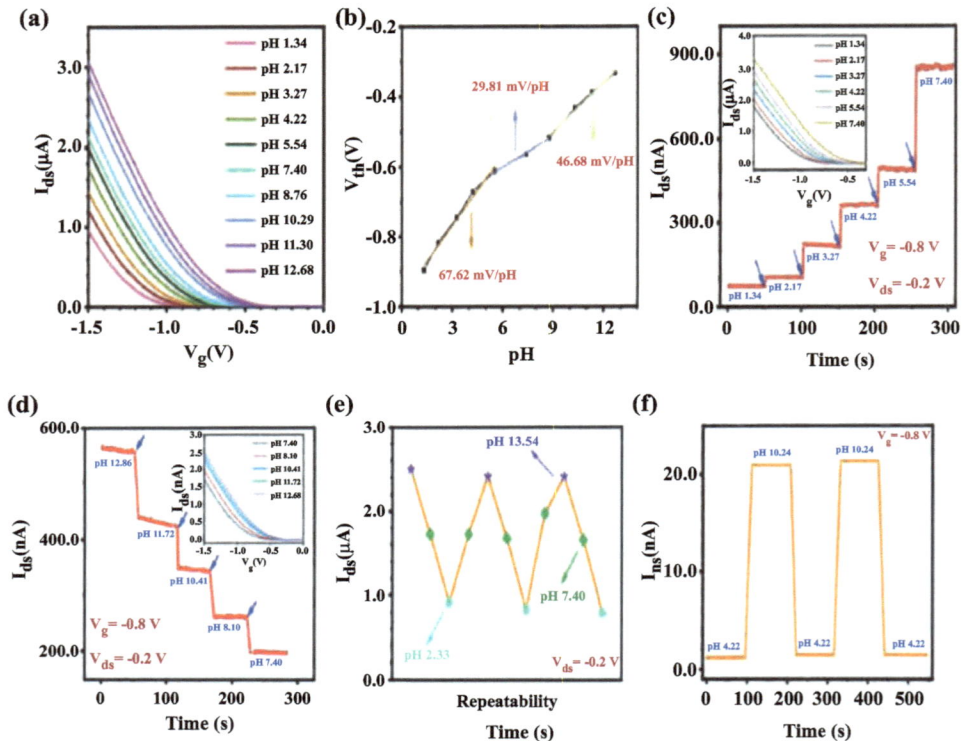

Fig. (10). Displays the transfer characteristic curves (Ids - Vg) of several pH solutions (ranging from 1.34 to 12.68) during a static test. (**b**) The Vth of different pH values and sensitivity. (**c, d**) Real-time current responses with different pH solutions. The repeatability of biosensors using (**e**) repeatability and (**f**) hysteresis [36].

Overall, the FG carbon nanotube Uni-polar transistor biosensor exhibits promise for use in demanding test environments and compatibility with diverse biotechnologies. These advancements contribute to the progress of FET biosensors, particularly in the testing at the point of sale industry.

DETERMINATION OF CARBAMATE PESTICIDES

Worldwide, carbamate insecticides such as carbofuran and carbaryl are used to manage insect-homes and crops. If fungicides, herbicides, and insecticides found their way into food and drinking water, they might have major health consequences for humans. As a result, there is a critical need for the creation of quick, easy, accurate, and focused testing tools. It is critical to test food and environmental samples on-site for carbamate detection. Semiconductors have recently become popular. Unipolar transistors (FETs) based on single-walled carbon nanotubes have shown several advantages. High carrier mobility is a significant advantage of Unipolar transistors according to single-walled carbon nanotubes, a high on or off ratio, electron transport that is nearly ballistic, and label-free identification are among the advantages. Real-time response is another feature. In this work, s-single-walled CNTs decorated with cobalt ferrite (CFO) nanoparticles were synthesized and investigated [36]. The non-enzymatic detection of carbaryl and carbofuran was carried out using a cobalt ferrite oxide-s-SWCNT/FET system that was prepared beforehand. The cobalt ferrite oxide / s-SWCNT hybrid film exhibited outstanding sensitivity and selectivity in the sensing system, achieving the estimated lowest limits of detection for carbaryl at 0.11 fM and for carbofuran at 0.07 fM. The linear detection range spanned from 10 to 100 fM [36]. In order to demonstrate the sensor's usefulness, it was also used to find carbaryl in samples of tomatoes and cabbage samples. This efficiency might be attributed to carbamate oxidation by cobalt ferrite oxides' strong catalytic activity, which altered the charge transfer mechanism on the semiconductor-single-walled CNTs/ Uni-polar transistor conduction channel. This research describes a unique cobalt ferrite oxide/s- single-walled CNTs sensing device that can be employed to find traces of pesticides in food items.

Carbamate Detection on Cobalt Ferrite Oxide/s-Single-walled Buckytubes

For s-single walled-CNT/FETs to function as a chemical sensor, they need to be repeatable and operate at low voltages. Several electrical measurements were conducted in ambient conditions to validate the reliability of cobalt ferrite oxide/semi-SWCNTs-FETs. The experiment consisted of adjusting the gate voltage within the range of -3.99 to +3.99 V while keeping the source voltage and drain constant of 0.20 V. The saturation zone's field-effect mobility was assessed to be 1.6×10^{-3} cm^2/Vs, accompanied by roughly on or off voltage ratio of -2.4×10^7. These numbers were adequate for using cobalt ferrite oxide/semi-SWCNTs-FETs in chemical detection applications. The cobalt ferrite oxide/s-SWCNTs-FET's strong mobility and good on or off ratio indicated that it can be used reliably in carbamate detection.

Our study reveals that the introduction of carbaryl or carbofuran influences the field effect and electrical conductivity of the cobalt ferrite oxide/s-single-walled buckytubes Uni-polar transistor (FET) device. The transfer characteristics of the cobalt ferrite oxide/s-SWCNTs were analyzed using the FET device. The introduction of carbaryl and carbofuran (Fig. **11a** and **11c**) respectively in 0.10 M PBS (phosphate-buffered saline) is illustrated in Fig. (**11a** and **11c**). This illustrates how the transfer curve shifts negatively when carbamate concentrations rise. This change is explained by the attachment of carbaryl or carbofuran on the outer layer of s-single-walled buckytubes coated with cobalt ferrite oxide, leading to charge transfer to the s-single-walled buckytubes and inducing a negative charge on the CNTs. Consequently, this shifts the threshold voltage towards a more negative gate voltage. The link between the electrochemical and physical oxidation routes of carbaryl and carbofuran is shown in Fig. (**11b** and **11d**). Within the range of values of 10 to 100 fM, there is a linear link between changes in threshold voltage. Each data point in the figures represents the average of three evaluations or the average deviation of VTH fluctuations under the same conditions (n = 3). It was estimated that the samples' limits of detection for carbaryl and carbofuran were 0.12 fM and 0.067 fM, respectively, using the equation (LOD = 3SD/S), where SD is the standard deviation, and S is the slope of the calibration curve. The fact that these results are the lowest for carbaryl and carbofuran among previously published electrochemical detectors (see Table **1**) suggests that chemically detectable devices that are based on s-single-walled buckytubes -FETs have the potential to establish updated guidelines for these compounds' low-level detection in several materials, respectively, using the following equation (LOD = 3SD/S).

To show this, numerous typically occurring chemicals and metal ions prevalent in fruits and veggies were introduced, and their impact on interference was examined using the cobalt ferrite oxide/s-SWCNTs-Uni-polar transistor with a carbaryl concentration of 70 fM. Cobalt ferrite oxide/s-SWCNTs-Uni-polar transistor exhibited carbaryl selectivity, as shown in Fig. (**12a**). Despite introducing several additional molecules along with carbaryl, the current response to carbaryl remained constant (Fig. **12a**). These results imply that any modifications to the cobalt ferrite oxide/s-SWCNTs Uni-polar transistor (FET) device's electrical properties were solely attributed to an inert chemical process on the device, supporting our initial hypotheses [36]. However, considering the nascent stage of solid-state electrodes' development, further research is necessary for a comprehensive understanding of their operating mechanisms.

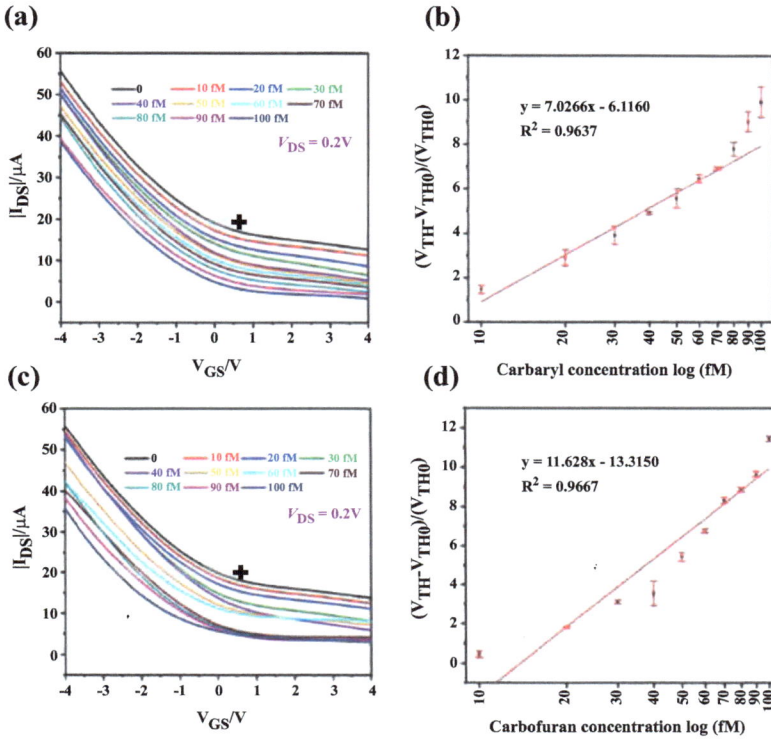

Fig. (11). (**a**) Transfer characteristics; (**b**) Changes in threshold voltage; (**c**) Transfer characteristics upon addition with carbofuran into 0.1 M PBS (pH 7.4) at VDS = 0.2 V. (**d**) Changes in VTH by adding carbofuran at various concentrations into a 0.1 M PBS (pH 7.4) at VDS = 0.2 V [37].

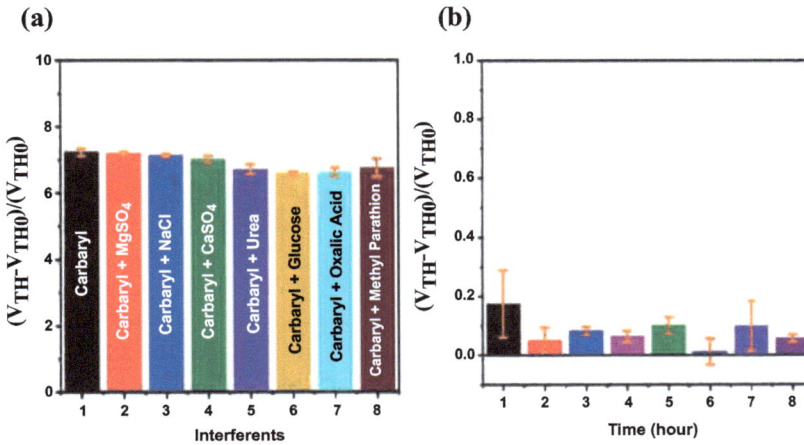

Fig. (12). (**a**) Changes in the VTH on addition Carbaryl; (**b**) Changes in the VTH on different time intervals [39].

Table 1. Analytical comparison.

Devices/Electrodes	Analytical Strategy	Technique	Target	Linear Range	Limit of Detection	References
AChE/PAMAM-Au/CNTs/ GCE	Enzymatic biosensor	DPV	Carbofuran	4.8×10^{-9} to 0.9×10^{-7} M	4.0×10^{-9} M	Qu et al. (2010)
AChE–TCNQ/SPE	Enzymatic biosensor	Chronoamperometry	Carbofuran	0.001–0.75 µM	0.001 µM	Nunes et al. (2004)
NF/AChE–CS/SNS–NF/ GCE	Enzymatic biosensor	CV	Carbofuran	10^{-12} to 10^{-10} M; 10^{-10} to 10^{-8} M	5.0×10^{-13} M	Yang et al. (2013)
Gd₂S₃/RGO/GCE	Non-Enzymatic sensor	DPV	Carbofuran	0.001–1381 µM	0.0128 µM	Mariyappan et al., (2021)
AChE-MWCNTs/GONRs/GCE	Enzymatic biosensor	Impedance	Carbaryl	5–5000 nM	1.7 nM	Liu et al. (2015)
AChE/SAMmix/Au	Enzymatic biosensor	Chronoamperometric	Carbaryl	0–1.75 µM	$3.45 \times$	Cancino et al. (2013)
PDDA-SWCNTs/AChE	Enzymatic sensor	Amperometric	Carbaryl	10^{-6} to 10^{-11} g/L	10^{-10} M	Firdosz et al. (2010)
GA/urease enzyme/graphene/Pt	IDS versus VGS	Carbaryl	2.58×10^{-7} to 2.58×10^{-2} µg/mL	10^{-12} g/L to 10^{-8} µg/mL	Thanh et al. (2018)	
rAChE-CNT/FET Ag–ZnO/s-SWCNTs-FET	Enzymatic sensor	IDS versus VGS	Acephate and Fenitrothion	1 pg/L to 1 ng/L; 1×10^{-16} to 1×10^{-4} M	-	Ishii et al. (2008)
-	Non-Enzymatic sensor	IDS versus VGS	Methyl parathion	0.27×10^{-16}	-	Kumar et al. (2020)
CFO/s-SWCNTs/FET	Non-Enzymatic sensor	IDS versus VGS	Carbaryl	10 fM to 100 fM	0.11 fM	This work
CFO/s-SWCNTs/FET	Non-Enzymatic sensor	IDS versus VGS	Carbofuran	10 fM to 100 fM	0.07 fM	This work

To determine the created sensor's capabilities for continually taking measurements at room temperature, we tested the inactive chemical procedure stability at every hour for a period of up to 8 hours, as shown in Fig. (**13**). The observed VTH variations were caused by a monotonic drift in the cobalt ferrite oxide layer. The potential output began to stabilize notably at 0.1 M PBS (pH 7.40), which is typically when the drift was most pronounced. This is why the first hour displayed a higher value compared to subsequent hours. The oxide layer undergoes gradual protonation/deprotonation as hydrogen ions penetrate it, leading to the formation of a continuous sequence of sharp turns, resembling the letter "Z" or a series of "zigzags" in VTH drifting over time [38]. In contrast, the cobalt ferrite oxide /s-SWCNTs-FET device showed minimal variations for up to 480 minutes. It was noted that the recently developed cobalt ferrite oxide /s-single-walled buckytubes-FET sensor can be used for continuous measurements throughout the day after being kept at ambient temperature.

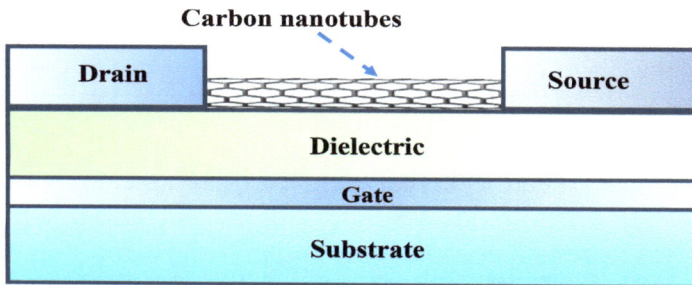

Fig. (13). Back-gated GFET.

Detecting Carbaryl in Genuine Samples with Cobalt Ferrite Oxide/s-Single-walled Buckytubes

To detect carbaryl in real-world samples, the researchers used Cobalt Ferrite Oxide/s-SWCNTs-FETs. Samples of tomatoes and cabbage were taken from a nearby market, cleaned, and then sprayed with a carbaryl pesticide solution. They were then left to dry for 120 minutes at room temperature. After that, the samples were submerged for a day in 30 milliliters of acetone [37].

Using carbaryl samples made from tomato and cabbage extracts, VTH alterations were investigated by calculating IDS-VGS transfer curves. The results showed that carbaryl was detected in tomato and cabbage samples using the cobalt ferrite oxide/s-Single walled CNT-Uni-polar transistor sensor. Carbaryl recovery values for 80 fM were 101% and 104% for tomato samples (a) and (b), respectively, while 99.50% was recovered for cabbage samples (Table **2**). The sensor's measured concentration is higher than predicted when the recovery value is more than 100%. Many things can cause this, such as the existence of matrix effects that amplify the response of the sensor. Nonetheless, the carbaryl recovery outcomes in the samples of cabbage and tomatoes were acceptable. These findings showed that the cobalt ferrite oxide/s-single walled CNT Uni-polar transistor device could serve as a carbamate pesticide detection sensor [37].

Table 2. Tomato and cabbage samples are taken.

Samples	Carbaryl added	Total found	RSD (n=3) in Percentage	Recovery in Percentage
sample - I	81 fm	82.5 fm	3.37	101.8
sample-II	81 fm	84.8 fm	1.96	104.7
Cabbage sample-I	81 fm	80.6 fm	2.47	99.5

Adjusting the V_{TH} revealed that carbaryl and carbofuran additions had a favorable effect on the cobalt ferrite oxide/s-Single walled CNT Uni-polar transistor. The created sensor was extremely responsive to carbaryl and carbofuran between 10 and 100 fM., with projected LODs of 0.07 fM and 0.11 fM, in that order. To further show that this non-enzymatic sensor may be used to track the nutritional value of food, it was also tested for the presence of carbaryl in samples of cabbage and tomatoes. With findings available in 15 minutes, our non-enzymatic cobalt ferrite oxide/s-Single walled CNT Uni-polar transistor assay provides a quick and easy substitute for conventional antibody-based assays. Furthermore, by combining metal nanoparticles with s-SWCNT/FET, this FET sensing model can be used to create a variety of FET-based sensors for many biological substances. Our results amply demonstrated the potential of the technology. The Metal Oxide Semiconductor Uni-polar transistor plays a crucial role in modern integrated circuits (ICs) and has seen significant advancements due to complementary metal oxide semiconductor (CMOS) technology. This progress has greatly contributed to the rapid evolution of information technology over the past few decades. The continuous downscaling of MOSFETs has led to substantial performance improvements, largely driven by innovations in materials science. Innovative materials such as high-Q gate dielectrics, metals, silicides, and nitrides have played a key role in advancing CMOS technology. Despite these material innovations, the active channel material in MOSFETs has remained relatively stable due to the scalability and processability of silicon (Si) technology.

However, as device dimensions, particularly channel lengths, approach, and fall below 10nm, new challenges arise. Direct tunneling between the source and drain and the emergence of severe short channel effects become significant obstacles for scaling Si devices further. Consequently, numerous academic and corporate research endeavors have focused on exploring novel semiconductors as potential channel materials. The objectives include enhancing carrier mobility and improving electrostatics at the nanoscale level, addressing the limitations faced by traditional Si-based devices in ultra-small dimensions.

CNTFET Type

So far, carbon nanotube field effect transistors have been classified as follows:

a) Back gate carbon nanotube-FET

b) Top gate carbon nanotube-FET

c) Wrap-around gate carbon nanotube-FET

d) Suspended carbon nanotube-FET

Back Gate Carbon Nanotube-FET

Carbon nanotube- Uni-polar transistors were first created by pre-patterning metal strips in parallel over a silicon dioxide substrate, followed by the random coating of CNTs on top [40]. The semiconducting carbon nanotubes positioned between the two metal strips fulfill all necessities for a basic Uni-polar transistor (FET) setup. One metal strip serves as the "drain" contact, while the other acts as the "source" contact. A metal contact on the back allows for the gating of the semiconducting CNT, and the SiO_2 substrate can function as the gate oxide as shown in Fig. (**13**). The transconductance ($gm = dI/dVG$) of back-gated CNTFETs investigated thus far is relatively low, around 10-9 A/V, due to their high contact resistances (1MΩ). This high resistance is attributed to the devices' weak Vander Waals binding to the noble metal electrodes in the chosen "side-bonding" configuration. Source and drain electrodes composed of transition metals compatible with silicon, such as Ti or Co, are produced on top of the SWNT after it has been distributed over the SiO2 layer. At the source and drain electrodes, low resistance Co connections or TiC contacts are created by annealing the material again at 400°C (for Co) or 820°C (for Ti) in an inert atmosphere. The Figure (displays the current-voltage characteristics of a p-type CNTFET with metallic Co or TiC contacts.

Top gate carbon nanotube-FET

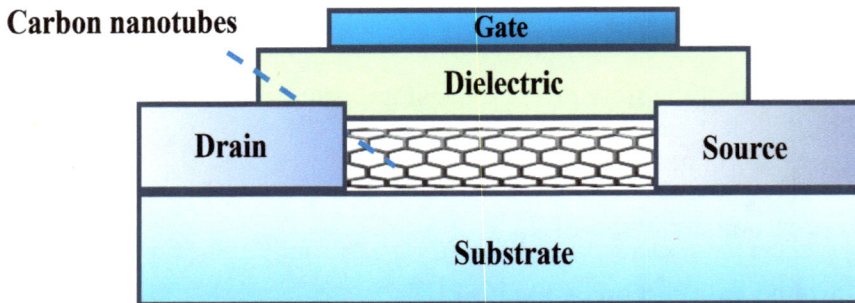

Fig. (14). Top-gated GFET.

The back-gate strategy was later replaced with a more sophisticated top-gate production technique by researchers (Fig. **14**) [41]. In the first stage, a silicon oxide substrate is a solution deposited with single-walled carbon nanotubes. Next, individual nanotubes are found using a scanning electron or atomic force microscope. Excellent quality lithography via electron beams is used to precisely define and pattern the source as well as the drain connections after isolating a single tube. Contact resistance is reduced by enhancing adhesion between the contact and CNT through a high-temperature annealing process. Subsequently, a thin top-gate dielectric layer is formed atop the nanotube using atomic layer deposition or evaporation techniques. Lastly, the procedure is finished when the top gate contact is placed onto the gate dielectric. Top-gated CNTFET arrays can be produced on the same wafer as back-gated CNTFET arrays due to the gate contacts' electrical isolation from one another. Moreover, a smaller gate voltage can create a larger electric field relative to the nanotube because of the thin gate dielectric. Despite their more involved manufacturing procedure, top-gated CNTFETs are frequently selected over back-gated CNTFETs due to these advantages.

Wrap-around gate carbon nanotube-FETs

Gate-wrapping CNTFETs (Fig. **15**), also referred to as gate-all-around CNTFETs, were introduced in 2008 [42] and represent an enhancement over top-gate device design. This device gates the entire length of the CNT rather than just the section closest to the metal gate contact. This feature is expected to increase the on or off ratio of the device and decrease the current loss in order to enhance the CNTFET's electrical efficiency.

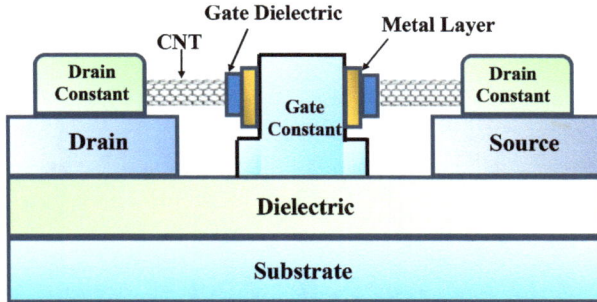

Fig. (15). Wrap-gated GFET.

Suspended carbon nanotube-FETs

To minimize interaction with the substrate and gate oxide, designers of CNTFET devices suspend the nanotube above a trench as shown in Fig. (**16**) [44]. This technique reduces spreading at the buckytube-substrate contact, thereby humanizing device performance [44 - 46]. Fabrication methods include growing suspended CNTFETs over trenches with catalyst particles [44, 46], transferring them onto a substrate and under-etching the dielectric beneath [46], and transfer-printing onto a trenched substrate [45]. One challenge with suspended CNTFETs is the limited choice of materials for the gate dielectric, typically limited to air or void. Applying a gate bias pulls the nanotube closer to the gate, limiting the extent of gating. Moreover, this approach is more suitable for shorter nanotubes, as longer ones may stretch and bow towards the gate, risking contact with the metal and device shorting. While suspended CNTFETs may not be ideal for commercial applications, they are valuable for studying the fundamental properties of pure nanotubes.

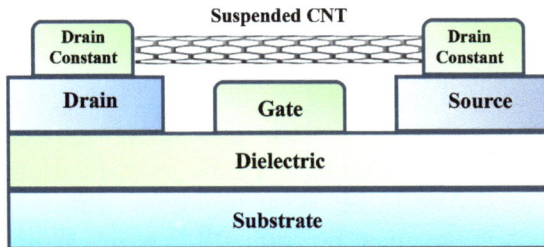

Fig. (16). Suspended GFET.

Structure

CNTs are graphene structures that have been rolled up and are made up of a hexagonal network of carbon atoms bound together by covalent connections.

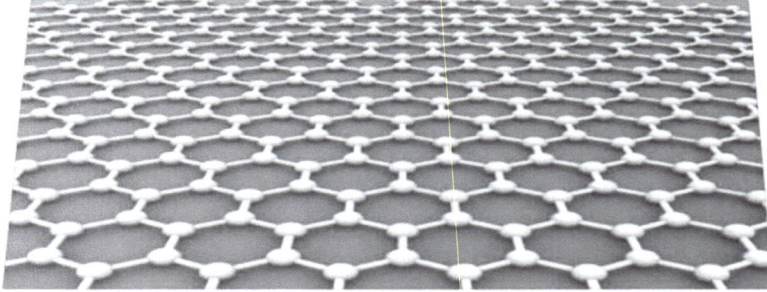

Fig. (17). Graphene's lattice structure.

Fig. (**17**) depicts the lattice structure of graphene. In a hexagon of carbon atoms, those in close proximity are not lattice-equivalent, while adjacent ones are. By selecting a carbon atom as the origin (0,0), the vectors extending from this origin to the nearest neighbouring atoms exhibit lattice equivalence. Within a hexagon, carbon atoms are non-lattice-equivalent when close together but become so when neighbouring. The vectors from the origin atom (0,0) to its adjacent atoms, which are lattice-equivalent, are labelled accordingly. The primitive unit cell of graphene is defined as a rhombus formed by vectors oriented at 60 degrees to each other. Rolling a graphene sheet into a tube results in a carbon nanotube [47].

The chiral vector of a labelled CNT corresponds to the circumference of the tube and is at a 90-degree angle to the tube axis T. This chiral vector can be mathematically expressed in terms of lattice vectors using the following equation:

$$\overrightarrow{C_h} = n\,\overrightarrow{a_1} + m\,\overrightarrow{a_2}$$

Process steps for creating complementary CNTFETs that are CMOS-compatible have been developed by recent research. The deposition techniques for carbon nanotubes must fulfill certain criteria to be viable for industrial adoption. They need to be scalable at the wafer level, ensuring high throughput at a low cost. Additionally, these techniques must be compatible with existing hardware, without introducing hazardous substances or particulates that are prohibited in industrial processes. A solution-based deposition method for carbon nanotubes, as described by Bishop et al., involves immersing a substrate in a solution containing carbon nanotubes. This allows the CNTs to adhere to the surface of the substrate, offering a practical approach to meet the aforementioned requirements. This 'incubation' method has several benefits for the first deployment of CNTFETs in these plants. First of all, the integration barrier is low (200 mm of homogeneous CNT deposition) [47].

The Similarities and Differences between Graphene and CNT

Both graphene and carbon nanotubes (CNTs) possess remarkable strength due to their sp^2 bonds, which are even more robust than the sp^3 bonds present in diamond. Both also exhibit incredibly high levels of chemical reactivity, electron mobility, and thermal conductivity.

Carbon nanotubes are thin films that are rolled like 3D tubes or cylinders, whereas graphene is a single thin layer 2D film. As an allotrope of carbon, graphene is classified as a semi-metal.

Fabrication of Carbon Nanotube-FETS

Steps in the process for creating complementary CNTFETs that are CMOS-compatible have been developed by recent research. Before being introduced to the industrial sector, this CNT deposition method needs to meet three crucial requirements: it needs to be compatible, efficient, and manufacturable (scalable at the wafer level, with high throughput at a low cost). Since the suggested CNTFETs were created using the same process as silicon wafers, they serve as a specific example of how CNT-based device fabrication can be done in a way that is compatible with CMOS technology [47]. The manufacture of pure CNT semiconducting solutions that fulfill the rigorous constraints for chemical and particle contamination has facilitated the development of these capabilities.

The study demonstrated the successful and uniform manufacturing of Carbon Nanotube Uni-polar transistors (CNTFETs) on industrially standardized 200 mm wafers. A total of 14,400 CNTFETs were produced across multiple wafers, ensuring uniformity and consistency with all outliers eliminated. The technology utilized for CNTFETs was at a 130 nm technical node. To evaluate the performance of CNTFET circuits, widely used Electronic Design Automation (EDA) tools and methods in the industry were employed. These tools were utilized to measure the total energy consumption and clock frequency of CNTFET circuits. The results indicated that CNT processing capabilities were not only viable but also showed potential for enhancing energy efficiency in CNTFET applications [47]. This achievement marks a significant step forward in the practical application and scalability of CNTFET technology in industrial settings.

Carbon Nanotube FET Applications

Bio Sensing and Healthcare

Using local electrolyte-gated CNTFETs, BSA (Bovine serum albumin) molecules have been effectively identified through the manipulation of channel conductance. High performance in electrical characteristics and sensing operation was demonstrated by local electrolyte-gated CNTFETs. Second, by employing CNTFETs functionalized with PNA probes on the Au back-gate surfaces of the devices, ultrasensitive real-time DNA hybridization detection was carried out. It was reported how aptamer-modified CNTFET biosensors were used to detect IgE label-free. In comparable circumstances, aptamers outperformed monoclonal antibodies in IgE detection. The sensitivity (signal-to-noise ratio) of CNTFET sensors was finally shown to have significantly improved. It was shown that pH sensor noise might drop without affecting signal strength [50]. Additionally, a modest amount of BSA (250 pM) was successfully detected using AC measurement, although DC measurement proved challenging. For upcoming, extremely sensitive sensors, such CNTFET devices hold great promise. It is anticipated that quick and high-throughput format CNTFET sensors will enable the sensitive multiplexed detection of many clinically significant biomolecules [50].

An external haemodialysis medium, an oxygen-based electrode, an interior partially permeable oxygen obstacle, and a thin layer of glucose oxidase (G-OX) make up a glucose-based biosensor. An enzyme electrode was created by placing the G-OX enzyme on top of an electrochemical detector [47]. The drop in oxygen concentration that was seen coincided with the concentration of glucose. Traditional chemical sensors and biosensors are different in two respects: Biosensors use biological substances as sensing components, such as proteins, polynucleotides or oligosaccharides, microbes, or specific biological tissues, to recognize bio-molecules or detect biological activities. Since the discovery of carbon-based nanomaterials, there has been significant interest in developing innovative biosensors using graphene and carbon nanotubes (CNTs). These nanomaterials possess exceptional electrical, mechanical, optical, and structural properties, leading them to be hailed as a new generation of nanoprobes. The unique characteristics of CNTs, such as their remarkable aspect ratio, sensitivity, chemical stability, and high conductivity, make them highly suitable for biosensing applications.

One of the key advantages of CNTs is their high rate of electron transfer, which is beneficial for biosensing applications. By leveraging graphene and CNTs in biosensor development, there is a potential to enhance the speed, precision, and reliability of sensors. The combination of these advanced materials promises improvements in biosensing technology, offering opportunities for more efficient and accurate detection in various fields. [47]

High-speed Memory

Over time, there has been a gradual advancement in the technology of carbon nanotube-based electrical devices, including sensors, Uni-polar transistors, switches, and electron field emitters. After the first carbon nanotube electromechanical memory, there were numerous studies on CNTFETs with memory properties. Although its charge storage stability is limited to 14 days, there is an opportunity for improvement. It offers good mobility. Another essential element of high-performance non-volatile memory is the speed with which it completes the writing and erasing tasks [47, 48]. The CNTFET memory's working frequency was formerly listed as 10 seconds. To enable carbon nanotube uni-polar transistor (CNTFET) memory to compete effectively with standard silicon-based memories, there is a critical need to improve performance metrics such as write and erase speeds. Marcus et al. developed a CNTFET-based memory system that utilizes 100 ns pulses for write and erase operations. This achievement represents a significant advancement compared to earlier documented operation rates for CNTFET memories, with speeds approximately 100 times faster. The remarkable speed achieved in this memory system is attributed to the use of HfO_2 as a gate oxide, which plays a crucial role in enhancing performance. The study not only evaluated write and erase speeds but also considered other important device characteristics such as retention and endurance performance. Memory cells based on CNTFETs under back gate control were specifically analyzed in the study, showcasing the potential for rapid advancements in CNTFET memory technology [49].

CONCLUSION

This work provides a comprehensive review of Carbon Nanotube Uni-polar transistors (CNTFETs). It specifically delves into the application of CNTFETs in pH sensors, confirming their suitability for continuous pH sensing applications. The

CNTFET device exhibited a high sensitivity of 68 mV/pH, capable of detecting a broad pH range from 1.30 to 13. Additionally, it demonstrated low hysteresis of 600 pA and efficient pH switching loops between 4.22 and 10.22. Furthermore, the study explored the amount of Carbamate pesticides using a non-enzymatic cobalt ferrite/s-SWCNTs-FET test. This test proved to be a rapid and straightforward alternative to traditional antibody-based tests, providing results within 15 minutes. Such advancements highlight the versatility and practicality of CNTFET technology in various sensing and detection applications. This Uni-polar transistor sensing paradigm, which includes metal nanoparticles and s-single-walled- CNT can be used to develop different Uni-polar transistors-based sensors for diverse substances, and carbon nanotube-FET designs, functionality, and traits of several carbon nanotube-FET varieties. Analysis of the performance of many features has been reported together with the operation and DC characteristics of CNTFETs.

Learning Objectives/Key Points

(1) The use of carbon nanotubes as the channel in CNFETs versus heavily doped silicon in MOSFETs is the primary difference between the two types of transistors. Complementary devices, semiconductors, are essential to both technologies.

(2) A biosensor for pH sensing was developed using this floating gate CNT-FET technology. It has accomplished an extensive detection range, notable sensitivity, the ability for continuous monitoring, and minimal hysteresis.

(3) The goal of the investigation is to use numerous dynamic testing cycles; repeatability between pH 4.22 and pH 9.92 was used to understand the interplay between ions at the solid-liquid interface and potential sensing interface defects.

(4) The FG carbon nanotube Uni-polar transistor biosensor exhibits promise for use in demanding testing environments and compatibility with diverse biotechnologies.

(5) The efficiency might be attributed to carbamate oxidation by cobalt ferrite oxides' strong catalytic activity, which altered the charge transfer mechanism on the semiconductor-single-walled CNTs/ Uni-polar transistor conduction channel.

(6) The semiconducting carbon nanotubes positioned between the two metal strips fulfill all necessities for a basic Uni-polar transistor (FET) setup. One metal strip serves as the "drain" contact, while the other acts as the "source" contact.

(7) A silicon oxide substrate is a solution deposited with single-walled carbon nanotubes. Excellent quality lithography via electron beams is used to precisely define and pattern the source as well as the drain connections after isolating a single tube.

Multiple Choice Questions (MCQs)

1). How many terminals does the FET transistor have?

a) One

b) Two

c) Three

d) Four

Answer : c) Three

2). The field-effect transistors used in _____

a) Amplifiers

b) Analog switch

c) Oscillator

d) All of the above

Answer : d) All of the above

3). The field-effect transistors have _____

a) Very high input impedance

b) Small in size

c) Low power consumption

d) All of the above

Answer : d) All of the above

4). Which one is a unipolar device?

a) FET

b) BJT

c) Both a and b

d) None of the above

Answer : a) FET

5). The operation of the BJT relies on _____

a) Free electrons

b) Holes

c) Both a and b

d) None of the above

Answer :C

6) A carbon nanotube (CNT)-FET pH sensor with enhanced environmental stability by introducing _____ film into the gate insulator

a) HfO_2

b) SiO_2

c) Al_2O_3

d) Both a and b

Answer : a) HfO_2

7) The FG CNT-FET fabrication techniques using the method is known as_____.

a) Semiconductor

b) Non-Semiconductor

c) CMOS

d) Oxide film

Answer : c) CMOS

8) The induced _____ thin film creates a series capacitance with the electric double layer (EDL), which acts as an interface capacitance between the solution and sensing layer

a) Hf_2O_2

b) Y_2O_3

c) Y_2O_3/HfO_2

d) Y_2O_4

Answer : c) Y_2O_3/HfO_2

9) In an alkaline environment, _____ results in the accumulation of negative charges on the HfO_2 surface, forming a lower Schottky barrier to holes and enhancing the on-state current.

a) Protonation of $-O^-$ groups into OH

b) Deprotonation of -OH groups into O^-

c) Deprotonation of -COOH groups into COO^-

d) Protonation of - COO^- groups into COOH

Answer : b) Deprotonation of -OH groups into O^-

10) A unique CFO/s-SWCNT sensing device that may be utilized to assess pesticide residues in food samples.

a) Cobalt ferrite

b) Chromium ferrite

c) Cadmium ferrite

d) None

Answer : a) Cobalt ferrite

11) Carbamates are used as sprays or baits to kill insects by affecting their brains and nervous systems.

a) affecting their brains and nervous

b) affecting their brains

c) affecting their brains and respiratory

d) affecting their nervous

Answer: a) affecting their brains and nervous

12) Which one is not a carbon nanotube field effect transistor

a) Back gate CNTFET

b) Top gate CNTFET

c) Around gate CNTFETs

d) Suspended CNTFETs

Answer : c) Around gate CNTFETs

13)Wrap-around gate CNTFETs are otherwise called as

a) gate-all-around

b) Back gate CNTFET

c) Top gate CNTFET

d) Suspended CNTFETs

Answer : a) gate-all-around

14) The chiral vector $\overrightarrow{C_h}$ can be expressed mathematically in the form of lattice vectors using the equation

a) $\overrightarrow{C_h} = n\overrightarrow{a_1} + m\overrightarrow{a_2}$

b) $\overrightarrow{C_h} = n\overrightarrow{a_1}$

c) $C_h = na_1 + ma_2$

d) $C_h = na_1 + ma_2$

Answer : a) $\overrightarrow{C_h} = n\overrightarrow{a_1} + m\overrightarrow{a_2}$

15) Which sensor is used to sense the components, such as proteins, polynucleotides or oligo, microorganisms

a) Electronic Sensor

b) Biosensor

c) Both a and b

d) None

Answer : b) Biosensor

REFERENCES

[1]   S. Zhang, "Review of modern field effect transistor technologies for scaling", In: *Journal of Physics: Conference Series,* vol. 1617. IOP Publishing, 2020, no. 1, p. 012054.

http://dx.doi.org/10.1088/1742-6596/1617/1/012054

[2]   P. Liu, Y. Liu, B. Cai, J. Li, and G. J. Zhang, "Probe-screened carbon nanotube field-effect transistor biosensor to enhance breast cancer-related gene assay," *Green Anal. Chem.*, vol. 2025, Art. no. 100267.

[3]   A. Kanwal, "A review of carbon nanotube field effect transistors", *Auburn University Samuel Ginn College of Engineering-in Alabama,* 2003. Available from:https://www. eng. auburn. edu/~ vagrawal/TALKS/nanotube_v3

[4]   R. Martel, V. Derycke, J. Appenzeller, S. Wind, and P. Avouris, "Carbon nanotube field-effect transistors and logic circuits," in *Proc. 39th Annu. Design Automation Conf. (DAC)*, 2002, pp. 94–98.

[5]   A. Sedra, K.C. Smith, T.C. Carusone, and V. Gaudet, *Microelectronic Circuits*, 8th ed. New York: Oxford University Press, 2020, pp. 1235–1236.

[6]   D. Srivastava, M. Menon, and Kyeongjae Cho, "Computational nanotechnology with carbon nanotubes and fullerenes", *Comput. Sci. Eng.,* vol. 3, no. 4, pp. 42-55, 2001.

http://dx.doi.org/10.1109/5992.931903

[7]   N. Gao, W. Zhou, X. Jiang, G. Hong, T.M. Fu, and C.M. Lieber, "General strategy for biodetection in high ionic strength solutions using transistor-based nanoelectronic sensors", *Nano Lett,* vol. 15, no. 3, pp. 2143-2148, 2015.

http://dx.doi.org/10.1021/acs.nanolett.5b00133 PMID: 25664395

[8]   J. Liu, X. Chen, Q. Wang, M. Xiao, D. Zhong, W. Sun, G. Zhang, and Z. Zhang, "Ultrasensitive monolayer MoS2 field-effect transistor based DNA sensors for screening of down syndrome", *Nano Lett,* vol. 19, no. 3, pp. 1437-1444, 2019.

http://dx.doi.org/10.1021/acs.nanolett.8b03818 PMID: 30757905

[9]   Vahid A. Mohammadi, J. Rosen, and Y. Gogotsi, "The world of two-dimensional carbides and nitrides (MXenes)", *Science,* vol. 372, no. 6547, 2021.

[10] H. Altug, S.H. Oh, S.A. Maier, and J. Homola, "Advances and applications of nanophotonic biosensors", *Nat. Nanotechnol.,* vol. 17, no. 1, pp. 5-16, 2022.

http://dx.doi.org/10.1038/s41565-021-01045-5 PMID: 35046571

[11] M. Negahdary, and L. Angnes, "Application of electrochemical biosensors for the detection of microRNAs (miRNAs) related to cancer", *Coord Chem. Rev.,* vol. 464, p. 214565, 2022.

http://dx.doi.org/10.1016/j.ccr.2022.214565

[12] R. Rao, C.L. Pint, A.E. Islam, R.S. Weatherup, S. Hofmann, E.R. Meshot, F. Wu, C. Zhou, N. Dee, P.B. Amama, J. Carpena-Nuñez, W. Shi, D.L. Plata, E.S. Penev, B.I. Yakobson, P.B. Balbuena, C. Bichara, D.N. Futaba, S. Noda, H. Shin, K.S. Kim, B. Simard, F. Mirri, M. Pasquali, F. Fornasiero, E.I. Kauppinen, M. Arnold, B.A. Cola, P. Nikolaev, S. Arepalli, H.M. Cheng, D.N. Zakharov, E.A. Stach, J. Zhang, F. Wei, M. Terrones, D.B. Geohegan, B. Maruyama, S. Maruyama, Y. Li, W.W. Adams, and A.J. Hart, "Carbon nanotubes and related nanomaterials: critical advances and challenges for synthesis toward mainstream commercial applications", *ACS Nano,* vol. 12, no. 12, pp. 11756-11784, 2018.

http://dx.doi.org/10.1021/acsnano.8b06511 PMID: 30516055

[13] F. Yang, M. Wang, D. Zhang, J. Yang, M. Zheng, and Y. Li, "Chirality pure carbon nanotubes: Growth, sorting, and characterization", *Chem. Rev.,* vol. 120, no. 5, pp. 2693-2758, 2020.

http://dx.doi.org/10.1021/acs.chemrev.9b00835 PMID: 32039585

[14] R. Eivazzadeh-Keihan, E. Bahojb Noruzi, E. Chidar, M. Jafari, F. Davoodi, A. Kashtiaray, M. Ghafori Gorab, S. Masoud Hashemi, S. Javanshir, R. Ahangari Cohan, A. Maleki, and M. Mahdavi, "Applications of carbon-based conductive nanomaterials in biosensors", *Chem. Eng. J.,* vol. 442, p. 136183, 2022.

http://dx.doi.org/10.1016/j.cej.2022.136183

[15] T. Li, Y. Liang, J. Li, Y. Yu, M.M. Xiao, W. Ni, Z. Zhang, and G.J. Zhang, "Carbon nanotube field-effect transistor biosensor for ultrasensitive and label-free detection of breast cancer exosomal miRNA21", *Anal. Chem.*, vol. 93, no. 46, pp. 15501-15507, 2021.

http://dx.doi.org/10.1021/acs.analchem.1c03573 PMID: 34747596

[16] B. Dai, R. Zhou, J. Ping, Y. Ying, and L. Xie, "Recent advances in carbon nanotube-based biosensors for biomolecular detection", *Trends Analyt Chem.*, vol. 154, p. 116658, 2022.

http://dx.doi.org/10.1016/j.trac.2022.116658

[17] S. Shylendra, M. Wajrak, and J. J. Kang, "Advancements in solid-state metal pH sensors: A comprehensive review of metal oxides and nitrides for enhanced chemical sensing: A review," *IEEE Sensors Journal*, vol. 25, no. 5, pp. 7886–7895, 2025.

[18] M. Zea, A. Moya, M. Fritsch, E. Ramon, R. Villa, and G. Gabriel, "Enhanced performance stability of iridium oxide-based pH sensors fabricated on rough inkjet-printed platinum", *ACS Appl. Mater. Interfaces,* vol. 11, no. 16, pp. 15160-15169, 2019.

http://dx.doi.org/10.1021/acsami.9b03085 PMID: 30848584

[19] P. Salvo, B. Melai, N. Calisi, C. Paoletti, F. Bellagambi, A. Kirchhain, M.G. Trivella, R. Fuoco, and F. Di Francesco, "Graphene-based devices for measuring pH", *Sens. Actuators B Chem.*, vol. 256, pp. 976-991, 2018.

http://dx.doi.org/10.1016/j.snb.2017.10.037

[20] T. Sakata, S. Nishitani, A. Saito, and Y. Fukasawa, "Solution-gated ultrathin channel indium tin oxide-based field-effect transistor fabricated by a one-step procedure that enables high-performance ion sensing and biosensing", *ACS Appl. Mater. Interfaces,* vol. 13, no. 32, pp. 38569-38578, 2021.

http://dx.doi.org/10.1021/acsami.1c05830 PMID: 34351737

[21] H. Wang, P. Zhao, X. Zeng, C.D. Young, and W. Hu, "High-stability pH sensing with a few-layer MoS 2 field-effect transistor", *Nanotechnology,* vol. 30, no. 37, p. 375203, 2019.

http://dx.doi.org/10.1088/1361-6528/ab277b PMID: 31170702

[22] O. Knopfmacher, M.L. Hammock, A.L. Appleton, G. Schwartz, J. Mei, T. Lei, J. Pei, and Z. Bao, "Highly stable organic polymer field-effect transistor sensor for selective detection in the marine environment", *Nat. Commun,* vol. 5, no. 1, p. 2954, 2014.

http://dx.doi.org/10.1038/ncomms3954 PMID: 24389531

[23] R. Ren, Y. Zhang, B.P. Nadappuram, B. Akpinar, D. Klenerman, A.P. Ivanov, J.B. Edel, and Y. Korchev, "Nanopore extended field-effect transistor for selective single-molecule biosensing", *Nat. Commun,* vol. 8, no. 1, p. 586, 2017.

http://dx.doi.org/10.1038/s41467-017-00549-w PMID: 28928405

[24] S. Kim, G. Myeong, W. Shin, H. Lim, B. Kim, T. Jin, S. Chang, K. Watanabe, T. Taniguchi, and S. Cho, "Thickness-controlled black phosphorus tunnel field-effect transistor for low-power switches", *Nat. Nanotechnol.,* vol. 15, no. 3, pp. 203-206, 2020.

http://dx.doi.org/10.1038/s41565-019-0623-7 PMID: 31988502

[25] S. Ju, A. Facchetti, Y. Xuan, J. Liu, F. Ishikawa, P. Ye, C. Zhou, T.J. Marks, and D.B. Janes, "Fabrication of fully transparent nanowire transistors for transparent and flexible electronics", *Nat. Nanotechnol.,* vol. 2, no. 6, pp. 378-384, 2007.

http://dx.doi.org/10.1038/nnano.2007.151 PMID: 18654311

[26] X. Wang, C. Zhu, Y. Deng, R. Duan, J. Chen, Q. Zeng, J. Zhou, Q. Fu, L. You, S. Liu, J.H. Edgar, P. Yu, and Z. Liu, "Van der Waals engineering of ferroelectric heterostructures for long-retention memory", *Nat. Commun,* vol. 12, no. 1, p. 1109, 2021.

http://dx.doi.org/10.1038/s41467-021-21320-2 PMID: 33597507

[27] T. Carey, S. Cacovich, G. Divitini, J. Ren, A. Mansouri, J.M. Kim, C. Wang, C. Ducati, R. Sordan, and F. Torrisi, "Fully inkjet-printed two-dimensional material field-effect heterojunctions for wearable and textile electronics", *Nat. Commun,* vol. 8, no. 1, p. 1202, 2017.

http://dx.doi.org/10.1038/s41467-017-01210-2 PMID: 29089495

[28] H.W. Shi, L. Ding, D.L. Zhong, J. Han, L.J. Liu, L. Xu, P.K. Sun, H. Wang, J.S. Zhou, L. Fang, and Z.Y. Zhang, "Radiofrequency transistors based on aligned carbon nanotube arrays", *Nature,* 2021.

[29] G. Seo, G. Lee, M.J. Kim, S.H. Baek, M. Choi, K.B. Ku, C.S. Lee, S. Jun, D. Park, H.G. Kim, S.J. Kim, J.O. Lee, B.T. Kim, E.C. Park, and S.I. Kim, "Rapid detection of COVID-19 causative virus (SARS-CoV-2) in human nasopharyngeal swab specimens using field-effect transistor-based biosensor", *ACS Nano,* vol. 14, no. 4, pp. 5135-5142, 2020.

http://dx.doi.org/10.1021/acsnano.0c02823 PMID: 32293168

[30] Y. Liang, M. Xiao, D. Wu, Y. Lin, L. Liu, J. He, G. Zhang, L.M. Peng, and Z. Zhang, "Wafer-scale uniform carbon nanotube transistors for ultrasensitive and label-free detection of disease biomarkers", *ACS Nano,* vol. 14, no. 7, pp. 8866-8874, 2020.

http://dx.doi.org/10.1021/acsnano.0c03523 PMID: 32574035

[31] B. Zeng, C. Liu, S. Dai, P. Zhou, K. Bao, S. Zheng, Q. Peng, J. Xiang, J. Gao, J. Zhao, M. Liao, and Y. Zhou, "Electric field gradient-controlled domain switching for size effect-resistant multilevel operations in HfO2-based ferroelectric field-effect transistor", *Adv. Funct. Mater.,* vol. 31, no. 17, p. 2011077, 2021.

http://dx.doi.org/10.1002/adfm.202011077

[32] P. Luo, C. Liu, J. Lin, X. Duan, W. Zhang, C. Ma, Y. Lv, X. Zou, Y. Liu, F. Schwierz, W. Qin, L. Liao, J. He, and X. Liu, "Molybdenum disulfide transistors with enlarged van der Waals gaps at their dielectric interface via oxygen accumulation", *Nat. Electron.,* vol. 5, no. 12, pp. 849-858, 2022.

http://dx.doi.org/10.1038/s41928-022-00877-w

[33] A.K. Singh, K. Adstedt, B. Brown, P.M. Singh, and S. Graham, "Development of ALD coatings for harsh environment applications", *ACS Appl. Mater. Interfaces,* vol. 11, no. 7, pp. 7498-7509, 2019.

http://dx.doi.org/10.1021/acsami.8b11557 PMID: 30585719

[34]   L. Manjakkal, D. Szwagierczak, and R. Dahiya, "Metal oxides based electrochemical pH sensors: Current progress and future perspectives", *Prog Mater. Sci.,* vol. 109, p. 100635, 2020.

http://dx.doi.org/10.1016/j.pmatsci.2019.100635

[35]   H. Ren, K. Liang, D. Li, M. Zhao, F. Li, H. Wang, X. Miao, T. Zhou, L. Wen, Q. Lu, and B. Zhu, "Interface Engineering of Metal-Oxide Field-Effect Transistors for Low-Drift pH Sensing", *Adv. Mater. Interfaces,* vol. 8, no. 20, p. 2100314, 2021.

http://dx.doi.org/10.1002/admi.202100314

[36]   K. Wang, X. Liu, Z. Zhao, L. Li, J. Tong, Q. Shang, Y. Liu, and Z. Zhang, "Carbon nanotube field-effect transistor based pH sensors", *Carbon,* vol. 205, pp. 540-545, 2023.

http://dx.doi.org/10.1016/j.carbon.2023.01.049

[37]   T.H.V. Kumar, J. Rajendran, R. Atchudan, S. Arya, M. Govindasamy, M.A. Habila, and A.K. Sundramoorthy, "Cobalt ferrite/semiconducting single-walled carbon nanotubes based field-effect transistor for determination of carbamate pesticides", *Environ. Res.,* vol. 238, no. Pt 2, p. 117193, 2023.

http://dx.doi.org/10.1016/j.envres.2023.117193 PMID: 37758116

[38]   M. Kaisti, "Detection principles of biological and chemical FET sensors", *Biosens Bioelectron,* vol. 98, pp. 437-448, 2017.

http://dx.doi.org/10.1016/j.bios.2017.07.010 PMID: 28711826

[39]   J. Fenoll, P. Hellín, C. Martínez, M. Miguel, and P. Flores, "Multiresidue method for analysis of pesticides in pepper and tomato by gas chromatography with nitrogen–phosphorus detection", *Food. Chem.,* vol. 105, no. 2, pp. 711-719, 2007.

http://dx.doi.org/10.1016/j.foodchem.2006.12.060 PMID: 26059152

[40]   R. Martel, T. Schmidt, H.R. Shea, T. Hertel, and P. Avouris, "Single- and multi-wall carbon nanotube field-effect transistors", *Appl. Phys. Lett,* vol. 73, no. 17, pp. 2447-2449, 1998.

http://dx.doi.org/10.1063/1.122477

[41]   P. Singh, and D.S. Yadav, "Impact of work function variation for enhanced electrostatic control with suppressed ambipolar behavior for dual gate L-TFET", *Curr. Appl. Phys.,* vol. 44, pp. 90-101, 2022.

http://dx.doi.org/10.1016/j.cap.2022.09.014

[42]   S.J. Wind, J. Appenzeller, R. Martel, V. Derycke, and P. Avouris, "Vertical scaling of carbon nanotube field-effect transistors using top gate electrodes", *Appl. Phys. Lett,* vol. 80, no. 20, pp. 3817-3819, 2002.

http://dx.doi.org/10.1063/1.1480877

[43]   Z. Chen, D. Farmer, S. Xu, R. Gordon, P. Avouris, and J. Appenzeller, "Externally assembled gate-all-around carbon nanotube field-effect transistor", *IEEE Electron. Device Lett,* vol. 29, no. 2, pp. 183-185, 2008.

http://dx.doi.org/10.1109/LED.2007.914069

[44] D.B. Farmer, and R.G. Gordon, "Atomic layer deposition on suspended single-walled carbon nanotubes via gas-phase noncovalent functionalization", *Nano Lett,* vol. 6, no. 4, pp. 699-703, 2006.

http://dx.doi.org/10.1021/nl052453d PMID: 16608267

[45] P. Singh, and D.S. Yadav, "Impactful study of f-shaped tunnel fet", *Silicon,* vol. 14, no. 10, pp. 5359-5365, 2022.

http://dx.doi.org/10.1007/s12633-021-01319-6

[46] J. Cao, Q. Wang, and H. Dai, "Electron transport in very clean, as-grown suspended carbon nanotubes", *Nat. Mater.,* vol. 4, no. 10, pp. 745-749, 2005.

http://dx.doi.org/10.1038/nmat1478 PMID: 16142240

[47] V.K. Sangwan, V.W. Ballarotto, M.S. Fuhrer, and E.D. Williams, "Facile fabrication of suspended as-grown carbon nanotube devices", *Appl. Phys. Lett,* vol. 93, no. 11, p. 113112, 2008.

http://dx.doi.org/10.1063/1.2987457

[48] Y.M. Lin, J.C. Tsang, M. Freitag, and P. Avouris, "Impact of oxide substrate on electrical and optical properties of carbon nanotube devices", *Nanotechnology,* vol. 18, no. 29, p. 295202, 2007.

http://dx.doi.org/10.1088/0957-4484/18/29/295202

[49] F. Zahoor, M. Hanif, U. Isyaku Bature, S. Bodapati, A. Chattopadhyay, F. Azmadi Hussin, H. Abbas, F. Merchant, and F. Bashir, "Carbon nanotube field effect transistors: an overview of device structure, modeling, fabrication and applications", *Phys. Scr,* vol. 98, no. 8, p. 082003, 2023.

http://dx.doi.org/10.1088/1402-4896/ace855

[50] Y. Yamamoto, K. Maehashi, Y. Ohno, and K. Matsumoto, "Highly sensitive biosensors based on high-performance carbon nanotube field-effect transistors", *Sens. Mater.,* vol. 21, no. 7, pp. 351-361, 2009.

CHAPTER 4

Advancements in Nanomaterial Integration for Enhanced Biosensing Applications: Focus on Field Effect Transistor (FET)-Based Devices

S. Grace Infantiya[1,*], D. Anbuselvi[1], C. Kathiravan[2], N. Suthanthira Vanitha[3], and T. Narmadha[4]

[1]*Department of Physics, Muthayammal Engineering College, Rasipuram, Namakkal, India*

[2]*Department of Chemistry, Muthayammal Engineering College, Rasipuram, Namakkal, India*

[3]*Department of Electrical and Electronics Engineering, Muthayammal Engineering College, Rasipuram, Namakkal, India*

[4]*Department of Computer Science Engineering, JAIN (Deemed-to-be University), Bengaluru, India*

Abstract: The accelerating advancement of nanoscience and nanotechnology has established an explosion of potential opportunities for the fabrication of miniaturized nanostructured components with specialized applications in biology, electronics, chemistry, mechanics, and computational functions. This has a huge impact on the special field of biosensors, empowering the fabrication of extremely sensitive, compact, and effective diagnostic equipment. Notably, among these aforementioned advances, the nano-based Field-Effect Transistor (NFET) serves as an attractive candidate for biosensor applications owing to its remarkable attributes, including label-free detection, a high level of sensitivity, rapid response times, continuous measurement capabilities, low consumption of electricity, and potential for miniaturization into compact devices. Each of these traits combines to make nano-based FET biosensors, an interesting and robust technology for a variety of biomedical applications. In recent years, the integration of semiconducting materials, polymers, and carbon-based biocompatible nanomaterials has significantly revolutionized biosensing applications. These materials have been strategically incorporated into various nanostructures to elevate the efficacy and sensitivity of biosensing devices, particularly in the realm of field-effect transistor (FET)-based systems. This proposed book chapter aims to explore the burgeoning landscape of biocompatible nanomaterials and their role in the evolution of biosensing technologies. The utilization of nanomaterials, including metal nanoparticles, polymer nanocomposites, and carbon-based structures, has offered unique opportunities to

* **Corresponding author S. Grace Infantiya:** Department of Physics, Muthayammal Engineering College, Rasipuram, Namakkal, India; E-mail: graceinfantiya@gmail.com

Dharmendra Singh Yadav & Prabhat Singh (Eds.)

enhance the performance and reliability of biosensors. Overall, the chapter strives to deliver an inclusive examination of the advancements including potential future directions in the realm of biocompatible nanomaterials, specifically focusing on their integration into FET-based biosensing devices. It aspires to be an essential resource for researchers, scientists, and practitioners in the field of nanotechnology and biosensing

Keywords: Biosensors, and bioelectronics, Field-effect transistor, Nano biosensors, Nanomaterials.

INTRODUCTION

In the contemporary era, the seamless integration of science and nanotechnology has become a defining feature of our daily existence, profoundly shaping the way we interact with and navigate our physical environment. This transition may be seen in the variety of equipment and gadgets that have become essential in modern lifestyles, including computing devices, smartphones, microwaveable devices, refrigerators, cooling appliances, televisions, and remote controls. These technological wonders, on which we rely for numerous areas of our daily lives, owe their efficiency and functionality to the delicate applications of nanotechnology [1]. At the heart of many of these innovations lie sensors operating at the nanoscale, facilitating precision, automation, and responsiveness. Nanoscience plays a critical part in elevating the performance and interconnection of various technologies, ultimately contributing to the harmonious flow of our daily lives, from the precision of infrared (IR) thermometers to the convenience of remotely operating equipment. The fascination with nanomaterials is driven by the requirement to precisely control substances vital for both the human body and the atmosphere [2]. These materials, which are typically less than 100 nanometers (nm) in at least one dimension (1-D), exhibit exceptional features that have sparked the attention of various applications in medicine, environmental solutions, and different technological advancements. The increased interest reflects a commitment to utilizing nanomaterials to develop personalized solutions to challenging issues [1-15].

Nanomaterials are widely acknowledged for their varied size distributions, ranging from bulk to extremely fine particles, as well as distinguishing characteristics that incorporate a large surface-to-volume ratio (LSVR), quantum mechanics (quantum confinement) effect, enhanced intensity, as well as exceptional sensitivity, which offer an extensive range of advantages over conventional compounds in various domains such as microelectronic devices, healthcare, food processing, agriculture, and pharmaceutical manufacturing. The unique properties of nanomaterials pave opportunities for novel implementations and developments in a variety of sectors, influencing the panorama of current technology and industries. Nanomaterial synthesis refers to the techniques used to achieve nanomaterials, which are found in nature and are investigated in numerous disciplines comprising physical and chemical science, geology, biology, and microelectronics. These nanomaterials exhibit variations in dimensions, structure, and topology. To acquire its distinctive characteristics, the synthesis process must be carefully regulated, tailoring parameters such as reactant concentration, environmental conditions, and time optimization [3, 4]. A comprehensive knowledge of the components mentioned above is essential for achieving the desired outcomes and may pave the way for the fabrication of multifunctional and revolutionary nanotechnologies. The sustainable synthesis of nanomaterials is the key to uncovering their distinctive characteristics and enabling advancements in a broad spectrum of scientific and technological realms [5-15].

Nanotechnology, a revolutionary discipline at an amalgamation of science and engineering, involves the fabrication, manipulation, and investigation of various techniques, materials, different modes, structural components, and vast applications such as catalysis, fuel cells, quantum-effect lasers, photovoltaic cells, solar transistors, molecular electronic devices, supercapacitors, biosensors, nanoactuators, surface-enhanced Raman spectroscopy, and enhanced energy storage devices. In today's world, nanotechnology has become synonymous with technological advances, particularly in the emergence of biosensors. Researchers are actively exploring the incorporation of cutting-edge nanomaterials, including quantum dots (QDs), nanotubes (NTs), and nanowires (Nws) in the development of biosensor medical diagnostic devices. Nanotechnology has changed biosensor architecture and effectiveness by incorporating several types of nanomaterials and various synthesis modes. The numerous nanomaterials employed in these applications have distinct properties characterized by chemical, physical, mechanical, and surface effects. The quantitative precision enables nanomaterials to unlock an endless number of possibilities, paving opportunities for significant improvements in new devices, notably smart biosensors [16-20].

This chapter provides a meticulous exploration of the current significant advances in biosensor technology, spanning from nanosensors to Field-Effect Transistor (FET) based innovations. The narrative navigates through the details of nanotechnology, highlighting the transformative impact of biosensors operating at the molecular and nanoscale levels. It also delves into the merging of electronics and biology through FET-based technologies [21, 22].

NANO BIOSENSORS

Introduction

Sensors and Nanosensors

Sensors and nanosensors stand at the forefront of innovation as pivotal detection tools, each contributing distinct advantages to our analytical capabilities. Traditional sensors, omnipresent in everyday applications, operate on a macro scale, detecting environmental changes and specific parameters across various industries [5]. They are instrumental in processes, such as quality assurance, system performance evaluation, and precise manufacturing control. Sensors at this scale contribute significantly to diverse sectors, including healthcare, automotive, and manufacturing. In contrast, to traditional sensors, nanosensors harness the unique attributes of nanomaterials, operating at the nanoscale to offer unparalleled sensitivity, particularly in biological and chemical detection. Exploiting surface effects and integrating with biomolecules, nanosensors exhibit enhanced precision and hold promise for real-time monitoring. While traditional sensors continue to possess a vital role, the evolution of nanosensors, fueled by ongoing advancements in nanotechnology, opens new frontiers in detection, with potential instances spanning from medical diagnostics to in vivo monitoring, ushering in a transformative era in sensing technologies [6].

In today's technologically advanced world, sensors serve as indispensable components across various industries and applications. These devices are essential for collecting data and monitoring diverse parameters such as temperature, pressure, humidity, and motion. The measurements acquired by sensors are essential for ensuring process control, maintaining quality assurance, and optimizing overall system performance [7]. By accurately measuring and controlling these dynamic variables, sensors facilitate precise manufacturing processes and effective machine control. Moreover, the continuous evolution of sensor technology has given rise to nanosensors, operating at the nanoscale. Nanosensors represent a significant advancement, offering the potential for heightened sensitivity, efficiency, and miniaturization compared to their traditional counterparts. At the forefront of this revolution are remarkable complexes comprising carbon nanotubes (CNTs) and quantum dots (QDs), which serve as the building block for an emerging class of biosensors [8].

Carbon nanotubes, with their exceptional electrical, thermal, and mechanical properties, have emerged as versatile components in nanosensors design. Their massive surface area and their unique electronic structure are ideal for capturing and transducing signals from biological molecules. Similarly, quantum dots (QDs), and semiconducting nanocrystals, demonstrate size-dependent luminescent properties, enabling precise detection and imaging at the nanoscale. These materials form the basis for nanosensors that can operate at unprecedented levels of sensitivity and specificity [9].

Single-Molecule Detection

One of the most remarkable capabilities of nanosensors is their capacity for Single-Molecule Detection. Traditional biosensors often grapple with the limitations of ensemble measurements, where the collective behavior of a group of molecules is observed. Nanosensors transcend this limitation, offering the ability to detect and analyze individual molecules. This extraordinary sensitivity has profound implications for fields such as medical diagnostics and drug discovery. In the realm of medical diagnostics, nanosensors capable of single-molecule detection open new possibilities for early disease diagnosis. Detecting minute concentrations of biomarkers allows for the identification of diseases at their nascent stages, facilitating timely and targeted interventions. Moreover, in drug discovery, the ability to study individual molecules enables a more precise understanding of drug interactions and mechanisms, resulting in improved and personalized therapies [9].

Challenges and Opportunities

While the nanoscale presents a realm of wonders, it also introduces unique challenges and opportunities in the field of biosensors. Working at such tiny scales poses challenges related to reproducibility, scalability, and integration into practical applications.

Addressing these challenges requires innovative solutions. Researchers are exploring novel fabrication techniques to enhance the reproducibility of nanosensors. Additionally, scalable manufacturing processes are being developed to ensure the widespread adoption of these technologies. Integration into practical applications involves interdisciplinary collaboration, bringing together experts from fields such as nanotechnology, biology, and engineering [23, 24].

In navigating the challenges associated with nanosensors, there is an array of opportunities waiting to be seized. Nanosensors' precision and versatility allow for implementations in a wide range of sectors, encompassing medical services, environmental monitoring, etc. The continued exploration of nanoscale wonders promises not only advancements in biosensing technology but also transformative impacts on how we perceive and interact with the biological world at the molecular level.

BASICS OF SENSOR TECHNOLOGY AND CLASSIFICATION

To fully understand the importance and functioning of sensors and nanosensors, it is essential to grasp the fundamental principles of sensor technology. A sensor is a component in a measurement system that acquires a physiological parameter and switches it into an electrical signal. These parameters can comprise temperature, pressure, light intensity, chemical composition, and various other variables, depending on the type of sensor. Sensors interface the physical world and electronic systems, providing crucial information for process control and decision-making. The sensing component of a sensor comes into interaction with the object or process being measured, and its output depends on the variation in the process variables or environmental conditions. For example, in the case of temperature sensors, they monitor modifications in temperature and convert them into electrical signals. Once the physical parameter is converted into an electrical signal, the data can be further processed, analyzed, and transmitted for monitoring or control purposes [25-30].

The characterization of sensors is a multifaceted process, primarily rooted in the basic sensing principle. Various types of sensors emerge from this classification, each serving a distinct function in the realm of measurement and data acquisition (Fig 1). This diverse array enables the adaptation of sensors to a wide range of applications and industries. Furthermore, sensor classification can be addressed from multiple perspectives, particularly depending on the physical characteristics they are developed to measure. The aforementioned perspective provides rise to an extensive variety of sensor types, each tailored to unique measurement requirements. Notable examples include, capacitive displacement sensors (CDS), temperature sensors (TS), inductive displacement sensors (IDS), resistance displacement sensors (RDS), laser interferometer displacement sensors (LIDS), chemical ingredient sensors (CIS), bore gauging displacement sensors (BGDS), ultrasonic displacement sensors (UDS), piezoelectric pressure sensors (PPS), optical beam deformation sensors (OBDS), optical encoder displacement sensors (OEDS), optical fiber displacement sensors (OFDS), imaging sensors, flow sensors, and intelligent sensors [31-35].

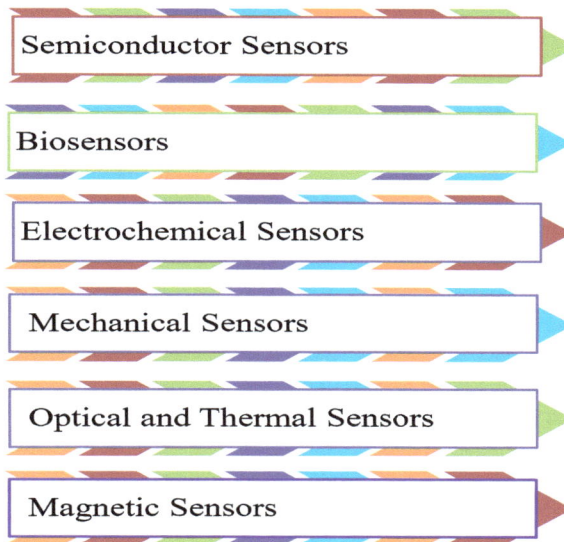

Fig. (1). Classification of sensors.

In the framework of biomedical applications, sensors perform a significant role in accumulating information about the functioning of the body in addition to pathological situations. Sensors possess a substantial element in biomedical applications by accumulating information about the functioning of the body and pathological situations. This particular branch of biomedical engineering emphasizes the development, implementation, and incorporation of sensors for monitoring physiological processes for medicinal purposes. Biomedical sensors make significant advancements to diagnostics, patient monitoring, and medical research, enabling a deeper comprehension of the human body's complexity and advancing healthcare technologies [36].

Biomedical sensors are categorized into three main categories: physical sensors, chemical sensors, and biosensors. Each type serves a distinct purpose in capturing and analyzing vital information related to the human body.

i) **Physical Sensors:** Physical sensors are instrumental in measuring various physiological parameters. Examples include blood pressure, flux, and viscosity, body temperature, as well as the biological magnetic field. These sensors provide quantitative data about the body's physical conditions, offering valuable insights into overall health and specific bodily functions [37].

ii) **Chemical Sensors:** Chemical sensors are designed to detect specific ingredients and concentrations within bodily fluids. These sensors serve an important function in investigating the chemical composition of biological fluids, especially the pH level, calcium ion concentration, and glucose concentration. By providing detailed chemical information, these sensor devices contribute to diagnostic and monitoring processes in healthcare.

iii) **Biosensors:** Biosensors are sophisticated sensors that detect biological molecules and signals. They can sense a wide range of substances, including enzymes, antigens, antibodies, hormones, DNA, RNA, and microorganisms. Biosensors are chemical sensors with a specific focus on biological signals, constituting them invaluable for applications in healthcare diagnostics, research, and monitoring [38-45].

Importance of Sensor Packaging and Biocompatibility in Biomedical Applications

The importance of sensor packaging and biocompatibility in the domain of biomedical applications is paramount for ensuring the efficacy, safety, and reliability of sensors utilized in healthcare applications. The factors mentioned above are foundational to seamlessly integrating sensors into medical equipment, and systems thereby optimizing patient care and precision in diagnostics [45-51].

- **Preservation and Protection:** Sensor packaging acts as a protective shield, insulating the sensitive components of biomedical sensors from environmental factors. This is particularly significant in biomedical applications where sensors may be exposed to human fluids and altering ambient conditions. Effective packaging ensures the long-term viability and functioning of the sensor.

- **Biocompatibility:** Biomedical sensors are frequently employed in applications involving direct contact with biological fluids and tissues. Biocompatible packaging components are required to ensure that the sensors can be employed properly within the human body while restricting the potential of unwanted responses or damage to the biological environment.

- **Durability for Real-world conductions:** The dynamic nature of biomedical applications demands robust sensor packaging. Durable packaging materials contribute to the sensors' ability to withstand the challenges of real-world usage, enhancing their longevity and reducing the need for frequent replacements.

- **Maintaining Signal Accuracy:** The packaging of biological sensors is crucial in maintaining the accuracy and dependability of the signals they produce. Signal distortion could affect the efficacy of examination or monitoring. Well-designed packaging ensures that sensor signals remain precise, delivering accurate information to medical professionals.

- **Integration with Medical Devices:** Medical Device Integration: Biomedical sensors are frequently incorporated into medical equipment and systems. For seamless incorporation, sensor packaging and the wider medical device infrastructure must be compatible. It enables the effective implementation of sensors into various healthcare solutions.

Key Parameters in Sensors

In sensor design, several key parameters are crucial for ensuring optimal performance and functionality. These parameters guide the engineering process to meet specific application requirements. Here are some key parameters in sensor design:

- **Measurement Range:** The series of values within the sensor can precisely compute the target parameter. It identifies the minimum and maximum values that the sensor is intended to handle.

- **Resolution:** The sensor detects even the smallest variations in the quantity being measured. Higher resolution allows for more precise measurements.

- **Precision:** The intensity to which the sensor's measurements align with the true or expected values. Precision is essential for reliable data acquisition.

- **Sensitivity:** Indicates how responsive the sensor reacts to changes in the measured parameter. It evaluates the interaction that exists between the sensor's input and output.

- **Response Time:** The time required for the sensor to identify a shift in the observed characteristic after it happens. Rapid response times are essential in dynamic applications.

- **Linearity:** Describes how well the sensor's output follows a straight line in response to changes in the measured parameter. Linearity ensures a predictable and consistent response.

- **Repeatability:** The proficiency of the sensor is to fabricate consistent output readings when exposed to the same input under the same conditions. Repeatability is essential for reliability.

- **Hysteresis:** The phenomenon in which the sensor's output fluctuates with the input, depending on whether it is increasing or decreasing. Minimizing hysteresis is important for accuracy.

- **Temperature Stability:** The sensor's ability to maintain consistent performance across a range of temperatures. Temperature stability is critical in applications with varying environmental conditions.

- **Power Consumption:** Specifies the amount of electrical power required for the sensor to operate. Reduced consumption of energy is frequently desired, particularly with battery-powered gadgets.

- **Environmental Conditions:** Describes the conditions under which the sensor is designed to operate effectively. This includes factors such as humidity, pressure, and exposure to specific chemicals.

- **Physical Size and Form Factor:** The physical dimensions and shape of the sensor, impact its integration into different devices or systems.

- **Durability and Robustness:** The facility of the sensor to tolerate physical stress, external factors, and other challenges in its operating environment.

- **Calibration:** Some sensors may require periodic calibration to maintain accuracy. Calibration procedures and intervals should be well-defined.

- **Cost:** The overall cost of manufacturing the sensor, which includes materials, production processes, and any additional features.

Understanding and optimizing these key parameters are essential for developing sensors that meet the specific requirements of diverse applications, ensuring accuracy, reliability, and efficiency.

AN OVERVIEW OF NANO BIOSENSORS

Nano biosensors have evolved at the intersection of two pivotal realms: nanotechnology and biosensors. Unlike conventional sensors solely designed for nano-scale measurements and activities, nano biosensors are a distinct category, primarily crafted from nanomaterials. What makes these sensors fascinating is their unique ability to harness the extraordinary physiochemical properties inherent in nanomaterials, attributes that markedly differ from those experientials in their bulk-scale counterparts [10].

Nanotechnology provides humanity with access to materials at the nanoscale, allowing for the creation of nano biosensors with exceptional sensitivity and specificity. These nanomaterials, characterized by their minute size and distinct properties, play a pivotal role in the intricate detection mechanisms employed by biosensing technology. The exclusive qualities of nanomaterials become particularly evident when compared to their counterparts produced on a larger scale.

In the realm of biosensing, nanomaterials contribute significantly to the detection mechanism. Their unique physiochemical properties enhance the sensor's ability to interact with biological entities, enabling precise and efficient detection of molecular targets. This interaction typically encompasses surface plasmon resonance (SPR), quantum effects (QE), and enhanced surface area, which are inherently associated with nanomaterials. Furthermore, the combination of nanomaterials and electrical components leads to the creation of microelectromechanical systems (MEMS) [11]. These integrated devices serve as transducers, switching biological signals into quantifiable electrical impulses. The synergy between nanomaterials and MEMS enhances the overall performance of nano biosensors, allowing for real-time monitoring and detection with remarkable accuracy. In essence, nano biosensors represent a convergence of nanotechnology and biosensing, capitalizing on the distinctive properties of nanomaterials to revolutionize detection mechanisms. The integration of these materials with MEMS showcases the interdisciplinary nature of this field, offering a glimpse into the transformative potential of nano biosensors across diverse applications, from healthcare to environmental monitoring.

In recent years, the exploration of nanomaterials has become pivotal in advancing biosensor technologies, harnessing their unique electrical and mechanical properties for refined biological signal processing. Noteworthy nanomaterials, including nanowires, nanotubes, nanorods, and thin films, characterized by their crystalline composition, have demonstrated exceptional efficacy in sensing applications. These nanomaterials have been harnessed across a spectrum of applications, showcasing their versatility and efficacy in diverse biological signal-processing tasks [12].

Nanowires (NWs), Nanotubes (NTs), Nanorods (NRs), and Thin Films

- **Crystalline Composition:** Crystalline nanomaterials, owing to their crystalline nature, exhibit exceptional electrical and mechanical properties that formulate them well-suited for biosensing applications.

- **Amperometric Devices:** Nanowires as well as nanotubes, for instance, find utility in amperometric devices employed for enzyme-based glucose sensing.

Quantum dots (QDs), with their distinguishing optical characteristics, serve as luminous reagents for binding detection, while bioconjugated nanomaterials enhance targeted biomolecular detection [48]. Colloidal nanomaterials, when coupled with antibodies, play a crucial role in immunosensing and immunolabeling applications. Furthermore, metal-based nanomaterials, leveraging their electronic and optical characteristics, prove particularly effective in electronic and optical biosensing applications, especially in the detection of nucleic acid sequences. Response time, selectivity, sensitivity, and linearity are essential properties for increased biosensor efficiency [13].

Research investigations highlight a continuous and notable surge in the utilization of various nanomaterials, demonstrating a growing interest in employing them as either transducers or receptors to enhance multi-detection sensitivity and capabilities. Nanoparticles (NPs), quantum dots (QDs), nanotubes (NTs), and other biological nanomaterials (BNMs) exemplify this trend, with the potential to support either the transducer, bio-recognition component, or both in biosensing applications. This evolution has significantly impacted biochemical and biological investigation disciplines, revolutionizing the rapid investigation of numerous chemicals in vivo through the development of nanosensors, nanoprobes, and other nanodevices [48]. Nanomaterials have emerged in recent years exhibiting a broad range of attributes, notably smaller dimensions, accelerated electron traversal, shorter distances, reduced power requirements, and lower voltages. The advanced nanoparticles (NPs) (Fig. **2**) contribute to the enhancement of electrochemical signals at the electrode-electrolyte interface in biocatalytic processes.

Multifunctional nanomaterials, especially combined with biological compounds which include protein molecules, peptides, and DNA, have been engineered for use in biosensors, expanding the versatility of these devices. Nanofabrication techniques encompass both top-down and bottom-up approaches, with the prior employing technologies that consist of the electrochemical etching process, ion milling laser ablation, and nano-lithography to downsize macroscopic materials to the nanoscale. Bottom-up nanofabrication incorporates techniques such as molecular beam epitaxy (MBE), evaporation, physical or chemical vapor deposition (CVD/PVD), and the synthesis of protein-polymer nanocomposites (PPN), all of which contribute to significant advances in the field of nanotechnology [14].

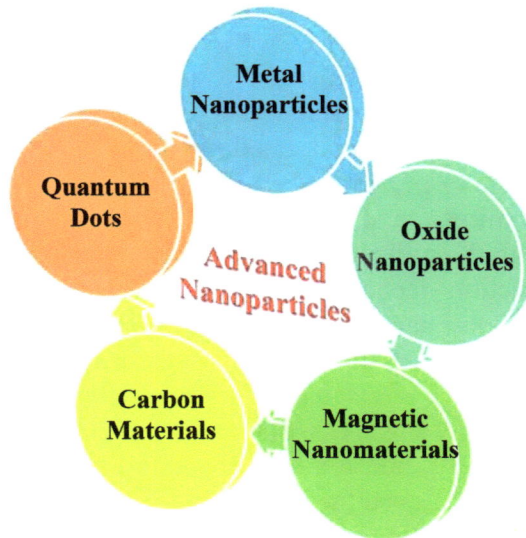

Fig. (2). Advanced Nanomaterials (NPs).

The Growth and Development of Nanobiosensors

In 1996, the International Union of Pure and Applied Chemistry (IUPAC) established a commission dedicated to the discovery and definition of biosensors, which was a significant step toward standardization. In accordance with IUPAC, a biosensor is a device that utilizes isolated enzymes, tissues, organelles, immune systems, or whole cells to assist certain biochemical reactions. These reactions serve the purpose of detecting chemical substances, with the interpretation of signals in electrical, thermal, or optical forms. This definition provides a comprehensive framework, encompassing a diverse array of biological components and signal detection mechanisms employed by biosensors in their analytical processes. At the heart of this concept lies the fundamental role of biosensors in converting biological interactions into measurable signals, underlining their significance in accurate and sensitive analytical assessments [14]. Moreover, this definition has been refined in accordance with recent technological and scientific advancements, particularly with respect to the emergence of bio-microchips. These bio-microchips offer the potential for creating smaller, more cost-effective, and highly precise devices compared to conventional microelectronics. They serve as the foundation for biosensor systems equipped with diverse sensing mechanisms, encompassing capabilities such as vision, hearing, olfaction, taste, and touch. This evolution underscores the continuous innovation in biosensor development, emphasizing enhanced functionalities and the integration of cutting-edge technologies [15].

Nano-biosensors represent a revolution in sensor technologies, offering the capability to analyze multiple samples at a desired time and location rapidly. These sensors exhibit outstanding performance in terms of selectivity, biocompatibility, non-toxicity, reversibility, and rapid response, enhancing sensitivity in determinations through the utilization of nanomaterials and introducing numerous novel signal transduction technologies. Various morphologies of nanotubes, nanowires, nanofibers, and nanorods have proven effective in the transduction of analytes [16]. Nano-biosensors represent a transformative revolution in sensor technologies, providing the remarkable ability to swiftly analyze multiple samples at specific times and locations. These sensors excel in performance, boasting selectivity, biocompatibility, non-toxicity, reversibility, and rapid response. This heightened sensitivity is achieved through the integration of nanomaterials and the incorporation of innovative signal transduction technologies. Various morphologies, including nanotubes, nanowires, nanofibers, and nanorods, have proven highly effective in transducing analytes, contributing to the versatility of nano-biosensors [17].

Furthermore, the utilization of biological materials, such as aptamers, as alternative sensing molecules adds another layer of innovation to this field. The amalgamation of nanomaterials with cutting-edge biological components exemplifies the multifaceted advancements that drive the capabilities and versatility of nano-biosensors. In this context, a "biosensor" is signified as a sophisticated and innovative quantitative instrumentation that incorporates a biological detecting component and therefore provides an extensive spectrum of applications (Fig. **3**). The convergence of nanotechnology and biosensing has propelled the development of nano-biosensors into a pivotal role across diverse fields, offering solutions to a broad spectrum of analytical challenges.

The promise of nano-biosensors in harnessing biological specimens for rapid identification of pathogenic organisms and infections has received widespread attention and acknowledgment. This interest has intensified following the discovery and utilization of nanomaterials in biosensor development. Notably, nanobiosensors can be broadly categorized into three generations, each representing a distinct stage in their evolution, functionality, and applications. Clark & Lyons created the first biosensor in 1962 to evaluate glucose levels in biological specimens, pioneering the groundbreaking technology for determining the electrochemical presence of oxygen (O_2)/hydrogen peroxide (H_2O_2). This innovative approach involved an immobilized glucose oxidase electrode [18-20]. In this setup, glucose served as the target substance, initiating an oxidation reaction catalyzed by glucose oxidase.

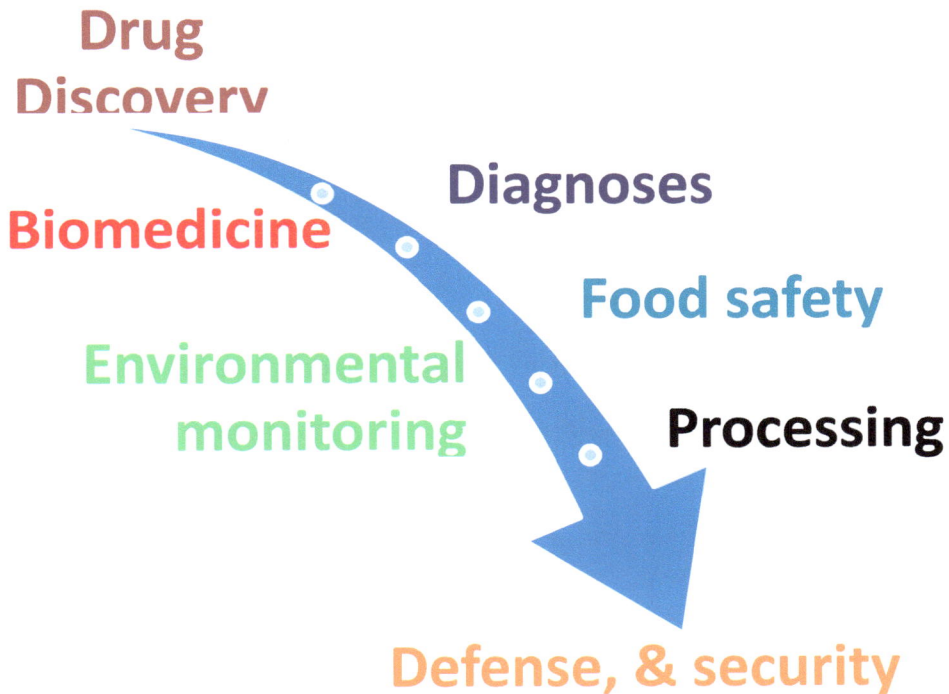

Fig. (3). Applications of Biosensors.

The functioning of this biosensor relied on the oxidation reaction of glucose, triggering the reduction of O_2, and the consequential electric current was quantified based on the rate of O_2 concentration. The reduction in electric current was directly correlated with the concentration of glucose. This breakthrough represented a foundational approach to biosensing, establishing a link between specific biological reactions and measurable electrical signals. Furthermore, an important innovation in biosensor development was the introduction of the potentiometric urea electrode, which is classified as a first-generation biosensor [21, 22]. A hallmark example of Leyland C. Clark's biosensor concept was the immobilization of glucose oxidase. The immobilized glucose oxidase played a pivotal role in converting the electrode into an effective tool for the precise uncovering of glucose ($C_6H_{12}O_6$) levels. The utilization of immobilized glucose oxidase not only demonstrated the versatility and power of biosensors but also marked a crucial milestone in the practical application of biosensing technology for addressing real-world health challenges [23]. This application held particular significance in the context of human samples, especially for individuals with diabetes. This innovative technique revolutionized a standard platinum electrode into a powerful analytical tool. These electrodes, designed to detect urea levels, contributed significantly to the early stages of biosensor evolution and paved the way for subsequent advancements in biosensing technology [24, 25].

In the realm of second-generation biosensors, a notable advancement involves the co-immobilization of auxiliary enzymes and/or co-reactants alongside the analyte-converting enzyme [26, 27]. This strategic approach aims to enhance the analytical quality of biosensors while concurrently simplifying their performance. By introducing auxiliary components, such as additional enzymes or co-reactants, the biosensor gains improved efficiency and functionality, allowing for more precise and reliable analytical outcomes. This evolution in biosensor design reflects a commitment to refining and optimizing the sensing capabilities of these devices for enhanced accuracy and broader applicability [28].

In the context of third-generation biosensors, biomolecules play a pivotal role in the interaction with biosensing materials, exemplified by technologies like SPR (Surface Plasmon Resonance) biosensors. These advanced biosensors leverage the involvement of biomolecules to enhance the sensitivity and specificity of the sensing process, providing a sophisticated platform for a diverse range of applications. Looking ahead to the fourth generation, the integration of developments in Micro-Electro-Mechanical Systems (MicroEMS), Nano-Electro-Mechanical Systems (NanoEMS), and BioNano Electro-Mechanical Systems (BioNEMS) is anticipated to bring forth a multitude of features. This convergence of nanotechnology along with bio-technology encompasses significant promise, with contributions from engineers and scientists across diverse fields. The evolution of biosensor technology in this generation opens up new and creative opportunities, showcasing versatility in their design and functionality [20, 29].

The applications of these advanced biosensors extend beyond traditional domains. They are increasingly utilized for the determination of biological and chemical substances in agricultural production, food analysis, and environmental monitoring. Additionally, biosensors play essential roles in areas such as mining, bioprocessing, bio-warfare detection, and homeland security. The collaborative efforts of experts from various disciplines continue to drive the occurrence of biosensors, offering ground-breaking elucidations to address complex challenges in diverse fields [30, 31].

Nanobiosensors are being constructed by integrating cutting-edge methodologies from both nanotechnology and biotechnology, resulting in non-intrusive and sensitive sensors that use unique materials and manufacturing techniques [32]. These sensors provide real-time highly sensitive indicators that may be collected and investigated easily. A typical nano biosensor is composed of three primary components:

Genetic Probe Module

This Fig. (**4**) provides a simplified visualization of the genetic probe module and its key components and interactions [33, 34].

- Materials with affinities such as enzyme-substrate, antibody-antigen, DNA, and cell-based interactions are encompassed [35-38].

- Nucleic acids, antibodies, and enzymes are all examples of biological components [39, 40].

- Bioanalytes are identified using biological probes that are coupled to nanomaterials such as magnetic (Fe_3O_4, $MnFe_2O_4$, Co_3O_4, $ZnFe_2O_4$, NiO), metallic (Ag, Au, Cu, Co, Ni, Za etc), graphene oxide (GO), quantum dots, as well as carbon nanotubes (CNTs) [41].

- From Fig. (**5**), we can see that multiple transduction methods have been employed in the transducer module in order to transform biological information into electrical impulses [42].

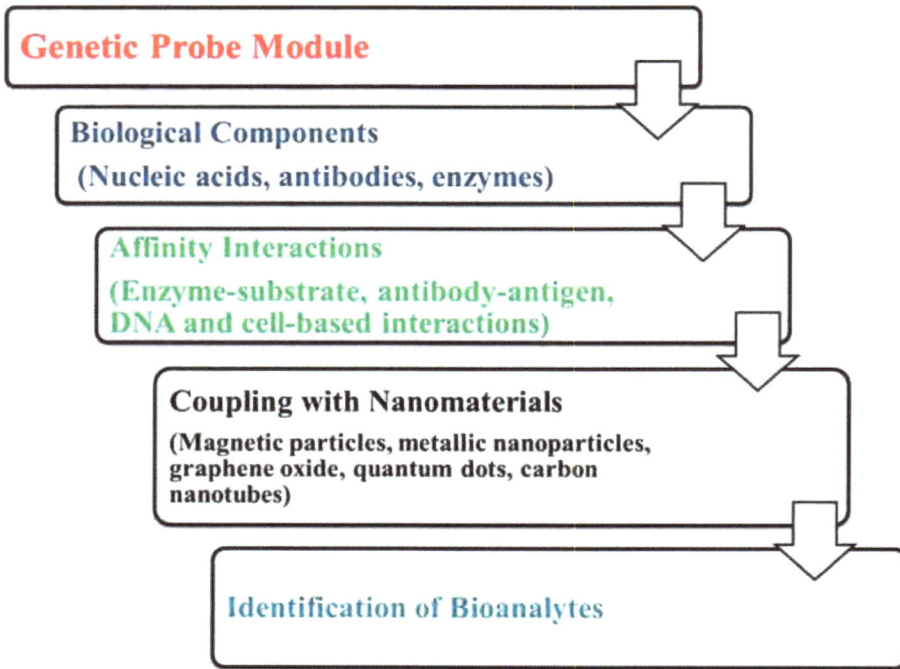

Fig. (4). Genetic Probe Module.

Transducer Module

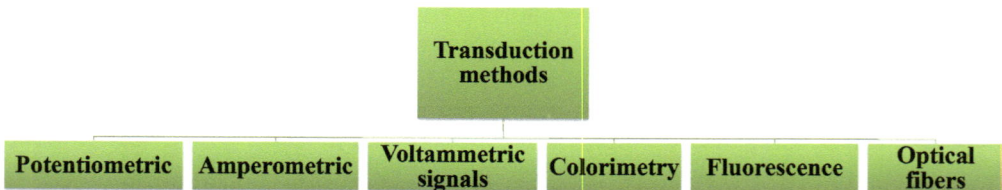

Fig. (5). Different Transduction Techniques.

Data Footage Unit

The Data Footage Unit facilitates the process of loading and distribution of sensor-collected information. This unit is critical to ensuring that data obtained from sensors are effectively processed and transferred for further examination or storage.

Advanced nanomaterials, ranging from sophisticated to cost-effective, are incorporated into the design and development of robust nano biosensors. Various nanomaterials, notably gold (Au), silver (Ag), diamond (C), and platinum (Pt) nanoparticles, showcase multiple features as well as are used in nano biosensor development [12, 46, 47]. Transducers are composed of various components, including fibers, ceramics, metals, silicon, plastics, and glasses, and they transform intermediate responses into particular electrical signals such as optoelectronic, electrochemical, piezoelectric, and thermal signals. The optimization constraints in nano biosensor development encompass a dynamic range of concentration and sensitivity, sensory nanomaterials, temperature, pH, detection time, volume of the reaction sample, and also rheological factors. The nano biosensor development aims at detecting various biological and electrochemical products such as cholesterol, choline, dopamine, vitamins, blood glucose, creatinine, albumin, drugs, enzymes, proteins, and nucleic acids [48]. These sensors efficiently convert biological or chemical reactions into electrical outputs, providing information on composition or concentration, rheological properties, amplitude, energy, pH, polarization, and decay time within a few seconds [15]. The nano biosensor development process encompasses the architecture, advancement, optimized performance, characterization, selection of materials, production, and testing of analytes on a miniaturized platform. Nano-biosensors are categorized and divided into multiple varieties, each requiring specific considerations for nanostructured components and biological detection mechanisms [48]. The nanoparticles utilized in these sensors, including carbon nanotubes, nanowires, nanoparticles, and quantum dots, possess exceptional qualities that enhance the sensing system. Table **1** provides a summary of different types of nanomaterials used in creating nano biosensors.

In the production of nanobiosensors, a critical aspect is the careful selection of substrates for efficient fabrication. The chosen substrate should allow the applied analyte on the nanomaterial surface to react effectively. Initially, silicon and glass were the primary materials for nano biosensor fabrication. However, as advancements in technology occurred, cutting-edge components such as polymer compounds, polyimide, paper, and aluminium foil have been developed.

Table 1. Different kinds of nanomaterials commonly used in the creation of nano biosensors.

Nanomaterials	Advantages	Disadvantages
Carbon Nanotubes (CNT)	• Excellent electrical conductivity • High surface area • Chemical stability • Biocompatible	• Production challenges • Potential toxicity concerns • Lack of selectivity
Nanowires	• High surface-to-volume ratio • Sensitive detection of biomolecules • Good charging conduction • Semiconductor properties • Extremely flexible. • Excellent electrochemical • & distinguishing properties. [10]	• Fabrication complexity • Potential biofouling • Minimal luminescence. • Significant proportion of emulsifier.
Nanoparticles	• Unique optical and catalytic properties • Tunable properties • High surface area	• Aggregation issues • Stability concerns • Size control challenges • Low optical signal

(Table 1) cont.....

	• Biocompatible	
Quantum Dots (QDs)	• Size-dependent fluorescent properties • Excellent photostability • Tunable emission spectra	• Toxicity concerns • Potential environmental impact
Graphene/Graphene Oxide	• Exceptional electrical conductivity • Biocompatibility • Large surface area.	• Limited functional groups on pristine graphene • Potential toxicity concerns.
Magnetic Nanomaterials	• Easy manipulation and separation of bioanalytes. • Magnetic-based detection methods.	• Potential agglomeration issues • Limited functionalization options.
Polymeric Nanomaterials	• Versatile platform for immobilizing biomolecule • Enhanced stability • Biocompatible.	• Limited electrical conductivity • Challenges in precise control of properties.
2D Nanomaterials	• Unique electronic properties • Large surface area • Tunable properties.	• Limited stability in biological environments • Production challenges

(Table 1) cont.....

Metal-Organic Frameworks (MOFs)	• High surface area for immobilizing biomolecules • Porous structure. • Better electrical conductivity • UV absorption	• Stability concerns in biological fluids • Limited options for certain applications.
Silica Nanoparticles	• Biocompatible • Ease of functionalization. • High surface area.	• Limited electrical conductivity • Potential toxicity at high concentrations.
Nanorods	• Size-tunable optimization of energy • Integrated with microelectromechanical systems.	• Metabolism is distinct with various compounds.

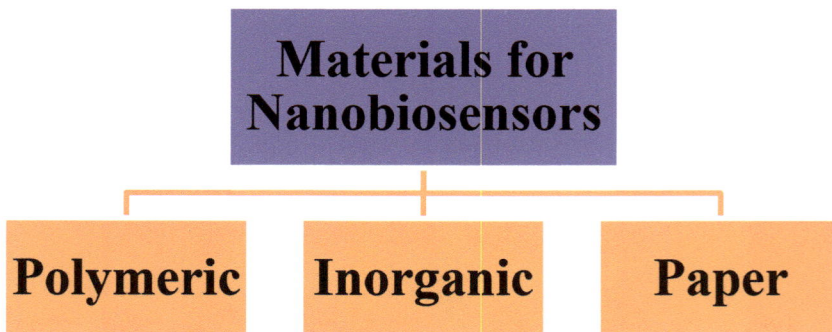

Materials for Nanobiosensors

Polymeric   **Inorganic**   **Paper**

Essential properties of these materials include exceptional optoelectric, thermal, and mechanical conductivity, as well as rapid melting ratio. Choosing the appropriate material is crucial for engineering nano biosensors, especially in Point-of-Care Testing (POCT) applications. Various progressive techniques, including laser ablation, 3D printing technology, and lithographical techniques, are employed in developing nanobiosensors. These techniques address considerations such as vast surface devices, low power consumption, a high degree of resolution, consequences for response samples, thermal parameters, crystallization, substrate damage, and carrier versatility competencies [48]. Conventional methods like non-impact printing as well as screen-printing technologies, which operate on a descent-on-demand feature, may lack uniformity and precision [10]. Table **2** provides an overview of materials used in nanobiosensor development. The selection of materials and fabrication methods plays a key function in heightening the effectiveness and applicability of nano-biosensors in various diagnostic combined with analytical applications.

Table 2. Overview of materials used in nanobiosensor development.

Materials: Polymeric Materials [20-23].
Role: Polymeric materials are known for their flexibility, biocompatibility, and ease of functionalization. They are often employed to enhance the stability of the sensor and facilitate interactions with biomolecules.
Advantages: Flexibility, biocompatibility, and the ability to modify surfaces for specific biomolecular interactions.

Examples of Polymeric Materials

Substances	**Properties**
Polyimide	•Polyimide is known for its high thermal stability, excellent mechanical strength, and flexibility. •All of these features make it attractive for flexible and wearable biosensor implementation.
Polyethylene Glycol (PEG)	•Polyethylene Glycol is a biocompatible polymer often used to modify surfaces and reduce non-specific binding in biosensors. •Its hydrophilic nature helps prevent biofouling.
Polydimethylsiloxane (PDMS)	•Polydimethylsiloxane is a silicone-based polymer with properties such as flexibility and biocompatibility.

(Table 2) cont.....

	•It is commonly used in microfluidic systems integrated with biosensors.
Polyethylene Terephthalate (PET)	•Polyethylene Terephthalate is a thermoplastic polymer with outstanding mechanical attributes. •It is widely employed in the fabrication of biosensor substrates and microfluidic devices.
Polystyrene	•Polystyrene is widely used in biosensors due to its ease of modification and compatibility with various biofunctionalization techniques. •It is often employed as a substrate for immobilizing biomolecules.
Polyvinyl Chloride (PVC)	•Polyvinyl Chloride is a versatile polymer with applications in biosensors, particularly in electrochemical sensors. It is known for its electrical insulating properties.
Polyvinyl Alcohol (PVA)	•PVA is a water-soluble polymer with good film-forming properties. •It is used in the development of sensors, especially those requiring hydrogel components.
Polyethylene (PE)	•Polyethylene is a widely used thermoplastic polymer known for its chemical resistance and versatility. •It can be employed in various biosensor components.

Materials: Inorganic Materials [24-26]

Role: Inorganic materials are chosen for their unique properties, including electrical conductivity, optical characteristics, and structural stability. They play an instrumental part in enhancing nanobiosensor reliability as well as efficiency.

Advantages: Enhanced conductivity, optical properties, and structural stability contribute to improved sensor performance.

(Table 2) cont.....

Examples of Inorganic Materials	
Substances	**Properties**
Gold Nanoparticles (AuNPs)	•Gold nanoparticles have unique optical features, such as surface plasmon resonance (SPR), which make them ideal for colorimetric as well as optical biosensors. •Gold nanoparticles are used for label-based detection.
Silver Nanoparticles (AgNPs)	•Silver nanoparticles possess excellent antimicrobial properties and are employed in biosensors for their catalytic and conductive features. •They find applications in electrochemical and optical sensing [50].
Quantum Dots (QDs)	•Quantum dots are semiconductor nanocrystals with size-dependent fluorescence properties. •They are used as fluorescent labels in biosensors, offering high sensitivity and multiplexing capabilities.
Iron Oxide Nanoparticles	•Iron oxide nanoparticles, such as magnetite (Fe_3O_4) and maghemite (γ-Fe_2O_3), are employed in magnetic biosensors. •They enable magnetic separation and enhance the sensitivity of detection.
Titanium Dioxide Nanoparticles (TiO_2)	•TiO_2 nanoparticles exhibit photocatalytic properties and are utilized in photoelectrochemical biosensors. •They can be integrated into sensors for enhanced light-induced electron transfer.
Platinum Nanoparticles (PtNPs)	•• Platinum nanoparticles exhibit catalytic activity, particularly in electrochemical sensors. •They are extensively utilized as catalysts for analyte detection.
Carbon Nanotubes (CNTs)	•CNTs are used in nanobiosensors for their high surface area, electrical conductivity, and ability to facilitate electron transfer in sensors.

(Table 2) cont.....

Silicon Nanowires (SiNWs)	•Silicon nanowires are used in the production of FET-based biosensors. •They offer high sensitivity and can be functionalized for selective biomolecular recognition.
Zinc Oxide Nanoparticles (ZnO)	•ZnO nanoparticles possess semiconducting properties and are utilized in various biosensors, particularly in piezoelectric and electrochemical sensors.
Graphene Oxide (GO)	•It is employed in biosensors for its large surface area, biocompatibility, and capability to adsorb biomolecules.

Materials: Paper-Based Materials [27]

Role: Paper-based materials are gaining attention for their simplicity, cost-effectiveness, and potential use in point-of-care applications. They offer a platform for developing portable and disposable biosensors.

Advantages: Low cost, simplicity, and disposability make paper-based materials suitable for applications in resource-limited settings.

Examples of Paper-Based Materials	
Substances	**Properties**
Filter Paper	• Filter paper is a commonly used porous substrate in paper-based biosensors. • Its ability to absorb liquids and maintain structural integrity makes it suitable for creating simple and disposable devices.
Cellulose Paper	• Cellulose paper, derived from wood fibers, is often used as a substrate for paper-based biosensors. • It provides a matrix for immobilizing biomolecules and allows capillary flow for sample transport.
Nitrocellulose	• Nitrocellulose is a modified form of cellulose that enhances the binding of biomolecules. • It is frequently used in lateral flow assays and other point-of-care applications due to its rapid fluid absorption.

(Table 2) cont.....

Paper Microfluidic Devices	• Paper microfluidic devices integrate paper with microfluidic channels to enable controlled fluid flow. • These devices can be designed for specific applications, such as sample separation or multi-step assays.
Paper-Based Electrodes	• Paper can be modified to serve as an electrode in electrochemical biosensors. The paper-based electrode allows for the immobilization of sensing components and facilitates electron transmission during electrochemical reactions.
Paper-Based Nanocomposites	• Paper can be combined with nanomaterials, such as nanoparticles or nanofibers, to enhance its properties. • This integration can improve the sensitivity, stability, and overall performance of paper-based biosensors.
Graphene-Infused Paper	• Paper can be infused with graphene or graphene oxide to enhance its electrical conductivity and sensing capabilities. Graphene-infused paper is utilized in advanced paper-based biosensors.

STRUCTURE OF BIOSENSORS

Biosensors are sophisticated analytical instruments that incorporate a biological or biologically induced detecting component and a physicochemical transducer. The combination of these components generates a signal that coincides with the concentration of a specific analyte. Biosensors are typically composed of three basic components: the biological detecting element, the transducer that transmits signals, and the signal refinement platform [49-51]. This amalgamation enables the detection and measurement of extremely low quantities of specific diseases, toxic substances, and variations in pH. This electronic module is pivotal in recognizing, recording, and transmitting information concerning physiological changes or the presence of diverse chemical or biological substances in the environment. The comprehensive components of a biosensor typically include:

(i) Analyte,

(ii) Bioreceptor,

(iii) Transducer,

(iv) Electronics,

(v) Display [51].

(i) Analyte

The analyte shows a momentous role in the functionality of biosensors. The analyte is a compound of interest and its components are identified or analyzed by the biosensor. Examples of analytes encompass a wide range of compounds, comprising ammonia, alcohol, lactose, glucose, and other specific chemicals or biological substances. The precise detection and quantification of the analyte are fundamental to the effectiveness of biosensors across miscellaneous applications, including therapeutic diagnostics, conservational monitoring, and beyond. The ability to accurately identify and measure the analyte allows biosensors to provide valuable information for numerous fields, contributing to advancements in research, healthcare, and environmental managing.

Examples of Analytes in Biosensors Include

1) **Glucose:** Widely detected in biosensors for applications in diabetes management and glucose monitoring.

2) **Ammonia:** Important for environmental monitoring, particularly in water quality assessments.

3) **Alcohol:** Detected in breathalyzer biosensors for alcohol level measurements.

4) **Lactose:** Found in biosensors used in the food industry for lactose intolerance testing.

5) **Cholesterol:** Measured in biosensors for cardiovascular disease risk assessment.

6) **DNA/RNA Sequences:** Used in genetic biosensors for detecting specific genetic material.

7) **Pathogens:** Such as bacteria or viruses, detected in biosensors for medical diagnostics and infectious disease monitoring.

8) **Environmental Pollutants:** Biosensors are capable of being utilized to uncover contaminants, which include metallic substances and synthetic pesticides during ecological surveillance.

9) **Neurotransmitters:** Biosensors can be employed to measure neurotransmitter levels in applications related to neuroscience and mental health.

These examples demonstrate the wide range of analytes that biosensors can detect, emphasizing their versatility in a variety of industries, including the health care sector environmental tracking, along food security. Analytes are chosen depending on each application's particular requirements and the insightful data that biosensor technology can deliver.

(ii) Biological Sensing Elements (Bioreceptor)

The bioreceptor is a fundamental component of biosensors, referring to a biomolecule or biological element capable of recognizing the target substrate, also known as the analyte [50]. Examples of bioreceptors include enzymes, cells, aptamers, deoxyribonucleic acid (DNA or RNA), as well as antibodies. The interaction between the bioreceptor and the substance being measured (Analyte) results in a signal, which can manifest in various forms, including changes in light, heat, modifications in pH, charge, or mass, as well as the release of plant or animal tissue and microbial metabolites. Biorecognition refers to the signal production process that takes place as a consequence of the interaction between the bioreceptor and analyte. The specificity and selectivity of the bioreceptor for the analyte are crucial for the accurate detection and measurement capabilities of biosensors in diverse applications [51].

Examples of Analytes in Biosensors Include

1. **Enzymes:** Biological catalysts that facilitate specific biochemical reactions, such as glucose oxidase for glucose detection.

2. **Cells:** Living cells or cellular components, used in applications like whole-cell biosensors for environmental monitoring.

3. **Deoxyribonucleic Acid (DNA) or Ribonucleic Acid (RNA):** Genetic material can be engineered or selected to bind to specific sequences, allowing for nucleic acid detection.

4. **Antibodies:** Immune system-produced proteins that bind to certain antigens and are widely employed in medical diagnostics.

These bioreceptors serve as the recognition elements in biosensors, interacting selectively with the target analyte and initiating the signal transduction process. The choice of bioreceptor depends on the nature of the analyte and the specific requirements of the biosensor application.

(iii) Transducer

The transducer, a critical element in biosensors, is a device designed to transmit information from one form of energy to another. In the framework of biosensors, the transducer serves an instrumental function in transforming the biorecognition process into a quantifiable indication, typically in the mode of electricity. This mechanism of energy transformation is often referred to as signalization. Transducers are capable of transmitting optical or electrical impulses according to the multitude of interactions that occur between the analyte (substance being studied) and the bioreceptor. Transducers are widely classified according to their underlying operating principles, ranging from electrochemical, optical, thermal, electrical, and gravimetric transducers. The selection of a suitable transducer is essential in shaping the sensitivity, precision, and application scope of the biosensor.

Various types of transducers are employed in nanobiosensors to convert the biorecognition events into measurable signals. These transducers play a crucial role in the detection and quantification of specific analytes [51]. The key types of transducers used in nanobiosensors include:

1. **Electrochemical Transducers:** These transducers measure fluctuations in electrical properties, including current/voltage, resulting from biorecognition events. Common examples include amperometric, potentiometric, and impedance-based transducers.

2. **Optical Transducers:** Optical transducers detect changes in light intensity, wavelength, or fluorescence during biorecognition. Surface Plasmon Resonance (SPR) and fluorescence-based transducers fall into this category.

3. **Thermal Transducers:** These transducers measure changes in temperature resulting from biorecognition events. Thermistor-based or thermoelectric transducers are examples.

4. **Electronic Transducers:** Electronic transducers convert biorecognition events into electronic signals. Field-Effect Transistors (FETs) and other semiconductor-based transducers are common in this category.

5. **Gravimetric Transducers:** Gravimetric transducers measure changes in mass, such as Quartz Crystal Microbalance (QCM) devices.

Each type of transducer offers specific advantages and is selected based on the requirements of the nano biosensor applications.

(iv) Electronics

The electronics component in biosensors acts as essential for interpreting and preparing the transduced signal for output. The electrical impulses obtained by the transducers are amplified and transferred to computational form. This processing stage enhances the quality and reliability of the signals. The signals are processed and interpreted by the visualization unit. The visualization unit then provides the information in a comprehensible format, enabling users to properly comprehend and critically evaluate the results. The integration of advanced electronics ensures accurate and efficient signal processing, contributing to the overall effectiveness of biosensors in various applications.

The electronic processing involves several key functions, including:

1. **Amplification:** The electronic signal is amplified to increase its strength. Amplification is essential to expand the signal-to-noise ratio, ensuring that the relevant information is distinguishable from background noise.

2. **Filtering:** The process of filtering minimizes undesirable distortion or interference from the transmitted signal. This step helps in isolating the relevant information and enhances the accuracy of the analysis.

3. **Digitization:** The analog signal is converted into digital form. Digitization enables more precise measurements and facilitates further processing and analysis using digital techniques.

These electronic processing tasks collectively contribute to refining the electronic signal, making it suitable for accurate analysis and interpretation. The effectiveness of biosensors in providing reliable and meaningful data heavily relies on the efficiency of the electronic processing stage.

(v) Display Unit

The display unit in biosensors serves as the interface through which users interpret and understand the generated output. Comprising systems like computers or printers, the display unit transforms the processed signals into a comprehensible and understandable format for the user. Depending on the end-user requirements, the output can take various forms, such as numerical values, graphical representations, tabular data, or figures. This versatility in output formats allows users to choose the representation that best suits their needs, making the information easily accessible and interpretable. The display unit is a crucial component that bridges the gap between the biosensor's analytical capabilities and the user's understanding, facilitating effective utilization in diverse applications.

CATEGORIZATION OF BIOSENSORS

A biosensor categorization is a broad and interdisciplinary field that requires a variety of characteristics to build a comprehensive framework (Fig. **6**). Several factors influence the classification of biosensors, and the proposed classification framework incorporates a variety of parameters for categorizing these analytical devices based on their different qualities and applications. Classification parameters may encompass the type of bioreceptor utilized, the transduction principle applied, the nature of the target analyte, the application domain, and the particular technology or material used in the biosensor design. This comprehensive approach enables a sophisticated and broad classification system that reflects the diverse range of biosensors and their uses in many domains [37-40].

As previously discussed, biosensors are classified based on various criteria, and one key classification is grounded on the type of bioreceptor employed. Enzymatic biosensors, the most common class, utilize enzymes as bioreceptors. Immunosensors, known for their high specificity and sensitivity, are particularly valuable in diagnostic applications, and rely on the immune system's recognition elements. Another important classification is recognized on the transduction principle Fig. (**7**).

Field Effect Transistor

The field-effect transistor (FET) is a crucial component in modern integrated circuit technology, where it is used for an assortment of applications such as amplification, switching, and signal processing. One of the key advantages of FETs is their high input impedance, which makes them suitable for use in many different circuit configurations. There are several diverse types of FETs, including the metal-oxide-semiconductor FET and the junction FET, each with its unique characteristics and applications.

In recent years, FET technology has continued to evolve, with advancements in materials and manufacturing techniques leading to improved performance and efficiency. As a result, FETs are now found in a wide range of electronic devices, from smartphones and laptops to high-power industrial applications. The future of FET technology looks promising, with ongoing research and development aimed at further enhancing their capabilities and expanding their potential uses. In today's rapidly changing world, the significance of accurate weather forecasts cannot be overstated. Moreover, the increasing integration of renewable energy sources, battery energy storage, and electric vehicle charging stations has led to a significant rise in the application of power electronics devices [29].

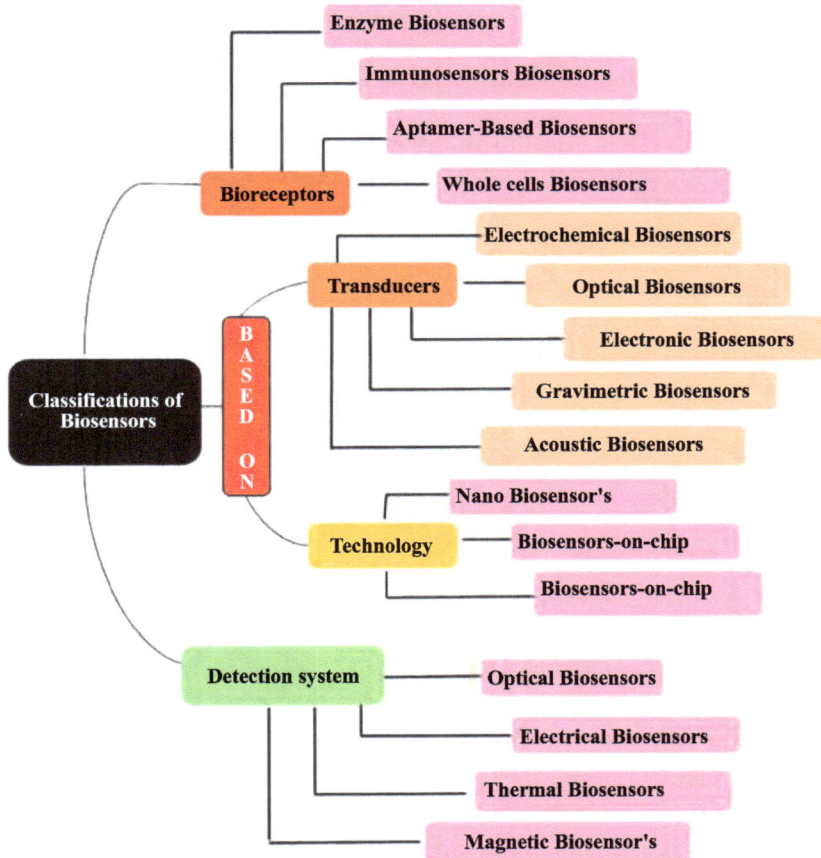

Fig. (6). Classification of Biosensors.

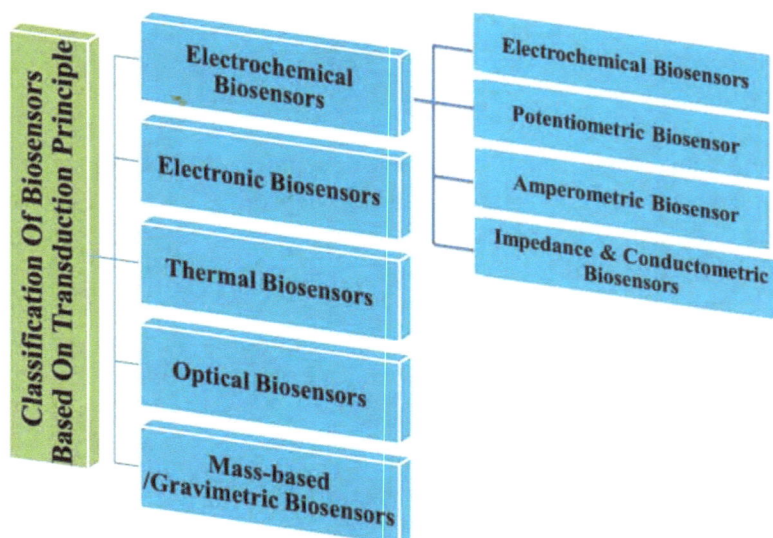

Fig. (7). Classification Based on Transduction Principle.

Among these devices, field-effect transistors play a vital role due to their ability to control the flow of electrical current in a precise and efficient manner a precise and efficient manner. This makes them indispensable in the development of advanced power electronics systems that are crucial for sustainable energy generation and storage. In conclusion, field-effect transistors are essential components in modern electronics, playing a crucial role in a wide array of applications such as amplification, switching, and signal processing. Their high input impedance makes them suitable for various circuit configurations, and advancements in FET technology continue to improve their performance and efficiency. As a result, FETs are widely used in electronic devices, from consumer electronics to high-power industrial applications.

The future of FET technology holds promise, with ongoing research aimed at further enhancing their capabilities and expanding their potential uses. In today's rapidly changing world, accurate weather forecasts are of utmost importance, and the increasing integration of renewable energy sources has led to a rise in the application of power electronics devices. Field-effect transistors, with their precise and efficient control of electrical current, are indispensable in the development of advanced power electronics systems crucial for sustainable energy generation and storage. Their ability to replace existing power-switching devices and contribute to global energy issues makes them a promising candidate in various applications such as microwave electronic devices and power amplifiers. In the field of biosensors,

FET-based biosensors are highly regarded for their speedy, low-cost, and easy detection capabilities.

Introduction

Biosensors are compact analytical tools that convert a biological response into a measurable electrical signal [1]. These miniaturized devices are designed to be sensitive, selective, rapid, cost-effective, and simple, operating independently of external factors like temperature and pH. Constructing efficient biosensors involves collaboration among scientists from various fields, including chemistry, biology, and engineering, due to the multidisciplinary nature of biosensor components. Researchers worldwide study various types of biosensors, including electrical, electrochemical, optical, photonic, magnetic, and piezoelectric biosensors [25]. Among electrical sensing devices, field-effect transistor (FET) biosensors have emerged as promising alternatives due to their speed, cost-effectiveness, and simplicity. The first FET biosensor was developed by Bergveld fifty years ago, and since then, this technology has evolved in different forms, proving to be an ideal approach for swift and precise detection of various analytes, including applications in drug discovery.

While conventional sensing techniques can specifically detect biomolecules, they often involve complicated instrumentations and complex protocols, making them expensive, labor-intensive, and time-consuming. To overcome these limitations, researchers focus on developing novel and reliable FET-based biosensors modified with specific probes on their conducting channels, enabling real-time and label-free analyses. This innovative approach holds great potential for advancing biosensing technologies and addressing various analytical challenges [26].

A field-effect transistor (FET) is a solid-state device where the electroconductivity of the semiconductor between the source and drain terminals is controlled by a third gate electrode through an insulator. These devices, often constructed using standard protocols, generally require minimal post-processing. However, to enhance their surface properties and increase sensitivity, additional materials like nanostructures are fabricated and integrated into their structure. FETs, being already commercialized, mass-produced, and widely utilized, hold significant promise for the future of medical diagnostics, especially in point-of-care (POC) applications. The design and fabrication of FETs using nanostructures have experienced remarkable development over the years. Their unique electronic characteristics, small size, dynamic range, and real-time biological detection capabilities, with limits of detection (LODs) reaching zeptomolar levels, make them one of the most powerful diagnostic platforms. In some cases, FETs based on nanostructures enable the evaluation of single molecules or particles, showcasing their potential for highly sensitive and precise diagnostic applications.

Field-effect transistor (FET) biosensors utilize biological receptors, including antibodies, nucleic acids, aptamers, enzymes, cells, microorganisms, or artificial biomaterials, immobilized on sensing channels connected to the source and drain electrodes. When exposed to target analytes, specific biological complexes such as antigen-antibody or enzyme-substrate are formed. The transducer system then converts these biochemical changes into a measurable signal. The attachment of charged biomolecules to the gate dielectric's surface is akin to applying a voltage through a gate electrode, leading to variations in the threshold voltage. This fundamental principle underscores FET biosensors, where conductance is contingent on the adsorbed species.

Field-effect transistor (FET) biosensors have emerged as the most advanced and widely adopted biosensing alternatives due to their numerous advantages. Among the various types of transistor-based sensing platforms, ion-sensitive field-effect transistors (ISFETs) and metal-oxide-semiconductor field-effect transistors (MOSFETs) stand out as the most utilized structures for biological applications. The key distinctions between ISFETs and MOSFETs lie in the technique of applying the gate voltage, the design, and the material of the gate and the channel region. The fundamental principle of regulating the channel conductance is similar across all FETs. However, the method of gate coupling can vary significantly, leading to the classification of ISFETs and MOSFETs. These biosensors offer a versatile platform for biological sensing, allowing for precise detection and measurement of various analytes. The advantages of FET biosensors include their high sensitivity, rapid response, label-free detection, and compatibility with miniaturization, making them suitable for a wide range of applications in medical diagnostics, environmental monitoring, and beyond.

The ISFETs, with their ion-sensitive gate, are particularly well-suited for detecting changes in ion concentrations, which is relevant in biological and environmental contexts. On the other hand, MOSFETs, with their metal-oxide gate, provide robust performance and versatility, enabling their use in diverse biosensing applications. As researchers continue to explore novel materials, fabrication techniques, and integration methods, FET biosensors are expected to witness further advancements. The versatility and adaptability of FET biosensors make them a cornerstone in the development of innovative and high-performance biosensing technologies, contributing to advancements in healthcare, diagnostics, and various other fields.

Ion-Sensitive Field-Effect Transistors (ISEFT'S) in Biosensing

Ion-sensitive field-effect transistors (ISFETs) stand out as integral components in microelectrochemical lab-on-a-chip (LOC) setups, particularly for biological applications. Their configuration closely resembles that of conventional metal-oxide-semiconductor field-effect transistors (MOSFETs), with a distinctive feature being the sensitive area or transistor gate, which transduces ion concentration into a measurable voltage [19].

In ISFETs, the metal gate and gate oxide of a MOSFET are represented by a sample solution with an immersed reference electrode and an insulating layer sensitive to ions. The nature of this insulating layer significantly influences the sensitivity and capability of the ISFET sensor. Materials like Si_3N_4, Al_2O_3, or Ta_2O_5 on SiO_2, with active groups on the oxide surface, selectively act on H^+ ions, making them particularly effective for pH sensing [20]. The pH-ISFET is a common category, and to enhance specificity for other analytes, ion-sensing or ion-blocking membranes like ionophores can be employed. These membranes provide selective interaction with analytes, reducing non-specific binding. For biomolecule sensitivity, a bio-recognition layer is introduced, transforming the ISFET into a biologically-sensitive field-effect transistor (BioFET). This layer facilitates selective interaction with analytes, reducing non-specific binding and enabling charge transfer [21].

The measurement in ISFETs involves regulating the current flowing between the source and drain terminals by a third electrode (gate terminal) directly in contact with the sample. Changes in conductance due to variations in surface charge, caused by biomolecular interactions, are measured. BioFETs have found applications in detecting various analytes, showcasing their potential in biosensing [22]. The development of ISFETs is facilitated through complementary metal oxide semiconductor technology (CMOS), commonly used in producing integrated circuits. CMOS technology allows the creation of inexpensive, extensive, and mass-produced devices capable of detecting minute amounts of analytes on miniaturized chips. This technology has been employed in the design of fully electronic Lab-on-Chip platforms for instantaneous DNA sensing, demonstrating the versatility and efficiency of ISFETs in modern biosensing applications [23].

Metal-oxide field-effect transistors (MOSFETs)

Metal-oxide-semiconductor field-effect Transistors (MOSFETs) are the most prevalent type of shielded gate FETs (IGFETs) and are widely utilized in numerous microelectronic circuits for the amplification and switching of electronic signals. They serve as the core of integrated circuits, owing to their small sizes, and are capable of being integrated into a single chip. A distinctive feature of MOSFETs is their metal oxide gate electrode [24]. This gate electrode is separated from the main channel, positioned between the drain and source, by an ultra-thin layer of insulating material, typically SiO_2. This insulating layer acts as a barrier, preventing current flow into the gate, and resulting in exceptionally high input resistance. MOSFETs, therefore, function like voltage-controlled resistors. It's essential to handle MOSFETs with care as they can be susceptible to damage from the accumulation of significant static charge due to their high input resistance. Despite this sensitivity, MOSFETs have found applications in biosensing devices. For instance, an industrial standard biological detection equipment based on MOSFET was industrialized for the recognition of C-reactive protein (CRP). The binding of CRP to its specific antibody was detected by quantifying the drain current of the MOSFET system, showcasing the potential for MOSFETs in creating cost-effective and simple FET-based biosensors. In another study, MOSFET-fixed cantilevers were designed for the detection of Bacillus thuringiensis. The attachment of the target to the cantilever, modified with gold, induced variations in the drain current of the MOSFET due to the flexing of the cantilever. The electronic attributes of the target also contributed to the alteration in the drain current, demonstrating the versatility of MOSFETs in biosensing applications with reported detection limits [25].

Nanotech in FET-based Biosensors

Nanotechnology has made significant advancements in the field of biosensors, revolutionizing the ways in which we detect and monitor various biomarkers and environmental factors. With the integration of nanomaterials, biosensors have become more efficient and versatile, expanding their assortment of applications in healthcare, environmental monitoring, and various additional industries. The emergence of graphene-encapsulated nanoparticle FET biosensors represents a significant advancement in the early identification of cancer biomarkers. These sensor devices have exhibited remarkable detection limits, enabling the identification of breast cancer biomarkers at concentrations as low as 100 pM. Such high sensitivity is crucial for the primary finding and nursing of diseases, facilitating timely interventions and enhancing patient outcomes [26]. Despite these achievements, current sensor developments primarily concentrate on detecting individual target biomarkers. To enhance diagnostic capabilities, there is a growing demand for biosensors capable of simultaneously measuring multiple biomarkers. This multifunctional approach holds promise for more comprehensive and effective disease diagnosis and monitoring.

The emergence of nanotechnology in the late 1990s has given rise to a groundbreaking category of biosensors known as nanobiosensors. These innovative devices capitalize on a diverse array of nanomaterials, including silicon and gold nanoparticles, to facilitate label-free and amplified biosensing. A noteworthy outcome of this development is the creation of the Advanced Wearable Biosensor Platform, representing a substantial leap forward in biosensor technology.

The Advanced Wearable Biosensor Platform stands at the intersection of nanotechnology and wearable technology, crafting a compact, non-invasive biosensor system. This platform has the capability to continuously monitor diverse physiological parameters, including heart rate, body temperature, glucose levels, and specific biomarkers in sweat or breath. By leveraging nanomaterials, this innovative system achieves real-time and personalized monitoring, providing intricate insights into an individual's health at the molecular level (Gao et al., 2016 [41]). The integration of nanomaterials has also conferred enhanced durability and flexibility to biosensors, making them well-suited for wearable applications (Yun et al., 2021 42]). These strides in nanomaterial-enabled wearable technology have the potential to reshape healthcare by enabling early disease detection, personalized medicine, and continuous health monitoring in real-time (Guk et al., 2019 [43]). Moreover, the versatility of nanomaterial-enabled wearable sensors extends beyond healthcare, with applications in environmental monitoring and other industries. In summary, nanotechnology's pivotal role in the evolution of wearable biosensors has brought about increased sensitivity, multiplexed biomarker detection, improved durability and flexibility, and real-time monitoring of diverse physiological parameters.

The rapid evolution of nanomaterials has positioned them as key players in biosensor manufacturing, finding applications across diverse fields like biochemistry, electrochemistry, agriculture, and biomedicine. Nanomaterials, renowned for their stability, potent electrocatalytic properties, and facile modifiability, offer a suite of advantages, making them highly suitable for biosensor development [35].

In the biosensor domain, nanotechnology has ushered in transformative changes, particularly in enhancing sensitivity, durability, and flexibility (Vaddiraju et al., 2010 [45]. This technological shift has given rise to miniaturized biosensors capable of integration into implantable devices, enabling continuous monitoring of metabolites and biomarkers within the human body (Vaddiraju et al., 2010 [36]). Nanotechnology's contributions have propelled biosensors to new heights, resulting in highly sensitive, durable, and flexible devices capable of continuous monitoring across various physiological parameters. Nanomaterials have further elevated the performance of electrochemical sensors and biosensors, particularly in the detection of biologically important analytes (Maduraiveeran et al., 2018 [46]). Notably, graphene-based electrochemical sensors, facilitated by nanomaterials, have demonstrated exceptional capabilities in biosensing applications. As a result, biosensors have found extensive applications in fields such as food, healthcare, environmental monitoring, water quality assessment, forensic medicine, drug development, and biology, transforming the way we approach detection and monitoring in various sectors. The integration of nanotechnology into biosensors has opened up endless possibilities for improved accuracy, sensitivity, and selectivity. This has also led to the development of wearable biosensors that can continuously monitor health parameters, allowing for personalized healthcare and early detection of diseases. In conclusion, the use of nanotechnology in FET-based biosensors has revolutionized the field, enhancing their sensitivity, durability, and flexibility [38]. It has also expanded their application potential, making them invaluable tools in various industries and fields. The integration of nanotechnology in FET-based biosensors has revolutionized the field, enhancing their sensitivity, durability, and flexibility. Additionally, the use of nanotechnology has allowed for the development of miniaturized and portable biosensors, enabling point-of-care testing and real-time monitoring in various settings. Overall, nanotechnology has significantly contributed to the advancement of FET-based biosensors by improving their performance and expanding their applications in various fields [38]. Nano biosensors, specifically those based on field-effect transistors, have emerged as a powerful tool in biosensing applications. Their ability to detect biomolecular analytes in a label-free, specific, and real-time manner makes them ideal for early biomarker detection and drug screening in biomedical research and clinical diagnostics. Additionally, the use of nanomaterials in FET-based biosensors has allowed for enhanced performance in terms of sensitivity, selectivity, and response time. Furthermore, the use of nanomaterials in FET-based biosensors has also addressed the challenges of low signal-to-noise ratio and non-specific binding, leading to improved accuracy and reliability in biosensing applications. Furthermore, the integration of nanomaterials in FET-based biosensors has also addressed the challenges of low signal-to-noise ratio and non-specific binding, resulting in improved accuracy and reliability in biosensing applications (Sung & Koo, 2021). Ultimately, the integration of nanotechnology into FET-based biosensors has revolutionized the field, enhancing their accuracy, sensitivity, and selectivity (Kulkarni et al., 2022). Overall, the integration of nanotechnology in FET-based biosensors has revolutionized the field by enhancing their accuracy, sensitivity, and selectivity. Nanotechnology has revolutionized the field of FET-based biosensors by enhancing their accuracy, sensitivity, and selectivity (Sedki et al., 2021). The integration of nanomaterials in FET-based biosensors has led to improved accuracy, sensitivity, and selectivity in biosensing applications (Sedki et al., 2021). In conclusion, the integration of nanotechnology in FET-based biosensors has revolutionized the field by enhancing their accuracy, sensitivity, and selectivity (Sedki et al., 2021). In conclusion, the integration of nanotechnology has revolutionized the field of FET-based biosensors, improving their accuracy, sensitivity, and selectivity in biosensing applications. In conclusion, the integration of nanotechnology has revolutionized the field of FET-based biosensors, significantly contributing to improved accuracy, sensitivity, and selectivity in biosensing applications. In conclusion, the integration of nanotechnology in FET-based biosensors has revolutionized the field by enhancing their accuracy, sensitivity, and selectivity in biosensing applications. The continued development and integration of nanomaterials in FET-based biosensors hold great promise for the future, offering the potential for further miniaturization, increased multiplexing capabilities, and enhanced portability. These advancements will undoubtedly drive the ongoing evolution of biosensor technology, opening up new frontiers in personalized healthcare, environmental monitoring, and biotechnology.

The escalating demand for real-time and highly sensitive biosensors across various industries underscores the significance of nanotechnology in advancing FET-based biosensors. The ongoing research and development efforts in this field promise to bring forth even more substantial innovations in biosensor design, leading to enhanced performance, broader applications, and an overall improvement in the quality of life. The integration of nanotechnology into FET-based biosensors has not only influenced the present state of biosensing technology but is poised to profoundly shape its future. This convergence solidifies nanotechnology's role as a cornerstone in the realm of advanced sensing capabilities. The field of biosensors encompasses various detection techniques, with a focus on electrochemical and optical biosensors due to their high specificity and low detection limits, making them well-suited for real-time processing. However, the emergence of simpler yet potent platforms like Field-Effect Transistors (FETs) has gained prominence, primarily attributed to their high sensitivity and rapid screening capabilities, which are explored in this section.

Biosensors can be further categorized based on detection techniques into two main types: (1) label biosensors and (2) label-free biosensors. In label-based detection, the phrase 'label' refers to a chemical or temporary attachment of a foreign molecule, effectively increasing the number of binding sites to facilitate the detection of the desired analyte. Label-based biosensing methods, especially those detectable by electrochemical sensors, offer accuracy and wide detection limits. However, they present challenges such as difficulties in multiplexing and the potential alteration of intrinsic properties of target molecules. For instance, an incorrect label can lead to the development of a drug with unintended side effects. Practical implementation of these methods demands a high level of assurance that the label does not interfere with the interaction between the target analyte and probe, potentially blocking important active sites. Whereas, label-free biological detection approaches do not encompass analyte labeling or specimen alteration. These methods investigate interfaces and physical or chemical properties of target analytes by monitoring their intrinsic physicochemical properties, comprising refractive index, affinity constants, dielectric permittivity, viscoelasticity, conductivity, and charge. Essentially, the molecular properties that link the target analyte to the biosensor are detected through the label-free approach. Label-free technology offers several advantages, including high throughput, high sensitivity, low sample/analyte consumption with minimal damage, high precision and simplicity, easy on-chip integration, the ability to use natural analytes and ligands, and cost-effectiveness compared to label-based methods (no reagent/label cost, no lab safety/waste disposal cost) [38]. Consequently, label-free biological sensing has evolved as a systematic and comprehensive strategy for constructing micro-scale assays that are extremely adaptable for distant diagnostics and field applications. Despite the industrial presence and uncovering accuracy of label-free detection, ongoing competition aims to enhance reliability, achieve a high signal-to-noise ratio (SNR), and reduce instrumentation costs. FET-based biosensors are a viable option for miniaturizing and commercializing biosensing platforms, among other label-free transducers [39].

Key Features and Advantages of FET-Based Biosensors

A field-effect transistor (FET)--based biological sensors indeed hold significant promise in the field of biosensing. These biosensors leverage the electrical characteristics of FETs to detect biological molecules, providing a label-free and highly sensitive approach. Here are some key features and advantages of FET-based biosensors:

1. **Label-Free Detection:** FET-based biosensors offer label-free detection, meaning that there's no need for additional labels or tags on the target molecules. This simplifies the sensing process and reduces the complexity of the assay.

2. **High Sensitivity:** FET-based biosensors can achieve high sensitivity, enabling the detection of low concentrations of target molecules. This is crucial in various applications such as medical diagnostics, environmental monitoring, and food safety.

3. **Real-Time Monitoring:** The real-time monitoring capability of FET-based biosensors allows for the continuous tracking of biological interactions. This is essential for understanding dynamic processes and kinetics of biomolecular binding events.

4. **Miniaturization:** FET-based biosensors can be miniaturized, allowing for the improvement of compact and portable gadgets. This is advantageous for point-of-care diagnostics, on-site environmental monitoring, and other applications where portability is essential.

5. **Low Detection Limits:** The inherent sensitivity of FETs enables the uncovering of low concentrations of analytes, formulating them suitable for applications requiring high precision and low detection limits.

6. **Versatility:** FET-based biosensor technology is capable of detecting multiple kinds of biological compounds, such as DNA, proteins, and cells other molecules. This adaptability makes them applicable in various domains including healthcare, biotechnology, and ecological examining.

7. **Commercial Potential:** The miniaturization and potential for mass production make FET-based biosensors commercially viable. The development of standardized and cost-effective biosensing platforms based on FETs could lead to widespread adoption in various industries.

It's important to note that while FET-based biosensors offer numerous advantages, there are also challenges such as optimizing sensor performance, ensuring stability, and addressing issues related to signal-to-noise ratios. Nevertheless, ongoing research and advancements in nanotechnology and materials science are likely to contribute to the continued development and improvement of FET-based biosensors for various applications.

Nanomaterials in FET-based Biosensors

Nanomaterials play a vital role in the expansion of FET-based biosensors. The incorporation of nanomaterials in FET-based biosensors offers abundant advantages, comprising enhanced sensitivity, improved detection limits, and increased surface area for biomolecule immobilization [48]. One of the most commonly used nanomaterials in FET-based biosensors is graphene, known for its high electrical conductivity and large surface area. Other nanomaterials, for example, carbon nanotubes (CNT), metal nanoparticles (MNPs), and quantum dots (QDs) have also shown promising results in improving the performance of FET-based biosensors [43]. The unique properties of these nanomaterials allow for the fabrication of incredibly sensitive and efficient biosensors for an assortment of biological and environmental applications. Furthermore, nanomaterials are capable of being enriched with distinct biomolecules instance protective antibodies or aptamers to target and detect individual analytes, making them extremely versatile for a widespread series of sensing applications.

In addition to their improved sensing properties, nanomaterials also contribute to the miniaturization of FET-based biosensors, making them ideal for point-of-care diagnostics and on-site environmental monitoring. The small size and efficient functionality of nanomaterial-integrated FET-based biosensors allow for easy integration into portable and handheld devices, enabling rapid and convenient detection of various analytes [44].

Furthermore, ongoing research in the field of nanomaterials continues to explore novel materials and their integration into FET-based biosensors, aiming to further enhance their performance, stability, and biocompatibility. With continued advancements, nanomaterial-integrated FET-based biosensors hold great potential for revolutionizing the field of biosensing, offering innovative solutions for real-time, accurate, and portable detection in diverse settings. Research in the field of nanomaterials continues to drive innovation in FET-based biosensors. One area of focus is the development of nanomaterials with specific properties tailored to the requirements of different biosensing applications. For example, the design of nanomaterials with enhanced biocompatibility and stability is crucial for long-term sensing applications, especially in medical diagnostics. Moreover, the exploration of hybrid nanomaterial systems, combining multiple types of nanomaterials to leverage their individual properties, is an exciting avenue for advancing the capabilities of FET-based biosensors [45]. These hybrid nanomaterials could potentially offer synergistic effects, leading to even greater sensitivity and selectivity in biosensing. Another intriguing prospect is the incorporation of nanomaterials with stimuli-responsive properties into FET-based biosensors. By integrating nanomaterials that respond to specific stimuli such as pH, temperature, or target analytes, the biosensors could exhibit dynamic and adaptive sensing capabilities, expanding their potential applications in complex biological and environmental environments[49][50]. As the understanding of nanomaterials and their interactions with biological systems continues to deepen, FET-based biosensors stand to benefit from the integration of increasingly sophisticated and tailored nanomaterials, paving the way for the next generation of high-performance biosensing technologies. The ability to tailor nanomaterials to specific sensing applications opens up a world of possibilities for the future of FET-based biosensors. By designing nanomaterials with properties that match the requirements of different environments and target analytes, researchers can create biosensors that are highly efficient and reliable across a wide range of scenarios [46,47].

Furthermore, the potential of hybrid nanomaterial systems is an exciting area of exploration. Combining the unique characteristics of different nanomaterials can lead to synergistic effects, enhancing the overall sensitivity and selectivity of FET-based biosensors. This methodology demonstrates the gigantic potential for refining the biosensor's performance in demanding detection tasks. The integration of nanomaterials with stimuli-responsive properties also represents a fascinating avenue for the advancement of FET-based biosensors. By incorporating nanomaterials that can respond to specific stimuli, for instance, changes in pH or temperature, the adaptability and dynamic nature of biosensors can be enhanced, opening up new possibilities for real-time monitoring in complex biological and environmental settings. In conclusion, the ongoing developments in the field of nanomaterials hold great potential for shaping the future of FET-based biosensors [51]. The ability to fine-tune the properties of nanomaterials to meet the specific demands of diverse sensing applications paves the way for highly effective and versatile biosensing technologies. With continued innovation and exploration in this field, we can expect to see significant advancements in the capabilities of FET-based biosensors, leading to groundbreaking solutions for real-time, accurate, and portable detection in various fields.

CONCLUSION

In conclusion, the integration of nanomaterials in biosensing applications, particularly in Field Effect Transistor-based devices, has exhibited promising potential for enhanced sensing capabilities. With their distinctive characteristics, such as high surface area, biocompatibility, as well as electronic properties, nanomaterials have revolutionized biosensing technologies, enabling sensitive and specific detection of biological analytes. As research in this field continues to progress, the development of novel nanomaterial-integrated FET biosensors delivers tremendous potential for diverse applications in healthcare, monitoring environmental conditions, and beyond. The intersection of nanotechnology and biosensing has the potential to contribute noteworthy enhancements in molecular diagnostics including personalized medicine, opening newfound avenues for addressing global health challenges. In addition, the integration of nanomaterials has laid the way for the advancement of portable, rapid, especially highly sensitive biological sensing devices. These devices have the proficiency to transfigure point-of-care testing, allowing for real-time monitoring of biomarkers in clinical settings and remote locations. Furthermore, the scalability and cost-effectiveness of nanomaterial-integrated FET biosensors make them a promising tool for widespread implementation in resource-limited settings, thereby improving access to diagnostic capabilities in underserved communities. The integration of nanomaterials in biosensing applications, particularly in Field-Effect Transistor-based devices, holds immense potential for enhancing sensing capabilities and addressing global health challenges. With their unique properties and vast application possibilities, nanomaterial-integrated biosensors have the capacity to revolutionize diagnostics, personalized medicine, and point-of-care testing on a global scale. Additionally, the ongoing advancements in nanomaterial synthesis and engineering will further expand the possibilities for tailoring biosensors with increased sensitivity, selectivity, and multiplexing capabilities. Overall, the future of nanomaterial-integrated biosensors looks extremely promising and has the potential to revolutionize various fields, ultimately improving human health and well-being.

Learning Objectives/Key Points

(1) Introduction to nano biosensors: Providing an overview of nanotechnology and biosensors, highlighting their combination for developing highly sensitive and selective detection platforms.

(2) Types of nano biosensors: Exploring various types of nano biosensors, including optical, electrochemical, and field-effect transistor (FET)-based sensors, and their applications in different fields such as healthcare, environmental monitoring, and food safety.

(3) Nanomaterials for biosensing: Discussing the use of nanomaterials such as nanoparticles, nanowires, and 2D materials like graphene and transition

metal dichalcogenides in the fabrication of nano biosensors to enhance sensitivity, selectivity, and stability.

(4) Design and fabrication of nano biosensors: Detailing the fabrication techniques and design considerations involved in developing nano biosensors, including surface functionalization, immobilization techniques, and signal transduction mechanisms.

(5) Applications and future prospects: Highlighting current applications of nano biosensors in areas such as disease diagnostics, environmental monitoring, and personalized medicine, and discussing future directions and challenges in the field, including commercialization and regulatory considerations.

(6) Field Effect Transistor: Field-effect transistor (FET)-based biosensors indeed hold significant promise in the field of biosensing. These biosensors leverage the electrical characteristics of FETs to detect biological molecules, providing a label-free and highly sensitive approach.

Nanomaterials in FET-based Biosensors: Nanomaterials play a crucial role in the development of FET-based biosensors. The integration of nanomaterials in FET-based biosensors offers several advantages, including enhanced sensitivity, improved detection limits, and increased surface area for biomolecule immobilization

Multiple Choice Questions (MCQs)

1. What is the advantage of nanosensors over conventional sensors?

a). Higher sensitivity and selectivity

b). Lower cost

c). Larger physical size

d). Higher power consumption

Ans: a) Higher sensitivity and selectivity

Explanation: Nanosensors have higher sensitivity and selectivity, allowing for more precise and accurate detection compared to conventional sensors. They can identify and distinguish between different molecules or analytes, making them highly valuable for applications where precision and accuracy are critical, such as in medical diagnostics, environmental monitoring, and scientific research.

2. What is the advantage of nanodevices over larger devices?

a). Improved performance and efficiency

b). Higher cost.

c). Less functionality

d). Reduced complexity

Ans: a). Improved performance and efficiency

Explanation: Nanodevices offer improved performance and efficiency due to their smaller size, enabling more advanced and precise functionalities compared to larger devices.

3. What is the main challenge of nanosensor fabrication?

a). Ensuring reproducibility and scalability

b). Reducing sensitivity and selectivity

c). Increasing cost

d). Avoiding miniaturization

Ans: a). Ensuring reproducibility and scalability

Explanation: One of the main challenges in nanosensor fabrication is ensuring reproducibility and scalability to achieve consistent sensor performance in large-scale production. While sensitivity, selectivity, and cost considerations are important in nanosensor development, ensuring reproducibility and scalability is a fundamental challenge that must be addressed to make nanosensors practical for various applications, including in healthcare, environmental monitoring, and industry.

4. Which property of nanodevices enables them to be used in various biological systems?

a). Rigidity

b). Toxicity

c). Opacity

d). Biocompatibility

Ans: d). Biocompatibility

Explanation: Nanodevices need to exhibit biocompatibility to interact safely with biological systems, making them a crucial property for their effectiveness in medicine and other applications. Nanodevices designed to be biocompatible can safely interact with cells, tissues, and biological molecules, making them suitable for applications in medicine, biotechnology, and other biological fields.

5. What is the role of the bioreceptor in a biosensor?

a). Convert biological signals into electrical signals

b). Amplify the detected signal

c). Recognize and bind to the target analyte

d). Convert electrical signals into biological signals

Ans: c). Recognize and bind to the target analyte

Explanation: The role of the bioreceptor in a biosensor is to recognize and bind specifically to the target analyte present in the sample. Bioreceptors can be various biomolecules such as enzymes, antibodies, nucleic acids, or whole cells, depending on the nature of the target analyte and the sensing mechanism employed in the biosensor. Once the bioreceptor binds to the target analyte, it initiates a biochemical or biophysical reaction that produces a detectable signal, which is then transduced by the biosensor for analysis.

6. Which biosensor type is commonly used for continuous glucose monitoring in diabetes management?

a). Amperometric biosensor

b). Optical biosensor

c). Potentiometric biosensor

d). Chemiresistor biosensor

Ans: a). Amperometric biosensor

Explanation: Amperometric biosensors are commonly used for continuous glucose monitoring in diabetes management. These biosensors typically employ an enzyme-based bioreceptor, such as glucose oxidase, which catalyses the oxidation of glucose. The resulting enzymatic reaction generates an electrical current proportional to the glucose concentration in the sample. By measuring this current, amperometric biosensors can provide continuous and real-time monitoring of glucose levels in diabetic patients, enabling timely adjustments in insulin dosage and dietary management.

7. What is a Field Effect Transistor (FET)

a). A type of light-emitting diode

b). A device that controls electrical current

c). A switch that converts digital signals

d). A component used in mechanical systems

Ans: b). A device that controls electrical current

Explanation: A Field Effect Transistor (FET) is an electronic device designed to regulate and amplify electrical current. It works based on the modulation of charge carriers in a semiconductor channel under the influence of an electric field. FETs are widely used in various applications, including amplifiers, switches, oscillators, and integrated circuits.

8. FET is a device which has:

a). High input impedance and is voltage-controlled

b). Low input impedance and current-controlled

c). Low input impedance and voltage-controlled

d). High input impedance and current-controlled

Ans: a). High input impedance and voltage-controlled

Explanation: FET (Field Effect Transistor)

- It is a voltage-controlled device.

- It has a high input impedance as the gate is Reverse for JFETs and because of the insulating layer in the case of MOSFETs.

- It has low gain bandwidth.

9. What is the purpose of the gate terminal in a FET-base biosensor?

a). To control the flow of current between the source and drain terminals.

b). To modulate the conductivity of the semiconductor channel.

c). To detect changes in analyte concentration.

d). To enhance the sensitivity of the biosensor.

Ans: b). To modulate the conductivity of the semiconductor channel.

Explanation: In a field-effect transistor (FET)-based biosensor, the gate terminal plays a crucial role in modulating the conductivity of the semiconductor channel. By applying a voltage to the gate terminal, the electrostatic field generated controls the flow of charge carriers within the semiconductor channel, which in turn affects the overall conductance of the device. This modulation of conductivity allows the biosensor to detect changes in the analyte concentration by measuring variations in electrical signals, such as current or voltage, resulting from interactions between the analyte and the biorecognition elements immobilized on the sensor surface.

10. In a FET-based biosensor, which component serves as the sensing element for detecting analytes?

a). Source terminal

b). Drain terminal

c). Gate terminal

d). Semiconductor channel

Ans: d). Semiconductor channel

Explanation: In an FET-based biosensor, the semiconductor channel serves as the sensing element for detecting analytes. When analyte molecules bind to the surface of the semiconductor channel, it induces changes in the conductivity or charge distribution within the channel, leading to variations in the electrical signals (such as current or voltage) passing through the device. These changes in electrical signals are then measured and correlated with the presence or concentration of the target analyte.

11. Which of the following is an advantage of FET-based biosensors?

a). Low sensitivity

b). Small detection range

c). Label-free detection

d). Limited scalability

Ans: c). Label-free detection

Explanation: An advantage of FET-based biosensors is label-free detection. Unlike some other biosensor technologies that require labeling of the target analyte with fluorescent or radioactive tags, FET-based biosensors can directly detect analytes without the need for additional labels or tags. This label-free detection approach simplifies the assay procedure, reduces assay time, and minimizes the risk of interference or false positives from labeling agents.

12. What type of analytes can be detected using FET-based biosensors?

a). Only proteins

b). Only nucleic acids

c). Various biomolecules (including proteins, and nucleic acids)

d). None of the above

Ans: c). Various biomolecules, including proteins, nucleic acids, and small molecules

Explanation: FET-based biosensors have the versatility to detect various types of biomolecules, including proteins, nucleic acids (such as DNA and RNA), and small molecules (such as ions, metabolites, and drugs). This broad range of analytes makes FET-based biosensors applicable in diverse fields such as medical diagnostics, environmental monitoring, food safety, and pharmaceutical development.

13. What is one advantage of using nanomaterials, such as carbon nanotubes or graphene, in FET-based biosensors?

a). Reduced sensitivity

b). Decreased detection limits

c). Increased specificity

d). Enhanced scalability

Ans: b). Decreased detection limits

Explanation: Nanomaterials like carbon nanotubes and graphene possess unique properties such as high surface area, excellent electrical conductivity, and biocompatibility, making them well-suited for biosensing applications. When incorporated into FET-based biosensors, these nanomaterials can enhance the sensitivity of the device, allowing for the detection of analytes at lower concentrations. This ultimately leads to decreased detection limits, enabling the biosensor to detect target molecules with higher precision and accuracy.

14.What makes nanosensors ideal for food packaging applications?

a). Ability to detect spoilage and contamination

b). Resistance to physical damage

c). Lower production cost compared to conventional packaging

d). Incompatibility with food products

Ans: a). Ability to detect spoilage and contamination

Explanation: Nanosensors can detect gases, chemicals, or indicators of spoilage and contamination, providing real-time information about the food's condition. This helps improve food safety, reduce food waste, and enhance the overall quality and shelf-life of food products, making nanosensors valuable tools for the food packaging industry.

15. What is the main challenge of nanosensor fabrication?

a). Ensuring reproducibility and scalability

b). Reducing sensitivity and selectivity

c). Increasing cost

d). Avoiding miniaturization

Ans: a). Ensuring reproducibility and scalability

Explanation: One of the main challenges in nanosensor fabrication is ensuring reproducibility and scalability to achieve consistent sensor performance in large-scale production. While sensitivity, selectivity, and cost considerations are important in nanosensor development, ensuring reproducibility and scalability is a fundamental challenge that must be addressed to make nanosensors practical for various applications, including in healthcare, environmental monitoring, and industry.

REFERENCES

[1] V.D. Krishna, K. Wu, D. Su, M.C.J. Cheeran, J.P. Wang, and A. Perez, "Nanotechnology: Review of concepts and potential application of sensing platforms in food safety", *Food. Microbiol.,* vol. 75, pp. 47-54, 2018.

http://dx.doi.org/10.1016/j.fm.2018.01.025 PMID: 30056962

[2] S. Sharma, and M. Madou, "2012. A new approach to gas sensing with nanotechnology", *Philos. Trans R Soc. Lond A,* vol. 370, pp. 2448-2473, 1967.

[3] D. Rawtani, N. Khatri, S. Tyagi, and G. Pandey, "Nanotechnology-based recent approaches for sensing and remediation of pesticides", *J. Environ. Manage,* vol. 206, pp. 749-762, 2018.

http://dx.doi.org/10.1016/j.jenvman.2017.11.037 PMID: 29161677

[4] G.A. Silva, "Introduction to nanotechnology and its applications to medicine", *Surg. Neurol.,* vol. 61, no. 3, pp. 216-220, 2004.

http://dx.doi.org/10.1016/j.surneu.2003.09.036 PMID: 14984987

[5] J. Riu, A. Maroto, and F. Rius, "Nanosensors in environmental analysis", *Talanta,* vol. 69, no. 2, pp. 288-301, 2006.

http://dx.doi.org/10.1016/j.talanta.2005.09.045 PMID: 18970568

[6] C. Yonzon, D. Stuart, X. Zhang, A. McFarland, C. Haynes, and R. Vanduyne, "Towards advanced chemical and biological nanosensors—An overview", *Talanta,* vol. 67, no. 3, pp. 438-448, 2005.

http://dx.doi.org/10.1016/j.talanta.2005.06.039 PMID: 18970187

[7] S.A. Perdomo, J.M. Marmolejo-Tejada, and A. Jaramillo-Botero, "Bio-nanosensors: Fundamentals and recent applications", *J. Electrochem Soc.,* vol. 168, no. 10, p. 107506, 2021.

http://dx.doi.org/10.1149/1945-7111/ac2972

[8] J.N. Anker, W.P. Hall, O. Lyandres, N.C. Shah, J. Zhao, and R.P. Van Duyne, "Biosensing with plasmonic nanosensors", *Nat. Mater.,* vol. 7, no. 6, pp. 442-453, 2008.

http://dx.doi.org/10.1038/nmat2162 PMID: 18497851

[9] S. Kurbanoglu, and S.A. Ozkan, "Electrochemical carbon based nanosensors: A promising tool in pharmaceutical and biomedical analysis", *J. Pharm. Biomed. Anal.,* vol. 147, pp. 439-457, 2018.

http://dx.doi.org/10.1016/j.jpba.2017.06.062 PMID: 28780997

[10] M.B. Kulkarni, N.H. Ayachit, and T.M. Aminabhavi, "Recent Advancements in Nanobiosensors: Current Trends, Challenges, Applications, and Future Scope", *Biosensors,* vol. 12, no. 10, p. 892, 2022.

http://dx.doi.org/10.3390/bios12100892 PMID: 36291028

[11] H. Becker, and W. Dietz, "Microfluidic devices for u-TAS applications fabricated by polymer hot embossing", In: *Microfluidic Devices and Systems,* vol. Vol. 3515. SPIE, 1998, pp. 177-182.

http://dx.doi.org/10.1117/12.322081

[12] R. Prakash, and K.V.I.S. Kaler, "An integrated genetic analysis microfluidic platform with valves and a PCR chip reusability method to avoid contamination", *Microfluid Nanofluidics,* vol. 3, no. 2, pp. 177-187, 2007.

http://dx.doi.org/10.1007/s10404-006-0114-7

[13] S. Gavrilaş, C.Ş. Ursachi, S. Perţa-Crişan, and F.D. Munteanu, "Recent trends in biosensors for environmental quality monitoring", *Sensors,* vol. 22, no. 4, p. 1513, 2022.

http://dx.doi.org/10.3390/s22041513 PMID: 35214408

[14] D.E.D.E. Sercan, and F. Altay, "Biosensors from the first generation to nano-biosensors", *International Advanced Researches and Engineering Journal,* vol. 2, no. 2, pp. 200-207, 2018.

[15] D. Grieshaber, R. MacKenzie, J. Vörös, and E. Reimhult, "Electrochemical biosensors-sensor principles and architectures", *Sensors,* vol. 8, no. 3, pp. 1400-1458, 2008.

http://dx.doi.org/10.3390/s80314000 PMID: 27879772

[16] K. Abu-Salah, S.A. Alrokyan, M.N. Khan, and A.A. Ansari, "Nanomaterials as analytical tools for genosensors", *Sensors,* vol. 10, no. 1, pp. 963-993, 2010.

http://dx.doi.org/10.3390/s100100963 PMID: 22315580

[17] S. Vigneshvar, C.C. Sudhakumari, B. Senthilkumaran, and H. Prakash, "Recent advances in biosensor technology for potential applications–an overview", *Front Bioeng. Biotechnol.,* vol. 4, p. 11, 2016.

http://dx.doi.org/10.3389/fbioe.2016.00011 PMID: 26909346

[18] L.C. Clark, and C. Lyons, "Electrode systems for continuous monitoring in cardiovascular surgery", *Ann N Y Acad. Sci.,* vol. 102, no. 1, pp. 29-45, 1962.

http://dx.doi.org/10.1111/j.1749-6632.1962.tb13623.x PMID: 14021529

[19] N. Fracchiolla, S. Artuso, and A. Cortelezzi, "Biosensors in clinical practice: focus on oncohematology", *Sensors,* vol. 13, no. 5, pp. 6423-6447, 2013.

http://dx.doi.org/10.3390/s130506423 PMID: 23673681

[20] A.P.F. Turner, "Biosensors: sense and sensibility", *Chem. Soc. Rev.,* vol. 42, no. 8, pp. 3184-3196, 2013.

http://dx.doi.org/10.1039/c3cs35528d PMID: 23420144

[21] B. Bhushan, "Nanotribology and nanomechanics of MEMS/NEMS and BioMEMS/BioNEMS materials and devices", *Microelectron Eng.,* vol. 84, no. 3, pp. 387-412, 2007.

http://dx.doi.org/10.1016/j.mee.2006.10.059

[22] G. Kaur, A. Kaur, and H. Kaur, "Review on nanomaterials/conducting polymer based nanocomposites for the development of biosensors and electrochemical sensors", *Polymer-Plastics Technology and Materials,* vol. 60, no. 5, pp. 504-521, 2021.

http://dx.doi.org/10.1080/25740881.2020.1844233

[23] M. Romero, M.A. Macchione, F. Mattea, and M. Strumia, "The role of polymers in analytical medical applications. A review", *Microchem J.,* vol. 159, p. 105366, 2020.

http://dx.doi.org/10.1016/j.microc.2020.105366

[24] M. Ramesh, R. Janani, C. Deepa, and L. Rajeshkumar, "Nanotechnology-enabled biosensors: A review of fundamentals, design principles, materials, and applications", *Biosensors,* vol. 13, no. 1, p. 40, 2022.

http://dx.doi.org/10.3390/bios13010040 PMID: 36671875

[25] L. Nie, F. Liu, P. Ma, and X. Xiao, "Applications of gold nanoparticles in optical biosensors", *J. Biomed. Nanotechnol.,* vol. 10, no. 10, pp. 2700-2721, 2014.

http://dx.doi.org/10.1166/jbn.2014.1987 PMID: 25992415

[26] H. Aldewachi, T. Chalati, M.N. Woodroofe, N. Bricklebank, B. Sharrack, and P. Gardiner, "Gold nanoparticle-based colorimetric biosensors", *Nanoscale,* vol. 10, no. 1, pp. 18-33, 2018.

http://dx.doi.org/10.1039/C7NR06367A PMID: 29211091

[27] C. Zhu, G. Yang, H. Li, D. Du, and Y. Lin, "Electrochemical sensors and biosensors based on nanomaterials and nanostructures", *Anal. Chem.,* vol. 87, no. 1, pp. 230-249, 2015.

http://dx.doi.org/10.1021/ac5039863 PMID: 25354297

[28] S. Chupradit, M. KM Nasution, H.S. Rahman, W. Suksatan, A. Turki Jalil, W.K. Abdelbasset, D. Bokov, A. Markov, I.N. Fardeeva, G. Widjaja, M.N. Shalaby, M.M. Saleh, Y.F. Mustafa, A. Surendar, and R. Bidares, "Various types of electrochemical biosensors for leukemia detection and therapeutic approaches", *Anal. Biochem.,* vol. 654, p. 114736, 2022.

http://dx.doi.org/10.1016/j.ab.2022.114736 PMID: 35588855

[29] C.S. Lee, S.K. Kim, and M. Kim, "Ion-sensitive field-effect transistor for biological sensing", *Sensors,* vol. 9, no. 9, pp. 7111-7131, 2009.

http://dx.doi.org/10.3390/s90907111 PMID: 22423205

[30] K. Nakazato, "An integrated ISFET sensor array", *Sensors,* vol. 9, no. 11, pp. 8831-8851, 2009.

http://dx.doi.org/10.3390/s91108831 PMID: 22291539

[31] A.M. Dinar, A.S.M. Zain, and F. Salehuddin, "Comprehensive identification of sensitive and stable ISFET sensing layer high-k gate based on ISFET/electrolyte models", *International Journal of Electrical and Computer Engineering (IJECE),* vol. 9, no. 2, pp. 926-933, 2019.

http://dx.doi.org/10.11591/ijece.v9i2.pp926-933

[32] K.B. Parizi, X. Xu, A. Pal, X. Hu, and H.S.P. Wong, "ISFET pH sensitivity: counter-ions play a key role", *Sci. Rep,* vol. 7, no. 1, p. 41305, 2017.

http://dx.doi.org/10.1038/srep41305 PMID: 28150700

[33] O. Synhaivska, Y. Mermoud, M. Baghernejad, I. Alshanski, M. Hurevich, S. Yitzchaik, M. Wipf, and M. Calame, "Detection of Cu2+ ions with GGH peptide realized with Si-nanoribbon ISFET", *Sensors,* vol. 19, no. 18, p. 4022, 2019.

http://dx.doi.org/10.3390/s19184022 PMID: 31540412

[34] I. Ferain, C.A. Colinge, and J.P. Colinge, "Multigate transistors as the future of classical metal–oxide–semiconductor field-effect transistors", *Nature,* vol. 479, no. 7373, pp. 310-316, 2011.

http://dx.doi.org/10.1038/nature10676 PMID: 22094690

[35] M. Sridhar, D. Xu, Y. Kang, A.B. Hmelo, L.C. Feldman, D. Li, and D. Li, "Experimental characterization of a metal-oxide-semiconductor field-effect transistor-based Coulter counter", *J. Appl. Phys.,* vol. 103, no. 10, pp. 104701-10470110, 2008.

http://dx.doi.org/10.1063/1.2931026 PMID: 19479001

[36] K. Girigoswami, and N. Akhtar, "Nanobiosensors and fluorescence based biosensors: An overview", *Int. J. Nanodimens,* vol. 10, no. 1, pp. 1-17, 2019.

[37] A.A. Nayl, A.I. Abd-Elhamid, A.Y. El-Moghazy, M. Hussin, M.A. Abu-Saied, A.A. El-Shanshory, and H.M.A. Soliman, "The nanomaterials and recent progress in biosensing systems: A review", *Trends in Environmental Analytical Chemistry,* vol. 26, p. e00087, 2020.

http://dx.doi.org/10.1016/j.teac.2020.e00087

[38] T. Wadhera, D. Kakkar, G. Wadhwa, and B. Raj, "Recent advances and progress in development of the field effect transistor biosensor: A review", *J. Electron. Mater.,* vol. 48, no. 12, pp. 7635-7646, 2019.

http://dx.doi.org/10.1007/s11664-019-07705-6

[39] N.N. Reddy, and D.K. Panda, "A comprehensive review on tunnel field-effect transistor (TFET) based biosensors: recent advances and future prospects on device structure and sensitivity", *Silicon,* vol. 13, no. 9, pp. 3085-3100, 2021.

http://dx.doi.org/10.1007/s12633-020-00657-1

[40] Y.C. Syu, W.E. Hsu, and C.T. Lin, "Field-effect transistor biosensing: Devices and clinical applications", *ECS J. Solid State Sci. Technol,* vol. 7, no. 7, pp. Q3196-Q3207, 2018.

http://dx.doi.org/10.1149/2.0291807jss

[41] W. Gao, S. Emaminejad, H.Y.Y. Nyein, S. Challa, K. Chen, A. Peck, H.M. Fahad, H. Ota, H. Shiraki, D. Kiriya, D.H. Lien, G.A. Brooks, R.W. Davis, and A. Javey, "Fully integrated wearable sensor arrays for multiplexed in situ perspiration analysis", *Nature,* vol. 529, no. 7587, pp. 509-514, 2016.

http://dx.doi.org/10.1038/nature16521 PMID: 26819044

[42] S.M. Yun, M. Kim, Y.W. Kwon, H. Kim, M.J. Kim, Y.G. Park, and J.U. Park, "Recent advances in wearable devices for non-invasive sensing", *Appl. Sci.,* vol. 11, no. 3, p. 1235, 2021.

http://dx.doi.org/10.3390/app11031235

[43] K. Guk, G. Han, J. Lim, K. Jeong, T. Kang, E.K. Lim, and J. Jung, "Evolution of wearable devices with real-time disease monitoring for personalized healthcare", *Nanomaterials,* vol. 9, no. 6, p. 813, 2019.

http://dx.doi.org/10.3390/nano9060813 PMID: 31146479

[44] W.A.D.M. Jayathilaka, K. Qi, Y. Qin, A. Chinnappan, W. Serrano-García, C. Baskar, H. Wang, J. He, S. Cui, S.W. Thomas, and S. Ramakrishna, "Significance of nanomaterials in wearables: a review on wearable actuators and sensors", *Adv. Mater.,* vol. 31, no. 7, p. 1805921, 2019.

http://dx.doi.org/10.1002/adma.201805921 PMID: 30589117

[45] S. Vaddiraju, I. Tomazos, D.J. Burgess, F.C. Jain, and F. Papadimitrakopoulos, "Emerging synergy between nanotechnology and implantable biosensors: A review", *Biosens Bioelectron,* vol. 25, no. 7, pp. 1553-1565, 2010.

http://dx.doi.org/10.1016/j.bios.2009.12.001 PMID: 20042326

[46] G. Maduraiveeran, M. Sasidharan, and V. Ganesan, "Electrochemical sensor and biosensor platforms based on advanced nanomaterials for biological and biomedical applications", *Biosens Bioelectron,* vol. 103, pp. 113-129, 2018.

http://dx.doi.org/10.1016/j.bios.2017.12.031 PMID: 29289816

[47] R. Ahmad, T. Mahmoudi, M.S. Ahn, and Y.B. Hahn, "Recent advances in nanowires-based field-effect transistors for biological sensor applications", *Biosens Bioelectron,* vol. 100, pp. 312-325, 2018.

http://dx.doi.org/10.1016/j.bios.2017.09.024 PMID: 28942344

[48] D. Sadighbayan, M. Hasanzadeh, and E. Ghafar-Zadeh, "Biosensing based on field-effect transistors (FET): Recent progress and challenges", *Trends Analyt Chem.,* vol. 133, p. 116067, 2020.

http://dx.doi.org/10.1016/j.trac.2020.116067 PMID: 33052154

[49] A. Panahi, and E. Ghafar-Zadeh, "Emerging Field-Effect Transistor Biosensors for Life Science Applications", *Bioengineering,* vol. 10, no. 7, p. 793, 2023.

http://dx.doi.org/10.3390/bioengineering10070793 PMID: 37508820

[50] P.C. Gupta, N. Sharma, S. Rai, and P. Mishra, "Use of Smart Silver Nanoparticles in Drug Delivery System", In: *Metal and Metal-Oxide Based Nanomaterials: Synthesis, Agricultural, Biomedical and Environmental Interventions.* Springer Nature Singapore: Singapore, 2024, pp. 213-241.

http://dx.doi.org/10.1007/978-981-99-7673-7_11

[51] M. Berkani, "MXene-based hybrid biosensors", In: *Mxene-Based Hybrid Nano-Architectures for Environmental Remediation and Sensor Applications.* Elsevier, 2024, pp. 327-349.

http://dx.doi.org/10.1016/B978-0-323-95515-7.00016-9

Advances in the Design and Application of Next-Generation Carbon-Based Field-Effect Transistor Biosensors

D. Anbuselvi[1],*, S. Grace Infantiya[1], N. Suthanthira Vanitha[2], T. Divya[3] and C. Kathiravan[4]

[1]*Department of Physics, Muthayammal Engineering College, Rasipuram, Namakkal, India*

[2]*Department of Electrical and Electronics Engineering, Muthayammal Engineering College, Rasipuram, Namakkal, India*

[3]*Department of Computer Science, Velalar College of Engineering Technology, Thindar, India*

[4]*Department of Chemistry, Muthayammal Engineering College, Rasipuram, Namakkal, India*

Abstract: The proposed book chapter aims to present a comprehensive review of nanomaterial-based biosensors utilizing field-effect transistors (FETs), exploring their diverse applications, advancements, and future potentials. The focus will be on examining the integration of carbon-based nanomaterials into FET-based biosensors and their role in revolutionizing biosensing technologies. Field effect-based biosensors (BioFETs) stand out among other biosensing technologies due to their unique features such as real-time screening, ultrasensitive detection, low cost, and amenability to extreme device miniaturization due to the convenient utilization of nanoscale materials. FET-based sensors operate on the principle that changes in the surrounding environment, such as alterations in temperature, pressure, gas concentration, or biological elements, modulate the electrical characteristics of the transistor. The integration of carbon-based nanomaterials into biosensing applications has emerged as a transformative development, significantly augmenting the efficacy and sensitivity of detection devices, particularly within the domain of field-effect transistor (FET) based technologies. The intent is to provide a holistic view of how these advancements have contributed to improving detection capabilities and to outline potential avenues for further research and applications in the field of biosensing.

Keywords: Bio-FETs, Biosensors, Biomolecules, Carbon nanotubes.

---

* Corresponding author D. Anbuselvi: Department of Physics, Muthayammal Engineering College, Rasipuram, Namakkal, India; E-mail: mailto:anbuselvivpm@gmail.com

INTRODUCTION

For one thousand millionths of a meter, the Greek word "nano," which means "dwarf" or "very small," is used (10 - 9 m). Making a distinction between nanotechnology and nanoscience is crucial. Nanotechnology is the application of knowledge about structures and chemicals at nanoscales, or between 1 and 100 nm, to real-world items like gadgets. Nanoscience is the study of these structures and chemicals [1].

It is significant to remember that a single human hair is 60,000 nm thick, while the radius of the DNA double helix is one nm (Fig. **1**) [2]. In the fifth century B.C., during the Greek and Democritus era, the area of nanoscience first emerged. The question of whether matter is continuous and so infinitely divided into smaller parts or if it is made up of tiny, indivisible, and unbreakable components known as atoms was up for debate among scientists.

Nanotechnology is one of the most fascinating emerging technologies of the twenty-first century. It is the ability to monitor, quantify, assemble, regulate, and produce materials at the nanoscale scale in order to put the theory of nanoscience into practice. The National Nanotechnology Initiative (NNI) in the United States defines nanotechnology as "a science, engineering, and technology conducted at the nanoscale (1-100 nm), where unique phenomena enable novel applications in a wide range of fields, from chemistry, physics, and biology, to medicine, engineering, and electronics" [3].

This definition states that two conditions must be met for nanotechnology to exist. First, there is the matter of scale. Utilizing objects by adjusting their proportions at the nanoscale is the main goal of nanotechnology. The second issue is novelty: because of the nanoscale, handling small items in a way that maximizes certain features is necessary for nanotechnology [4-15].

Making a distinction between nanotechnology and nanoscience is crucial. Nanoscience is the combination of physics, materials science, and biology, which deal with the manipulation of materials at the atomic and molecular levels. Nanotechnology is the study, measurement, manipulation, assembly, control, and production of matter at the nanoscale size. There are a few articles available that describe the history of nanoscience and technology, but none that offer an overview of the area's evolution from its beginnings to the present with an emphasis on breakthroughs in the field. Therefore, it is essential to assemble a synopsis of significant occurrences in this field in order to completely appreciate the progress of nanoscience and technology [16-25].

At the beginning of the twenty-first century, there was an increase in interest in the fields of nanoscience and nanotechnology. National scientific goals in the United States were greatly impacted by Feynman's hypothesis of atomic-level matter manipulation. In a speech at Caltech on January 21, 2000, President Bill Clinton made the case for funding research in nanotechnology. Three years later, George W. Bush signed the 21st Century Nanotechnology Research and Development Act into law. The Act also made nanotechnology research a national priority and established the National Technology Initiative (NNI) [26-35].

BRIEF HISTORY OF BIOSENSOR FIELD EFFECT TRANSISTOR (FET)

An Introduction to FET Transistor

Field-Effect transistors are referred to as FET transistors. An electric field is used by the field-effect transistor (FET) to regulate the current flowing through a semiconductor. FETs have a source, a gate, and a drain on their three terminals. The current flow in FETs is regulated by applying a voltage to the gate, which changes the conductivity between the drain and the source. Julias Edgar submitted the first FET transistor patent application in 1926. Numerous developments have occurred since then. In 1934, Oskar Heil applied for another patent. William Shockley invented the junction gate utilized in field-effect transistors at Bell Laboratories. Over time, FET transistors have seen several more advances [33-35].

Working of FET Transistor

The voltage applied controls the current flowing in a FET transistor, which is a voltage-operated device. Because they operate using a single carrier, they are often referred to as unipolar transistors. All FET shapes and sizes have a high input impedance. The field-effect transistor's terminal applies voltage to control the conductivity in all situations. Furthermore, conductivity is influenced by the carrier charge density. The three main parts of a FET transistor are the source, drain, and gate. The majority of the carriers enter the bar from one of the FET transistor's terminals, which is the source. Most carriers lead the bar via the second terminal, which is called the drain. Two terminals on the Gate are internally linked to one another [35-55].

Field-effect transistor (FET)

The field-effect transistor also referred to as (FET) is a specific kind of transistor that modulates the flow of electricity in a semiconductor material. Additionally, there are two distinct categories such as MOSFET (metal-oxide-semiconductor FET) and junction-gate FET (JFET). A FET comprises three distinct terminals: drain, gate, and source. FETs regulate the rate of flow of electrical power through applying voltage to the gate, thereby influencing the electrical conductivity between the drain and the source. Fig. (**1**) shows the schematic diagram of FET[56, 57].

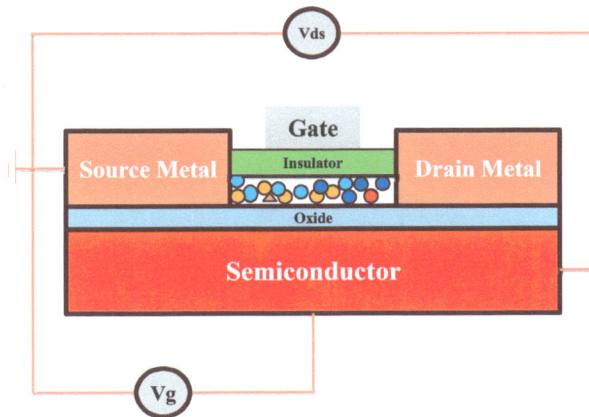

Fig. (1). FET Schematic Diagram.

FETs are commonly mentioned as unipolar transistors because they only have one terminal. Field-effect transistors (FETs) operate by employing a combination of holes (in p-channel FETs) or electrons (in n-channel FETs) as the predominant charge transporters, but not both. Field-effect transistors are accessible in a wide range of types and configurations. Field-effect transistors have an extraordinarily high input impedance, especially at lower frequency ranges. The MOSFET, or metal-oxide-semiconductor field-effect transistor, was one of the earliest and most widespread kinds of field-effect transistors [58-78].

Metal-oxide-semiconductor FET (MOSFET)

The MOSFET (MOS field-effect transistor) was created in 1959 by Mohamed Atalla (on the left) and Dawon Kahng (on the right). The work of Egyptian engineer Mohamed Atalla in the late 1950s was a milestone in FET research. He demonstrated experimentally in 1958 that the growth of thin silicon oxide on a clean silicon surface neutralizes surface states. Surface passivation is the term for this technique, which proved essential to the semiconductor industry by enabling silicon-integrated circuit mass production [79-81].

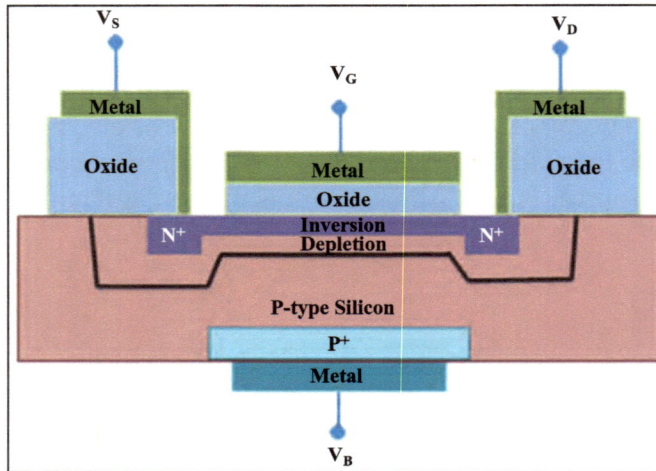

Fig. (2). Metal-oxide-semiconductor FET (MOSFET).

The MOSFET had an important influence on the advancement of modern digital technological devices, largely replacing both the JFET and the bipolar transistor (as illustrated in Fig. (**2**). Its flexibility significantly reduced energy consumption, and higher density has enabled the development of high-density integrated circuits (ICs). Moreover, the MOSFET can handle higher power levels compared to *the* JFET. The MOSFET was the initially extremely tiny semiconductor that was capable of being produced in large quantities for a wide range of purposes [80, 110].

CNT-BIOFET SENSING MECHANISM

Among the most prominent biochemical communications used in field-effect transistor-based biosensors are enzyme-substrate complex formation, antibody-antigen binding, and DNA probe-target oligo hybridization. Field effect biosensors, which are biochemically and charge-sensitive devices, are capable of detecting any type of charge or potential change within or close to the conducting channel triggered by a molecular interaction (such as molecule adsorption or binding) or biological reaction [101, 102]. Carbon nanotubes (CNTs) are able to react to biological processes and small biomolecules that become trapped on their surface because of their nanoscale size [109, 110]. The biochemical reaction that occurs between the probe biological species and the targeted species modifies the electrical charge distribution on the surface of the carbon nanotubes. This results in a shift in the CNT's threshold voltage, which is a manifestation of field effect behavior. Depending on the analyte concentration, conductivity can be utilized to monitor changes in resistance both before and after the biorecognition process[111].

Providing a voltage equal to the gate voltage to a semiconducting network in a conventional field effect transistor alters channel conductance and amplifies or diminishes the effects by causing the majority of charge carriers to accumulate or deplete depending on the biasing. In a traditional field-effect transistor, providing a gate voltage to a semiconductor channel causes the majority of the charge carriers to accumulate or deplete depending on the biasing, which changes the channel conductance and amplifies or reduces the effects [112-120].

Biomolecules can bind or adsorb on the active surface of CNT channels to create one or more of the following electrical properties and adjust charge: There are now three known mechanisms: There are three types of gating: (1) surface charge-induced gating or electrostatic gating; (2) charge transfer between the biomolecule and the nanotube (nanotube doping); and (3) a hybrid gating that combines electrostatic gating with the Schottky barrier effect. Using I-V characteristic curves, a previous investigation examined the effect of protein adsorption on SWCNT devices with ambipolar conduction. The vast majority of the experiments demonstrated that the mechanism of protein biosensing was driven by a combination of Schottky barrier effects and electrostatic gating. In comparison to conventional electrochemical sensors, FET-based biosensors offer a multitude of potential benefits, including low output impedance, label-free assay, fast reaction times, on-chip integration of biosensor arrays, small size and light weight, inexpensive mass manufacturing, and small, portable microanalysis. Due to their inherent shrinking and compatibility with microfabrication techniques, FET devices are very desired for inclusion into microfluidics and micro-analytical devices. FET-based technologies are currently in considerable demand for applications such as microfluidics and micro-analytical devices owing to their extremely small dimensions and the flexibility of microfabrication techniques[110, 111].

However, one challenge with BioFETs is Debye screening, which is the process of filtering the analyte charge by the electrolyte ions large molecules, macromolecules, and enzymes are challenging to identify because the standard dimension of a protein or antibody receptors molecule is 10-15 nm, which exceeds the Debye length of the optimum range for charge identification, which is approximately one nm. To address this issue, attempts have been made to use single-chain variable antibody fragments or short nanobody receptors with detection limits as low as a sub-picomolar regime. The ionic concentration of the solution plays a crucial role in controlling the Debye length to ensure that a certain macromolecule binding contributes to the sensor response [120-132].

CARBON NANOTUBES AND WATER PURIFICATION

Single-walled CNTs are typically created by wrapping a single graphite layer, known as a graphene layer, into an uninterrupted cylindrical. Viewed as a sequence of compressed SWCNTs with varying diameters, multi-walled carbon nanotubes (MWCNTs) are made up of many layers of graphite wrapped into a tube form. MWCNTs may reach sizes of up to 100 nm, whereas SWCNTs have diameters of just 0.3 to 3 nm. Fig. (**3**) illustrates the use of CNTs in various water purifying procedures[111].

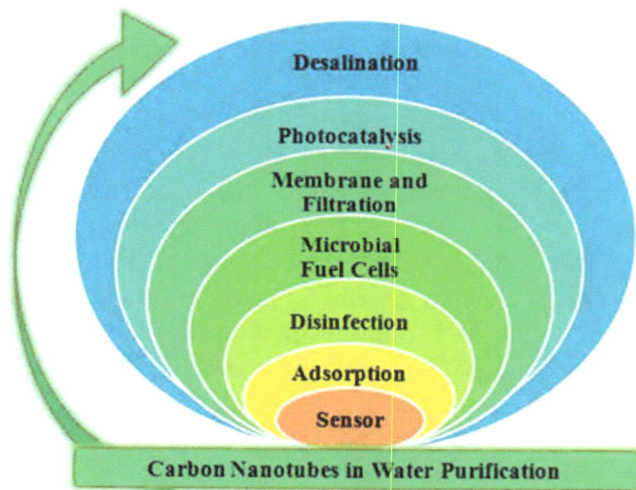

Fig. (3). Applications of carbon nanotubes (CNTs) in water purification.

It has been shown that carbon nanotubes are a superior adsorbent for taking impurities out of water, leaving behind a clean environment. While CNTs may absorb pollutants on their own, they can also be combined with other materials to create composites that have a higher potential for adsorption. Typically, the carbon nanotube surface is functionalized in order to maximize the adsorbent's adsorption capabilities and investigate them [7]. Through improved dispersion in the adsorption system, this method enhances the interaction between the adsorbent and the pollutant. The synthesized carbon nanotubes (CNTs) do not always function at their best because of the contaminants that are added throughout the process. As a result, to enhance their functionality [112-133], CNTs can be functionalized in the presence of acidic and alkaline solutions.[112]

All of these impurities would alter the fundamental characteristics of the carbon nanotubes if they were to appear on their surface. To avoid changing the properties, impurities will be extracted from the carbon nanotube using an acidic or basic solution. As a result, the functionalization process, which involves introducing another group of functions to the surface of carbon nanotubes, will change the surface of the nanoparticles [8]. An additional negatively charged functional group at the open end or sidewall of the carbon nanotubes would increase their solubility in any kind of solvent and strengthen their surface by substituting strong electrostatic forces for weak Van der Waals interactions[113-134].

Carbon nanotubes have been effectively coated with other carbon-based compounds to boost their efficacy in removing pollutants from water. At optimized reaction conditions of 0.2g adsorbent (functionalized CNTs), pH 5.0, 90° C, and a temperature of 100 minutes, the result for the removal of Cu + from aqueous solution showed that CNTs which are functionalized using HNO 3 and H 2 SO 4 display a higher adsorption capacity as compared to the non-functionalized CNTs [9,114,115]. As a result, carbon nanotubes (CNTs) were chosen as an adsorbent to give a more promising and efficient substitute. $Pb^{2+}>Ni^{2+}>Zn^{2+}>Cu^{2+}>Cd^{2+}$ was the order of metal ion adsorption to different kinds of carbon nanotubes (CNTs), with surface-functionalized CNTs via NaOCl, HNO_3, and $KMnO_4$ having a greater adsorption capacity than non-functionalized carbon nanotubes (CNTs) [10]. Pb^{2+}, Ni^{2+}, Zn^{2+}, Cu^{2+}, and Cd^{2+} was the sequence in which metal ions were adsorbed to various carbon nanotube (CNT) types; surface-functionalized CNTs from NaOCl, HNO_3, and $KMnO_4$ had a higher adsorption capacity than CNTs that were not functionalized [40]. The adsorption mechanism is defined by the interaction between the metal ions and the ions located on the surface of CNTs, such as the functional groups that remain after the functionalization process. Consequently, the treatment of divalent metal ions in water and wastewater makes extensive use of both functionalized and non-functionalized carbon nanotubes. Consequently, carbon nanotubes (CNTs) were selected as an adsorbent to provide a more effective and promising substitute [135].

Medical Application of Carbon Nanotube

Carbon Nanomaterials in Diagnosis and Therapy: Clinical Applications

Carbon nanoparticles have the power to alter patient care and diagnosis in medical settings, as shown in Fig. (**4**). Potential applications for carbon-based nanoparticles include treating chemo-resistant cancer, promoting tissue regeneration, stem cell banking, and a host of other therapeutic applications. Examples of these applications include carbon nanotubes, nanodiamonds, graphene oxides, carbon nanofibers, and graphene [11].

Biomedical Applications of Nanodiamonds

A particular class of carbon nanoparticles called a nanodiamond shows significant promise for therapeutic uses as agents in medication transport and imaging. The most common nanodiamonds utilized in therapeutic systems are fluorescent or tiny detonation nanodiamonds. Nanodiamonds have several advantages as a drug delivery and imaging vehicle, including unique facet-dependent surface electrostatic potentials, a chemically inert core that improves biocompatibility, and a surface studded with changeable functional groups. When used alone as a medication delivery mechanism or in combination with other elements, nanodiamonds have the power to completely transform medicine [12].

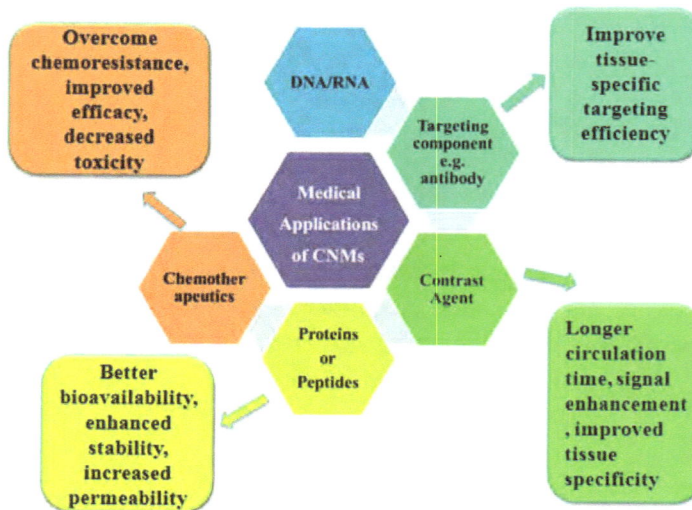

Fig. (4). Medical application of carbon nanoparticles.

Chemo-resistant Cancers and Nanodiamond Drug Delivery

One of the most intriguing potential applications for nanodiamonds is enhanced chemotherapy drug delivery, particularly for chemoresistant tumors. Chemotherapy drugs known as anthracyclines work by forming complexes with DNA and preventing topoisomerase II, which results in the death of cancer cells. While anthracyclines can be used to treat a number of cancers, one common mechanism of anthracycline resistance is ABC transporter-mediated chemo-resistance, which occurs because many ABC transporters have the ability to efflux anthracyclines. While several anthracyclines and other chemotherapeutic medications have been administered via nanodiamonds, the majority of the most comprehensive scientific work that has been published has utilized eirenicon and doxorubicin. Because the modified doxorubicin-nanodiamond complex is kept for a significantly longer period of time in cells with overexpressed ABC transporters, it prevented the occurrence of active efflux of doxorubicin, in contrast to the regular unaltered form [116]. This higher cellular retention led to improved survival and smaller tumors in mice models of liver and breast cancer, as well as increased death of chemo-resistant malignant cells [13, 117]. It was found that the efficacy of nanodiamonds–anthracycline complexes can be improved while keeping the low side effects that are characteristic of nanodiamonds–anthracycline combinations. This is achieved by surface functionalizing the complexes with selected elements. When it comes to the efficacy and safety of anthracyclines, nanodiamonds are an effective technique for improving drug delivery in cancer [11].

Tissue Engineering and Regeneration using Functionalized Carbon Nanotubes

Carbon-based nanomaterials, such as carbon nanotubes, have special characteristics that make them useful as frameworks for tissue engineering and regeneration. The ability to induce precursor cells in the area of the defect to undergo differentiation into particular kinds of connective tissue according to particular growth factors and the surfaces on which they emerge is essential to the process's efficacy [11]. Stem cells are essential in this sector due to their vast ability for differentiation. As such, in order for them to develop into a particular cell lineage, they still require biochemical inputs. Carbon nanotubes have been shown to support normal cellular development and accelerate stem cell differentiation into several lineages when used to treat skeletal muscle injuries [12, 118].

Carbon Nanofibers for Electrochemical Sensors and Biosensors

Processes such as carbonization and electrospinning enable the manufacturing of CNFs at far larger lengths than CNTs. Additionally, CNFs built in this way may have substantially larger hollow internal dimensions. Because of their wider width and the polymer-carbon composite that is formed during production—which controls the characteristics of their electrical conductivity—CNFs are appealing to electrochemical devices. CNFs provide new possibilities in neuroscience since they are biocompatible, flexible, and electrically conductive. For example, they may be used to create electrodes for biosensors that track electroactive neurotransmitters [13].

Electrochemical Sensors Based on Carbon Nanoparticles

Graphene, crystalline diamond, carbon nanotubes, and diamond-like carbon are examples of carbon nanoparticles with exceptional electrochemical properties that have led to their widespread application. Since the distinct carbon-derived nanostructures exhibit properties that are unthinkable in normal materials, the potential applications of these nanomaterials in sensing systems are evident. As an outcome, they can operate with increased sensitivity along with selectivity spanning an expanded temperature and dynamic range, as well as in antagonistic environments.

Carbon Nanotube-Based Electrochemical Sensors

In view of their unique structure, which includes mechanical deformations, the capacity to form extremely thin wires that combine the toughness of diamond with the conductivity of graphene, and high surface-to-volume ratios, carbon nanotubes are a good material to use in the development of nanoscale sensors that appear to be sensitive to mechanical, physical, and chemical environments. The electrical properties of carbon nanotubes are influenced by the curvature of graphene sheets [14, 119].

Chemicals that are either electron donors or acceptors like NH_3, O_2, and NO_2, when dopped with single-wall carbon nanotubes either increase or reduce their conductivity by providing the nanotubes with more holes or charge carriers. Not only do carbon nanotubes improve the electrochemical reactivity of vital biomolecules, but they also enhance electron-based transfer reactions of proteins as well [119]. The capacity of modified electrodes with carbon nanotubes to reduce surface contamination has been established, as in the event of primary oxidation of

NADH due to high over-potentials which lead to oxidation products fouling the electrode surface. Nucleic acid is an example of a vital biomolecule that is concentrated by carbon nanotubes which help in improving the sensitivity and selectivity of the probe.For usage in electrochemical sensing applications, carbon nanotubes must be suitably functionalized and immobilized. The electrochemical functionalization of carbon nanoparticles was also carried out with metallic nanoparticles, and the application of the resulting CNTs-metal nanocomposite for catalysis and sensing has grown recently [14,119].

DNA Sensors Based on Carbon Nanotubes

Since carbon nanotubes can form bonds between their conjugated systems and nucleobases, they are good candidates for application in DNA and RNA sensing. Since carbon nanotubes are intrinsically electrical, they also amplify the DNA/RNA detecting signal. The intrinsic conductivity of CNTs is important because methods of amplification, such as the addition of Nanoparticles or enzymes that promote the passage of electrons, are frequently needed to enhance the typically weak DNA/RNA sensing signal. Albumin, amino acids, and glucose are detected and quantified using glassy carbon electrode electrodes that have been modified with multi-walled carbon nanotubes. Moreover, DNA is immobilized by carbon nanotubes by the use of a single data point attachment to bond DNA to a solid surface.[119]

This suggests that the detection limit and manufacturing time of the DNA sensing technology were improved. An underlying glassy carbon electrode was modified using multiwalled carbon nanotubes to provide a well-known biomarker for oxidative DNA damage sensors with good electrochemical performance to the oxidation of 8-OHdG [15].

BIO –CNTFET APPLICATIONS-REVIEW

An analytical tool known as a biosensor consists of a biological recognition element that is in close spatial proximity to a transduction element. The quick and easy conversion of biological processes into observable signals is ensured by this integration.[16,63]. In the past 20 years, new directions in the field of biosensors have been made possible by the discovery of abundant nanomaterials and the creation of sophisticated nanofabrication instruments such as focussed ion beams, electron beams, and nano impression lithography[17,18,63]. Specifically, researchers worldwide have been customizing a wide range of electrical biosensors based on nanomaterials and creating innovative approaches to use them in ultrasensitive biosensing. Fig. (**5**) shows the schematic illustration of Bio –FET employing nanomaterials, their immobilization techniques, and application[119].

Carbon nanotubes, [19-28] nanowires, [26-36] nanoparticles, [21– 40] nanopores, [41-42] nanoclusters, and graphene are a few examples of these nanomaterials[44–47]. Nanomaterial-based electronic biosensing presents a number of advantages over traditional optical, biochemical, and biophysical methods. These advantages include enhanced sensitivity and novel sensing mechanisms, high spatial resolution for localized detection, ease of integration with standard wafer-scale semiconductor processing, and label-free, real-time detection in a nondestructive way. Field-effect transistor (FET)-based devices have garnered significant interest among various electrical biosensing architectures due to their exceptional biosensor capabilities, which enable the direct conversion of biological interactions between target molecules and the FET surface into readable electrical signals [48–49]. Current travels over a semiconductor channel in a typical FET, which is coupled to the source and drain electrodes.

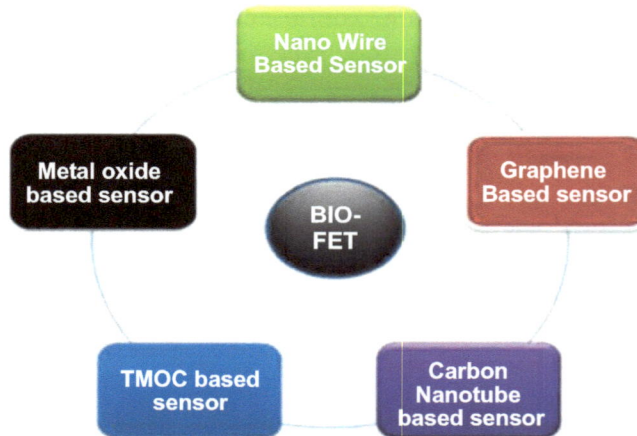

Fig. (5). Schematic Illustration of Bio –FET Employing Nanomaterial, their Immobilization Techniques, and Application.

A third (gate) electrode, capacitively connected through a thin dielectric layer, controls the channel conductance between the source and the drain. The conducting channel of a FET-based biosensor is in direct touch with the environment, providing better control over the surface charge than in typical CMO semiconductor-fabricated transistors, where the channel is buried inside the substrate. This suggests that surface FET-based biosensors may be more sensitive because biological activity at the channel surface may cause the semiconductor channel's surface potential to fluctuate, which in turn could change the channel conductance[63].

FET-based biosensors are appealing substitutes for current biosensor technologies due to their surface ultrasensitivity, ease of on-chip integration of device arrays, and low cost of device fabrication. We provide an overview of current developments in ultrasensitive biosensors made of FETs based on nanomaterials in this article. We focus especially on single-walled carbon nanotubes (SWNTs), a type of carbon nanomaterial. This is a broad and quickly expanding field of research, and we will only be able to cover a few of the most significant contributions due to space and reference constraints.

Since most earlier studies have overlooked these crucial elements, we will draw attention to them here. These include methods for boosting sensitivity, dynamic detection in cells and liquid environments, DNA hybridization, and single-molecule detection. Thankfully, there exist some outstanding prior review articles in the literature that address different facets of biosensors based on carbon nanomaterials, which can rectify these shortcomings [19–28, 45–49,63].

A particular recognition element can be biofunctionalized into either the gate or the semiconductor in EG-FET biosensors[121]. Due to the strong affinity of thiol ending molecules for gold gates and the ease of thiol functionalization techniques, gate functionalization has proven to be exceedingly durable and adaptable, even though integrating a bioreceptor into the semiconductor could become challenging.[51-53]. Solution-processed organic semiconductors have emerged as a viable option for the active material in the hunt for low manufacturing costs of EG-FET-based biosensors,[54-55] due to their compatibility with flexible, inexpensive substrates and large-area, low-temperature printing processes.[56–58,121]

BASIC INTRODUCTION TO CARBON NANOMATERIALS

Significant carbon nanoforms have recently been discovered, sparking intense interest in both research and specific scientific uses. Carbon nanotubes (CNTs) can be produced by a variety of techniques, including chemical vapor deposition(CVD), electric arc discharge, and laser ablation. The production of CNTs is facilitated by functionalization, chemical addition, doping, and filing, which enable detailed characterization and manipulation of the material. Their stability and reactivity are significantly influenced by the elasticity, electromechanical, chemical, and optical properties that are inherent in CNTs.

Perhaps because of their   strength and flexibility, they are able to validate prospective application in a variety of industries, ensuring that these CNTs will undoubtedly play a significant part in nanotechnology.

The application of carbon nanomaterials in novel nanoscale biosensors has sparked a wave of intense interest. Due to their distinct physicochemical characteristics, graphene and, more recently, SWNTs are leading this development. Made up of sp2-hybridized carbon atoms organized in a honeycomb lattice, graphene is a two-dimensional (2D) crystalline monolayer[59–61]. Indeed, the fundamental component of all other dimensionality's graphitic materials is graphene; for example, a well-ordered, hollow graphitic nanomaterial, known as an SWNT, is created by folding a graphene sheet into a cylinder along a specific lattice vector [62-63].

These two carbon allotropes, which maximize the surface-to-volume ratio, have the most straightforward chemical composition and atomic bonding configuration in two dimensions. Ultrasensitive biosensing is based on the fact that every carbon atom on a nanocarbon surface is exposed to the environment. Even slight changes in this environment can have a significant impact on the electrical characteristics of the nanocarbon device[63–72].

Biomolecular-Nanoparticle Interactions

Modification of Carbon Nanotubes

The complete fulfillment of CNTs' potential has been hampered by their processability due to their chemical inertness. It is widely established that the chemical modification of CNTs promotes significant unbundling, modifies their electrical properties, and provides mobility in an array of solvents. Due to bundle formation, CNTs have solubility issues that render them rarely soluble in typical solvents. Due to the restricted solubility of carbon nanotubes in water, flocculation and phase separation must be prevented with the right concentrations of stabilizers.[1] The fact that CNTs are extremely hydrophobic and typically form insoluble aggregates is a drawback to their usage in biochemistry and medicinal applications. It's also exceedingly hard to separate one carbon nanotube from the other because CNTs are less soluble in any of the solvents. Except for the nanotube caps, where the existence of dangling bonds makes them more reactive, carbon nanotubes are generally nonreactive, similar to graphite. The chirality or curvature of the tube can affect the reactivity of the sidewalls in the carbon nanotube system. Through - and/or van der Waals interactions, the hydrophobic surfaces of carbon nanotubes absorb a broad variety of compounds [73–75,1]. Thus, for technological applications, appropriate stability of CNT dispersions is necessary.

When functionalizing carbon nanotubes (CNTs), there are two major approaches that are typically taken: attaching organic moieties directly to the surface double bonds or attaching them to carboxylic groups that are produced when CNTs are oxidized with strong acids [77]. To create extremely soluble bundles of nanotubes, insert about one organic group for every 100 carbon atoms in the nanotube [78]. By combining the characteristics of the organic moiety and the nanotubes, the solubilization of the nanotubes creates a novel and intriguing class of materials that open up new possibilities for materials science applications, such as the creation of nanocomposites.[1]

Fluorination, introducing carbenes and nitrenes, with electrophiles, and peroxyl radicals were all successful processes for sidewall covalent functionalization of carbon nanotubes (CNTs) [77,78]. Since CNTs have intriguing sizes, shapes, and structures in addition to appealing optical and electrical capabilities, there has been a new push to investigate the possible biological applications of CNTs. Firstly, SWNTs have an ultrahigh surface area (theoretically 1300 m2/g) due to all of their atoms being exposed on the surface, which makes it possible to load numerous molecules efficiently along the sidewall of the nanotube. Secondly, aromatic molecules can be readily bound supramolecularly by stacking them onto the polyaromatic surface of nanotubes [79,80,81-83,1].

The complexity of biomacromolecules makes it challenging to comprehend the microscopic details of interactions between molecules, in addition to the dynamic mechanics of biomolecule-CNT systems. The biocompatibility, possible uses, and applications of new nanomaterials in biotechnological processes will be determined by their interaction with biological systems. This interaction is of both basic and practical interest. Comparatively speaking, nevertheless, the research on CNT-organic nanoparticle hybrid structures is still in its infancy. For instance, not enough research has been done on how the composition, shape, and surface chemistry of nanomaterials affect the structure and functionality of conjugated proteins.[1]

The selection of carbon nanotubes (CNTs) is a critical factor in these studies, influenced by several key factors:

1. Manufacturing and Preparation Process: The method used to manufacture and prepare the CNTs significantly impacts their properties and suitability for specific applications.

2. Structural Characteristics of CNTs: The structural features of CNTs, such as diameter, length, and chirality, play a role in determining their performance and behavior in various applications.

3. Surface Characteristics and Functional Groups: The surface properties of CNTs, including the presence and nature of functional groups, greatly influence their interaction with biological systems.

To ensure biocompatibility for cell interactions, it is essential to use CNTs that are compatible with biological environments. CNTs can be made water-soluble through covalent or noncovalent surface functionalization, which modifies their surface properties and enhances their biocompatibility for specific applications involving cells.[1]

The Noncovalent Immobilization

The non-covalent technique promises to be additional feasible planning because the conformational structure of the immobilized enzymes is less affected[85, 86]. Protein-CNT interaction is primarily triggered by the protein that has physically adsorbed onto the CNTs. In order for protein to adsorb onto CNTs, both hydrophobic and electrostatic interactions are required [1, 87, 88]. Because of their extreme hydrophobicity, pristine CNTs will work with proteins' side chains of hydrophobic amino acids to interact with the sidewall of CNTS, facilitating hydrophobic interactions. Regarding the electrostatic interaction, the CNTs' surface electrons will come into contact with the aromatic ring electrons of amino acids, such as tryptophan and phenylalanine.[1]

Other mechanisms that may promote enzyme adsorption onto carbon nanotubes (CNTs) include hydrophobic and electrostatic interactions, nonspecific adsorption, and hydrogen bonding [85]. To physically adsorb the enzyme onto the surface, the CNTs are normally soaked in an enzyme solution, agitated, and then any remaining enzyme is washed away. This technique may be used in two ways: selective adsorption, which entails adsorption with the help of substances like polymers, surfactants, and linking molecules, or direct physical adsorption onto CNTs. When proteins come into contact with carbon nanotubes (CNTs) in a solution, adsorption takes place spontaneously and displays a pseudo-saturation feature because CNTs naturally bind a range of proteins.

Enzyme adsorption onto carbon nanotubes (CNTs) may also be facilitated by hydrogen bonding, nonspecific adsorption, and electrostatic and hydrophobic interactions [85]. Usually, the adsorption process is soaking the CNTs in an enzyme solution, shaking the sample to give the enzyme time to physically adsorb onto the surface, and then washing out any enzyme that has not adsorbed. This method can be applied either by direct physical adsorption onto CNTs (direct physical adsorption) or by selective adsorption, which involves adsorption with the aid of materials like polymers, surfactants, and connecting molecules. Since CNTs naturally bind an extensive variety of protein molecules, when proteins get close to interaction with CNTs in a solution, adsorption occurs spontaneously and with a pseudo saturation characteristic. To physically adsorb the enzyme onto the surface, the CNTs are normally soaked in an enzyme solution, stirred, and then cleansed to eliminate any remaining enzyme.[1]

A CNT's hydrophobic, nanoscale environment can affect the structure and functionality of proteins and enzymes. The strong contact between the adsorbed protein and the CNTs resulted in a deformed or partially unfolded conformation, as observed by means of FT-IR and AFM techniques. Research has shown that during adsorption, enzymes undergo structural alterations, with certain enzymes experiencing a notable reduction in their alpha-helix composition. The degree to which a protein or enzyme is denatured during adsorption depends on its type.

Molecular dynamics simulations have demonstrated that immobilizing the lipase enzyme on the surface of the CNT can better retain the hydrogen bonds at the enzyme's active site in the organic solvent heptane [89]. CNTs can stabilize enzymes at high temperatures and in organic solvents more effectively than standard flat supports, even if the denaturation of proteins or enzymes still happens [87, 88]. It has been proposed that the sharply curved surfaces of CNTs inhibit lateral contacts between neighboring adsorbed enzymes and make them unsuitable for enzyme denaturation under severe circumstances.[1]

Enzyme steadiness on CNTs is higher than on smooth substrates because of the greater curvature of CNTs, which may help to lessen harmful interrelating energy between immobilized protein molecules. The interacting force between the protein or enzyme and CNTs in the direct physical adsorption technique is mostly a hydrophobic interaction. The kind of enzyme, the chemistry of the CNT's surface, and operational parameters are some of the elements that influence the quantities of enzymes adsorbed on CNTs. In addition to direct noncovalent immobilization, proteins or enzymes can also be more easily immobilized by the use of polymers and biomolecules affixed to carbon nanotubes (CNTs).

To functionalize CNTs, polymers and biomolecules have been used. The anchor molecule facilitates the formation of the enzyme-CNT complex, and the functionalized CNTs exhibit high water dispersibility. On the surface of CNT-polymer complexes, polymers coated on CNTs can supply functional groups that are negatively and positively charged. Numerous molecular characteristics, including hydrophobicity, electrostatic interactions, hydrogen bonding, and steric features, combine to give enzymes and proteins their binding specificity [90].

The Covalent Immobilization

Direct Covalent Linking Proteins onto CNTs

Protein or enzyme molecules have been bonded to carbon nanotubes (CNTs) via covalent bonds to stop them from leaching. There are two primary approaches for achieving the covalent functionalization of CNTs: esterification or amidation of oxidized tubes, and covalent attachment of functional groups to the side walls [1, 77, 78]. The sidewalls of carbon nanotubes may be covalently attached to alkyl groups ending with a carboxylic acid using an effective one-step process. By further functionalizing the carboxyl groups on the side chains of the nanotubes with terminal amines and diamines, better solubility and processing are achieved. This may enable the nanotubes to attach to polymer composite materials through chemical bonding with the terminal functional groups on the nanotubes. Additionally, these functional groups offer locations for the covalent attachment of physiologically active molecules, such as oligonucleotides, proteins, peptides, and amino acids. The process of allowing the free amine groups (on the protein surface) to react with the carboxylic acid groups produced by the sidewall oxidation of CNTs, which is aided by 1-ethyl-3-(3dimethylaminipropyl) carbodiimide, has also been used to covalently immobilize enzymes on CNTs [1, 87,88].

The technique has been frequently used to immobilize proteins covalently on carboxylated carbon nanotubes. Enzyme loadings over 1000 µg per mg of CNTs have been reported for some enzymes. The conjugates have a number of beneficial properties, including low mass transfer resistance, high activity and stability, and reusability. At high temperatures, they remain stable as well. Reports suggest that the exposed surface of carbon nanotubes (CNTs) may aid in heat transfer and improve the substrate's accessibility to the enzyme. In organic solvent catalysis, covalently linked lipase on carbon nanotubes (CNTs) offers advantages over free lipase. Immobilized lipase dramatically improves substrate conversion over native lipase. It has been demonstrated that when enzymes are linked to SWCNTs, their activity is much greater than when they are coupled to MWCNTs.

Applications of biofunctionalization to FET Biosensors

Biosensors

Different biomolecules, including DNA and protein, may be functionalized onto carbon nanotubes [123, 124]. It was discovered that compounds that alter the quantity of injected charge have a significant impact on the electrical transport and thermopower of CNT [124]. These special qualities, which include high electrocatalytic activity [125] and high electron transfer rate [124], have led to numerous studies showing that CNT-modified electrodes [125], polymer–CNT composites [126], CNT bundles [123], and CNT membranes [124] can all be employed as efficient electrochemical biosensors. The primary benefits of CNTs are the sensing element's nanosize and the correspondingly low material requirements for a detectable response. In order to function as ribonucleic acid (RNA), enzyme, DNA, glucose, and even protein sensors, well-aligned carbon nanotube arrays have been used. When used as a field effect transistor (FET) configuration, carbon nanotubes (CNTs) have the benefits of size compatibility, sensitivity to minuscule electrical perturbations, and potential biocompatibility in the identification of biological species. Owing to its distinct features, the CNTFET biosensor has found extensive use in the biological domain, encompassing protein, glucose, enzyme, antigen and antibody molecules, DNA molecules, bacteria, and hormones. The silicon substrate of the CNTFET biosensor is covered with an insulating layer of silicon dioxide, and metal electrodes are produced to serve as the source and drain electrodes. A gate electrode is introduced, acting as a conductive channel, by connecting a particular CNT between the two electrodes. This allows the gate's source voltage to be imported, controlling conduction. The CNTFET biosensor's mechanism is depicted in Fig. (**6**). It is possible to coat certain antibodies with CNT. An electric signal that is seen and recorded may result from the particular adsorption between the antigen and the anti-body.

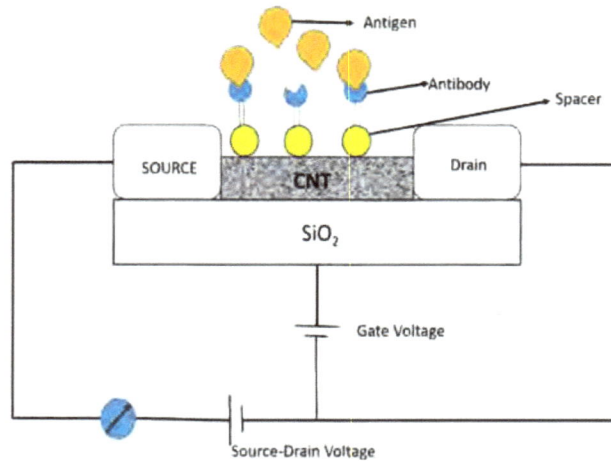

Fig. (6). Schematic diagram of the biosensor using CNT [125].

There are two types of biofunctionalization: covalent and non-covalent methods. The non-covalent functionalization is made possible by the direct adsorption of bioreceptors on the NWs via vdW and electrostatic forces. The direct biofunctionalization of bioreceptors (enzymes, antibodies, DNA probes, etc.) with low IEP via electrostatic contact is made possible by the high IEP of metal oxides like ZnO and TiO_2. Physical adsorption has been used to functionalize several metal oxides with biomaterials [91,122]. However, metal oxide NWs' surfaces are favorable in that they naturally provide oxygenated moieties for biofunctionalizations. Using a self-assembled monolayer (SAM) of hydroquinone (HQ) and its derivatives to supply the peptide ligand for cell adhesion is one method used to immobilize bioreceptors on In_2O_3 [93,122]. Biofunctionalization and applications to FET biosensors.

By utilizing 4-(1,4-dihydroxybenzene)butyl phosphonic acid (HQ-PA) to produce SAM on In_2O_3 NWs, it was further advanced to a selective technique of functionalizing DNA probes on In_2O_3 NWs. QPA, or oxidized HQ-PA, forms conjugations with functional groups such as amines, cyclopentadienes, thiols, and azides (94). Cheung et al. used mmaleimidobenzoyl-NHS as the thiol-amine linker to functionalize the thiolated DNA on In_2O_3 FETs. In addition, non-covalent linkers are frequently employed to immobilize bioreceptors. Organosilanes, like APTES, have the ability to conjugate with biomolecules that include active amine groups by attaching to their surface with –OH groups. Table **1** lists the most current FET biosensors with enhanced sensing capabilities that North American researchers have described utilizing these nanomaterials [122].

Table 1. List of ID FET biosensors.

1 DNanomaterial	Bioreceptor	Analyte	Refs.
GaN NWs	PEG-biotin	Streptavidin	[96]
GaN NWs	Biotin	Streptavidin	[97]
In_2O_3 NWs	Antibody	---	[98]
In_2O_3 NWs	Antibodies	cTnI, CK-MB,and BNP	[95]
In_2O_3Nanoribbons	Glucose/chitosan/SWCNT hybrid	Glucose	[95]

Single-walled Nanotube (SWNT) FET-Based Biosensors

SWNT dynamic detection in living cells are innovative nanoscale biosensors. It seems that graphene is becoming more and more popular than SWNTs for biosensing applications.

We will highlight eminent works in this review that have made significant contributions to the area of carbon nanotubes for sensing applications [63,1].We will specifically review the research on carbon nanotube-based pressure, light, bio, and gas sensors (Fig. **7**). We regret any omissions as it was not feasible to examine all of the excellent research on this issue due to their sheer number.

Numerous applications have been found for carbon nanotubes. They have realized gas sensors, biosensors, pressure sensors, and photodetectors among other sensing devices [63].

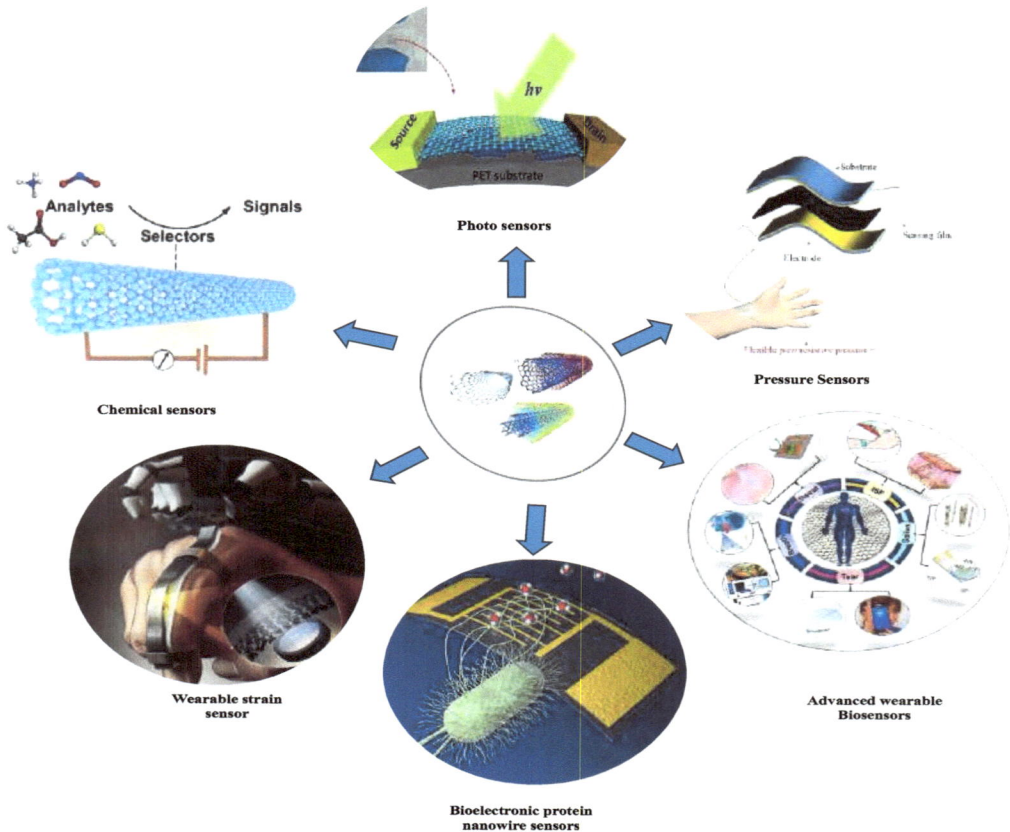

Fig. (7). Applications of sensors.

Sensors Based on the Interstitial Fluid of the Skin

The fluid that surrounds cells is called skin interstitial fluid (ISF), and it acts as a link between blood and cells by continuously maintaining equilibrium with blood capillaries through the process of diffusion [130]. ISF is the body fluid that is easiest to reach; it is mostly found in the subcutaneous tissue layer and makes up 70% of its volume [131]. ISF may be constantly and non-invasively collected, and it has been shown to correlate with blood, providing valuable information for monitoring physiological parameters [132]. For the past few decades, glucose has been the most often found biomarker for the diagnosis of diabetes. Microneedles are usually used to measure the concentration of chemicals in ISF. Without causing pain to the patient, these microneedles have shown to be extremely effective in monitoring salt levels in ISFs [133–135]. Conventional microneedles, on the other hand, frequently have stiff architectures, which means that they are not appropriate for use in devices that demand flexibility or elasticity. An extended-gate FET biosensor that can stretch to detect sodium, a biomarker for the minimally invasive diagnosis of dysnatremia, in order to overcome this constraint. The FET sensor has an expanded gate with skin-piercing microneedles that allow it to reach the ISF, which is used to detect sodium. The gadget under investigation demonstrated notable features such as high sensitivity, mechanical stability on the body, low detection limits, outstanding biocompatibility, and real-time monitoring. In order to identify C-reactive protein in ISF, recently a FET biosensor was created employing silicon nanowire arrays. The authors overcame Debye screening and made label-free detection possible by using SiNW arrays immobilized with anti-body fragments. To guarantee the identification of particular proteins, the reference subtraction approach was used. The FET sensors that were described enabled prompt diagnosis and identified CRP levels between 60 ng/mL and 100 µg/mL, which corresponds to physiological concentrations.

Advantages of CNT

The unique mechanical, electrical, and magnetic characteristics of carbon nanotubes are of essential scientific interest and may have important technological applications. One-dimensional conductors, fiber reinforcement in very strong carbon composite materials, and sorption material for gases like hydrogen are a few potential uses that have been proposed [5, 64]. Research revealed that the graphitic and tubular structures of carbon nanotubes provide them with great strength and stiffness [65]. With a quarter of the weight of steel and a strength that is 10–100 times greater than the strongest steel, carbon nanotubes (CNTs) have the potential to create a whole new class of sophisticated materials [26].

In a vacuum, CNT is likewise thermally stable to 2800 °C. CNT has a thermal conductivity of > 200 W/m K, which is almost twice that of diamond. Because CNTs have an electric current carrying capacity that is around a thousand times more than copper wire, they can improve electrical performance, solve the electromigration reliability issue, and are thus a good choice for VLSI interconnects. Electronic devices including field emission displays, scanning probe microscopy tips, and microelectronic devices have benefited from the special qualities of carbon nanotubes [13, 26, 66].

Disadvantages of CNT

Although carbon nanotubes (CNTs) have numerous advantageous qualities, their hydrophobicity and chemical inertness are two drawbacks that prevent their industrial use. Due to strong van der Waals forces between the nanotubes, it is challenging to suspend them in solution. The majority of applications are hindered by the aggregation of nanotubes in solution, which reduces the unique electrical and mechanical characteristics of each tube. Covalent or non-covalent surface modifications can be used to change the CNT's surface in order to solve these issues. Dissolution in water has been made easier by the use of surfactants (such as sodium dodecyl sulfate, sodium dodecylbenzenesulfonate, cetyltrimethylammonium bromide, polyethylene glycol, etc.), polymer wrapping, or chemical modifications. This is important for potential biomedical applications and biophysical processing schemes.

CNT flaws are another significant disadvantage. Realistic carbon nanotubes have a variety of structural flaws that are produced during growing and are not flawless. Defects in carbon nanotubes have an impact on the material's chemical and physical characteristics [67–69]. Three categories may be used to group the CNT defects:

(1) Topological flaws, which may be corrected, caused topological modifications in the graphene sheet by introducing ring sizes other than hexagons, such as pentagons and heptagons as local defects. Dissolution in water has been made easier by the use of surfactants (such as sodium dodecyl sulfate, sodium dodecylbenzenesulfonate, cetyltrimethylammonium bromide, polyethylene glycol, etc.), polymer wrapping, or chemical modifications. This is important for potential biomedical applications and biophysical processing schemes. CNT flaws are another significant disadvantage. Realistic carbon nanotubes have a variety of structural flaws that are produced during growing and are not flawless. Defects in carbon nanotubes have an impact on the material's chemical and physical characteristics [67–69]. Three categories may be used to group the CNT defects:

(1) Topological flaws, which may be corrected, caused topological modifications in the graphene sheet by introducing ring sizes other than hexagons, such as pentagons and heptagons.

(2) The flaws often linked with graphite may also exist in nanotubes, and they include incomplete bonding and other defects (such as vacancies, dislocations, attachment of impurities, *etc.*). For instance, the nanotubes may have point defects like vacancies or dislocations in the graphene sheet. While they are rare,

dislocations are occasionally noticed in the high-temperature-formed nanotubes. It could be very different for nanotubes that are generated catalytically. When analyzing the qualities of nanotubes that have been seen, the matter of flaws in nanotubes is crucial. The chemistry may be impacted, but not significantly because nanotubes, like graphite, are chemically quite inert. If possible, nanotubes should at least be annealed if not also purified, before physical measurements are made. Only then are the results likely to be consistent and unambiguous.

CONCLUSION AND PROSPECTS

Many customized synthesis techniques have been developed to produce CNTs on a wide scale for commercial usage.[1].The tremendous progress made in the field of carbon nanotube chemistry will surely propel carbon nanotube applications. Functionalizing CNTs, particularly those with specific length, diameter, and chirality, will provide more molecular control over CNT-based materials and devices. This research shows that their immense potential for biotechnology and biomedicine is just now starting to be realized. The contact and immobilization of various biomolecules (enzymes, proteins, or DNA/RNA) on the CNTs can lead to a wide range of applications.[1,120]

The promise of the Bio-FET device design for multiple analyte detection has been highlighted in this review study, as it has been employed with nanomaterials. To improve sensing performance, novel materials with dynamic morphologies, chemistries, and architectures have been employed in nanomaterial-based bio-FET sensing during the past 10 years. Generally speaking, bio-FETs have become more significant in the sensing industry since flexible sensors with low power requirements were introduced. These result from a number of factors, such as reproducible device performance and manufacturing as well as cost optimization that takes into account increases in device yield and mobility for POCT application. As the next evolutionary step with such development, implantable devices are crucial; in fact, it might be argued that they are inevitable given the rate of progress in the pertinent fields of study for a broad variety of applications. According to the authors, figuring out the main issues with nanomaterial Bio-FETs that still need to be fixed is essential to creating solutions[120].

Bio-FETs need to be both biocompatible and environmentally stable in order to be used to manufacture implanted devices. This will increase their durability and prolong their shelf life, especially for wearable tech. The sensor chips in implanted devices will eventually come into interface with complicated biological surroundings, potentially reducing the device's usefulness. In order to concentrate on this concern, comprehensive studies into the stability of 2-dimensional substances, including protection and storing techniques, are continuously established. One method to improve stability and reusability is to keep the electrode and sensor sections apart to allow connections to be made between them. Another barrier to better commercialization is understanding the nuances of the systems controlling the interactions between biological molecules and sensing materials. In this context, further understanding is needed regarding several phenomena, including the kinetics of adsorption and desorption when interfering molecules are present, the effects of contaminants and their tendency to reappear at various stages of material/device fabrication, and the dynamics of charge transfer between enzymes and the sensing layer. These phenomena include anomalous variations in biomolecule sensitivity. Given the introduction of promising technologies on several fronts, it is therefore plausible that nanomaterial-based Bio-FETs have immense potential for usage in commercial biomedical devices.[120]

Learning Objectives/Key Points

(1) CNTs are very promising materials for integration into electrochemical biosensors.

(2) Carbon Nanotube Field Effect Transistor (CNTFET) technology is the best replacement for CMOS technology because of its excellent properties such as high tensile strength, ballistic transport operation, and superior thermal conductivities.

(3) CNTFET Structure Carbon nanotube field effect transistors (CNTFETs) utilize semiconducting single-wall CNTs to assemble electronic devices.

(4) Carbon nanotubes have been applied to a wide variety of different types of biosensors, exploiting different aspects of their advantageous physical and chemical properties.

(5) Nano biochemical sensors play an important role in detecting the biomarkers related to human diseases, and carbon nanotubes (CNTs) have become an important factor in promoting the vigorous development of this field due to their special structure and excellent electronic properties.

Multiple Choice Questions (MCQs)

1. The lotus effect refers to

a. self-wetting property

b. self-cleaning property

c. self-drying property

d. self-oxidizing property

Ans: b. self-cleaning property

2. Quantum dots (QDs) are which type of nanomaterials?

a. 0-D

b. 1-D

c. 2-D

d. 3-D

Ans: 0-d

3. Which one of the following is an example of 1-D nanomaterial?

a. Thin films

b. Quantum well

c. Quantum wire

d. Quantum dots

Ans: c. Quantum wire

4. The properties of nanomaterials are different from the bulk. This is related to

a. Large surface-to-volume ratio of nanomaterials

b. Small surface-to-volume ratio of nanomaterials

c. Large surface area in bulk materials compared to nanomaterials

d. Absence of quantum confinement effect in nanomaterials

Ans: Large surface-to-volume ratio of nanomaterials

5. What are the terminals of FET?

a. Anode and cathode

b. Source, gate, and drain

c. Collector, emitter, and base

d. None of the above

Ans: b. Source, gate, and drain

6. The current in FET flows between _____ terminals.

a. Drain and gate

b. Drain and source

c. Both a and b

d. None of the above

Ans: b. Drain and source

7. The field-effect transistors used in _____

a. Amplifiers

b. Analog switch

c. Oscillator

d. All of the above

Ans: d. All of the above

8. The advantages of FET are _____

a. It has better thermal stability

b. It predicts less noise

c. It can be used at high frequency

d. All of the above

Ans: d. All of the above

9. In FET, the input resistance is higher due to _____

a. Forward bias

b. Reverse bias

Ans: b. Reverse bias.

10. In FET, the noise level is _____

a. High

b. Very low

c. Moderate

d. Low

Ans: d. Low

11. Who invented FETs?

a. Julius Edgar Lilienfeld

b. Shockley

c. Harris

d. None of the above

Ans: a. Julius Edgar Lilienfeld

12. In FET configuration, the voltage gain of the common gate is _____

a. High

b. Very low

c. Moderate

d. Low

Ans: a. High

13. Which of the following statements is not true for biosensors?

a. Biosensors convert a biological signal into an electrical signal.

b. Biosensors are used to determine the concentration of substances and other parameters of biological interest even where they do not utilize a biological system directly.

c. Biosensors utilize different biological systems such as enzymes, whole-cell metabolism, ligand binding, and the antibody-antigen reaction.

d. A biosensor consists of a vessel, or series of vessels, used to perform a desired conversion by enzymic means.

Ans: d. Biosensor consists of a vessel, or a series of vessels, used to perform a desired conversion by enzymic means.

14. Which of the following features is not possessed by biosensors?

a. The biocatalyst used in the biosensor must be highly specific for the purpose of the analyses.

b. The reaction occurring in the biosensor should be independent of such physical parameters.

c. The active site is mainly constituted by non-polar amino acids for catalysis to take place.

d. The response from the biosensors should be accurate, precise, reproducible and linear.

Ans: d. Active site is mainly constituted by non-polar amino acids for catalysis to take place.

15. The operation of flexure-FET relies on:

a. Change in the mass of the beam following the capture of biomolecules.

b. Change in the spring constant following the capture of biomolecules.

c. Change in the damping coefficient following the capture of biomolecules.

d. Change in surface reflection of the cantilever following the capture of a biomolecule.

Ans: b. Change in the spring constant following the capture of biomolecules.

16. A flexure FET provides the highest sensitivity when the transistor is operated in the:

b. Inversion mode.

b. Subthreshold mode.

c. Accumulation mode.

d. Percolation mode.

Ans: b. Subthreshold mode.

17.A flexure FET requires:

a. A reference electrode.

b. A counter electrode.

c. An auxiliary electrode.

d. None of the above.

Ans: d. None of the Above

18. The settling time of a cantilever-based biosensor is:

a. Identical to that of a potentiometric sensor.

b. Still limited by diffusion of molecules to the sensor surface.

c. Is dictated by the shape of the cantilever.

d. All of the above.

Ans: d. All of the above.

19. The very high sensitivity of a flexure-FET arises from the following feature of the sensor.

a. It does not suffer from screening limits.

b. It does not need a reference electrode.

c. There is no redox reaction on the sensor surface.

d. Displacement can be measured far more easily compared to charges or current.

A. Ans: a. It does not suffer from screening limits.

REFERENCES

[1] N. Saifuddin, A.Z. Raziah, and A.R. Junizah, "Carbon nanotubes: a review on structure and their interaction with proteins", *J. Chem.,* vol. 2013, no. 1, p. 676815, 2013.

http://dx.doi.org/10.1155/2013/676815

[2] A. Gnach, T. Lipinski, A. Bednarkiewicz, J. Rybka, and J.A. Capobianco, "Upconverting nanoparticles: assessing the toxicity", *Chem. Soc. Rev.,* vol. 44, no. 6, pp. 1561-1584, 2015.

http://dx.doi.org/10.1039/C4CS00177J PMID: 25176037

[3] M.C. Roco, "National Nanotechnology Initiative at 20 years: enabling new horizons", *J. Nanopart Res.,* vol. 25, no. 10, p. 197, 2023.

http://dx.doi.org/10.1007/s11051-023-05829-9

[4] T. Manimekala, R. Sivasubramanian, and G. Dharmalingam, "Nanomaterial-based biosensors using Field-Effect Transistors: A review", *J. Electron. Mater.,* vol. 51, no. 5, pp. 1950-1973, 2022.

http://dx.doi.org/10.1007/s11664-022-09492-z PMID: 35250154

[5] H. He, L.A. Pham-Huy, P. Dramou, D. Xiao, P. Zuo, and C. Pham-Huy, "Carbon nanotubes: applications in pharmacy and medicine", *BioMed Res. Int.,* vol. 2013, pp. 1-12, 2013.

http://dx.doi.org/10.1155/2013/578290 PMID: 24195076

[6] F.H. Hussein, B.S. Hasan, M.B. Mageed, Z.H. Nafaee, and G.J. Mohammed, "Pharmaceutical Application of Carbon Nanotubes Synthesized by Flame Fragments Deposition Method", *J. Environ. Anal. Chem.,* vol. 4, no. 4, 2017.

http://dx.doi.org/10.4172/2380-2391.1000e115

[7] S.P. Gautam, B. Kaur, and T. Gautam, "Carbon Nanotubes: Exploring Intrinsic Medicinal Activities and Biomedical Applications. Open Access J.", *Oncol. Med.,* vol. 3, pp. 230-232, 2019.

[8] M. Thiruvengadam, G. Rajakumar, V. Swetha, M. Ansari, S. Alghamdi, M. Almehmadi, M. Halawi, L. Kungumadevi, V. Raja, S. Sabura Sarbudeen, S. Madhavan, M. Rebezov, M. Ali Shariati, A. Sviderskiy, and K. Bogonosov, "Recent insights and multifactorial applications of carbon nanotubes", *Micromachines (Basel),* vol. 12, no. 12, p. 1502, 2021.

http://dx.doi.org/10.3390/mi12121502 PMID: 34945354

[9] K.P. Loh, D. Ho, G.N.C. Chiu, D.T. Leong, G. Pastorin, and E.K.H. Chow, "Clinical applications of carbon nanomaterials in diagnostics and therapy", *Adv. Mater.,* vol. 30, no. 47, p. 1802368, 2018.

http://dx.doi.org/10.1002/adma.201802368 PMID: 30133035

[10] N. Kumar, and A. Raman, "Design and analog performance analysis of charge-plasma based cylindrical GAA silicon nanowire tunnel field effect transistor", *Silicon,* vol. 12, no. 11, pp. 2627-2634, 2020.

http://dx.doi.org/10.1007/s12633-019-00355-7

[11] E. Perevedentseva, Y.C. Lin, M. Jani, and C.L. Cheng, "Biomedical applications of nanodiamonds in imaging and therapy", *Nanomedicine (Lond),* vol. 8, no. 12, pp. 2041-2060, 2013.

http://dx.doi.org/10.2217/nnm.13.183 PMID: 24279492

[12] B. Pei, W. Wang, N. Dunne, and X. Li, "Applications of carbon nanotubes in bone tissue regeneration and engineering: Superiority, concerns, current advancements, and prospects", *Nanomaterials (Basel),* vol. 9, no. 10, p. 1501, 2019.

http://dx.doi.org/10.3390/nano9101501 PMID: 31652533

[13] J. Huang, Y. Liu, and T. You, "Carbon nanofiber based electrochemical biosensors: A review", *Anal. Methods,* vol. 2, no. 3, pp. 202-211, 2010.

http://dx.doi.org/10.1039/b9ay00312f

[14] D.C. Ferrier, and K.C. Honeychurch, "Carbon nanotube (CNT)-based biosensors", *Biosensors (Basel),* vol. 11, no. 12, p. 486, 2021.

http://dx.doi.org/10.3390/bios11120486 PMID: 34940243

[15] V. Vamvakaki, and N.A. Chaniotakis, "Carbon nanostructures as transducers in biosensors", *Sens. Actuators B Chem.,* vol. 126, no. 1, pp. 193-197, 2007.

http://dx.doi.org/10.1016/j.snb.2006.11.042

[16] D.S. Yadav, P. Singh, and P. Roat, "Assessing the Impact of Source Pocket Length Variation to Examine DC/RF to Linearity Performance of DG-TFET", *Nano,* vol. 18, no. 4, p. 2350027, 2023.

http://dx.doi.org/10.1142/S1793292023500273

[17] J. Liu, Z. Cao, and Y. Lu, "Functional nucleic acid sensors", *Chem. Rev.,* vol. 109, no. 5, pp. 1948-1998, 2009.

http://dx.doi.org/10.1021/cr030183i PMID: 19301873

[18] N. L. Rosi, and C. A. Mirkin, "Nanostructures in biodiagnostics", *Chem. Rev.,* vol. 105, no. 4, pp. 1547-1562, 2005.

[19] W. Yang, K.R. Ratinac, S.P. Ringer, P. Thordarson, J.J. Gooding, and F. Braet, "Carbon nanomaterials in biosensors: should you use nanotubes or graphene?", *Angew Chem. Int. Ed,* vol. 49, no. 12, pp. 2114-2138, 2010.

http://dx.doi.org/10.1002/anie.200903463 PMID: 20187048

[20] J. Wang, "Nanomaterial-based electrochemical biosensors", *Analyst,* vol. 130, no. 4, pp. 421-426, 2005.

http://dx.doi.org/10.1039/b414248a PMID: 15846872

[21] B.L. Allen, P.D. Kichambare, and A. Star, "Carbon nanotube field-effect-transistor-based biosensors", *Adv. Mater.,* vol. 19, no. 11, pp. 1439-1451, 2007.

http://dx.doi.org/10.1002/adma.200602043

[22] E. Lahiff, C. Lynam, N. Gilmartin, R. O'Kennedy, and D. Diamond, "The increasing importance of carbon nanotubes and nanostructured conducting polymers in biosensors", *Anal. Bioanal Chem.,* vol. 398, no. 4, pp. 1575-1589, 2010.

http://dx.doi.org/10.1007/s00216-010-4054-4 PMID: 20706831

[23] Z. Liu, S. Tabakman, K. Welsher, and H. Dai, "Carbon nanotubes in biology and medicine: In vitro and in vivo detection, imaging and drug delivery", *Nano Res.,* vol. 2, no. 2, pp. 85-120, 2009.

http://dx.doi.org/10.1007/s12274-009-9009-8 PMID: 20174481

[24] S. Liu, Q. Shen, Y. Cao, L. Gan, Z. Wang, M.L. Steigerwald, and X. Guo, "Chemical functionalization of single-walled carbon nanotube field-effect transistors as switches and sensors", *Coord Chem. Rev.,* vol. 254, no. 9-10, pp. 1101-1116, 2010.

http://dx.doi.org/10.1016/j.ccr.2009.11.007

[25] S. Roy, and Z. Gao, "Nanostructure-based electrical biosensors", *Nano Today,* vol. 4, no. 4, pp. 318-334, 2009.

[26] G. Gruner, "Carbon nanotube transistors for biosensing applications", *Anal. Bioanal Chem.,* vol. 384, no. 2, pp. 322-335, 2005.

http://dx.doi.org/10.1007/s00216-005-3400-4 PMID: 16132132

[27] D.R. Kauffman, and A. Star, "Electronically monitoring biological interactions with carbon nanotube field-effect transistors", *Chem. Soc. Rev.,* vol. 37, no. 6, pp. 1197-1206, 2008.

PMID: 18497932

[28] J.L. Arlett, E.B. Myers, and M.L. Roukes, "Comparative advantages of mechanical biosensors", *Nat. Nanotechnol.,* vol. 6, no. 4, pp. 203-215, 2011.

PMID: 21441911

[29] S. Song, Y. Qin, Y. He, Q. Huang, C. Fan, and H.Y. Chen, "Functional nanoprobes for ultrasensitive detection of biomolecules", *Chem. Soc. Rev.,* vol. 39, no. 11, pp. 4234-4243, 2010.

PMID: 20871878

[30] Y. He, C. Fan, and S.T. Lee, "Silicon nanostructures for bioapplications", *Nano Today,* vol. 5, no. 4, pp. 282-295, 2010.

[31] K.I. Chen, B.R. Li, and Y.T. Chen, "Silicon nanowire field-effect transistor-based biosensors for biomedical diagnosis and cellular recording investigation", *Nano Today,* vol. 6, no. 2, pp. 131-154, 2011.

[32] A. Raman, K.J. Kumar, D. Kakkar, R. Ranjan, and N. Kumar, "Performance investigation of source delta-doped vertical nanowire TFET", *J. Electron. Mater.,* vol. 51, no. 10, pp. 5655-5663, 2022.

[33] E. Stern, J.F. Klemic, D.A. Routenberg, P.N. Wyrembak, D.B. Turner-Evans, A.D. Hamilton, D.A. LaVan, T.M. Fahmy, and M.A. Reed, "Label-free immunodetection with CMOS-compatible semiconducting nanowires", *Nature,* vol. 445, no. 7127, pp. 519-522, 2007.

PMID: 17268465

[34] P. Singh, and D. S. Yadav, "Design and investigation of F-shaped tunnel FET with enhanced analog/RF parameters", *Silicon,* pp. 1-16, 2021.

[35] B. Tian, T. Cohen-Karni, Q. Qing, X. Duan, P. Xie, and C.M. Lieber, "Three-dimensional, flexible nanoscale field-effect transistors as localized bioprobes", *Science,* vol. 329, no. 5993, pp. 830-834, 2010.

PMID: 20705858

[36] I.L. Medintz, H.T. Uyeda, E.R. Goldman, and H. Mattoussi, "Quantum dot bioconjugates for imaging, labelling and sensing", *Nat. Mater.,* vol. 4, no. 6, pp. 435-446, 2005.

PMID: 15928695

[37] R. Wilson, "The use of gold nanoparticles in diagnostics and detection", *Chem. Soc. Rev.,* vol. 37, no. 9, pp. 2028-2045, 2008.

PMID: 18762845

[38] J. M. Nam, C. S. Thaxton, and C. A. Mirkin, "Nanoparticle-based bio-bar codes for the ultrasensitive detection of proteins", *Science,* vol. 301, no. 5641, pp. 1884-1886, 2003.

[39] P. Alivisatos, "The use of nanocrystals in biological detection", *Nat. Biotechnol.,* vol. 22, no. 1, pp. 47-52, 2004.

PMID: 14704706

[40] C. Dekker, "Solid-state nanopores", *Nat. Nanotechnol.,* vol. 2, no. 4, pp. 209-215, 2007.

PMID: 18654264

[41] S. Howorka, and Z. Siwy, "Nanopore analytics: sensing of single molecules", *Chem. Soc. Rev.,* vol. 38, no. 8, pp. 2360-2384, 2009.

http://dx.doi.org/10.1039/b813796j PMID: 19623355

[42] L. Soleymani, Z. Fang, E.H. Sargent, and S.O. Kelley, "Programming the detection limits of biosensors through controlled nanostructuring", *Nat. Nanotechnol.,* vol. 4, no. 12, pp. 844-848, 2009.

http://dx.doi.org/10.1038/nnano.2009.276 PMID: 19893517

[43] K.R. Ratinac, W. Yang, J.J. Gooding, P. Thordarson, and F. Braet, "Graphene and related materials in electrochemical sensing", *Electroanalysis,* vol. 23, no. 4, pp. 803-826, 2011.

[44] Y. Ohno, K. Maehashi, and K. Matsumoto, "Chemical and biological sensing applications based on graphene field-effect transistors", *Biosens Bioelectron,* vol. 26, no. 4, pp. 1727-1730, 2010.

http://dx.doi.org/10.1016/j.bios.2010.08.001 PMID: 20800470

[45] X. Huang, Z. Yin, S. Wu, X. Qi, Q. He, Q. Zhang, and H. Zhang, "Graphene-based materials: synthesis, characterization, properties, and applications", *Small,* vol. 7, no. 14, pp. 1876-1902, 2011.

[46] Q. He, S. Wu, Z. Yin, and H. Zhang, "Graphene-based electronic sensors", *Chem. Sci,* vol. 3, no. 6, pp. 1764-1772, 2012.

http://dx.doi.org/10.1039/c2sc20205k

[47] Y. Cui, Q. Wei, H. Park, and C. M. Lieber, "Nanowire nanosensors for highly sensitive and selective detection of biological and chemical species", *Science,* vol. 293, no. 5533, pp. 1289-1292, 2001.

[48] F. Patolsky, G. Zheng, and C.M. Lieber, "Fabrication of silicon nanowire devices for ultrasensitive, label-free, real-time detection of biological and chemical species", *Nat. Protoc,* vol. 1, no. 4, pp. 1711-1724, 2006.

http://dx.doi.org/10.1038/nprot.2006.227 PMID: 17487154

[49] P. Romele, M. Ghittorelli, Z.M. Kovács-Vajna, and F. Torricelli, "Ion buffering and interface charge enable high performance electronics with organic electrochemical transistors", *Nat. Commun,* vol. 10, no. 1, p. 3044, 2019.

http://dx.doi.org/10.1038/s41467-019-11073-4 PMID: 31292452

[50] S. Casalini, F. Leonardi, T. Cramer, and F. Biscarini, "Organic field-effect transistor for label-free dopamine sensing", *Org Electron.,* vol. 14, no. 1, pp. 156-163, 2013.

http://dx.doi.org/10.1016/j.orgel.2012.10.027

[51] P. Singh, and D.S. Yadav, "Assessing the Impact of Drain Underlap Perspective Approach to Investigate DC/RF to Linearity Behavior of L-Shaped TFET", *Silicon,* vol. 14, no. 17, pp. 11471-11481, 2022.

http://dx.doi.org/10.1007/s12633-022-01814-4

[52] M.Y. Mulla, E. Tuccori, M. Magliulo, G. Lattanzi, G. Palazzo, K. Persaud, and L. Torsi, "Capacitance-modulated transistor detects odorant binding protein chiral interactions", *Nat. Commun,* vol. 6, no. 1, p. 6010, 2015.

PMID: 25591754

[53] A. Dey, A. Singh, D. Dutta, S.S. Ghosh, and P.K. Iyer, "Rapid and label-free bacteria detection using a hybrid tri-layer dielectric integrated n-type organic field effect transistor", *J. Mater. Chem. A Mater. Energy. Sustain.,* vol. 7, no. 31, pp. 18330-18337, 2019.

http://dx.doi.org/10.1039/C9TA06359E

[54] M. Berto, C. Diacci, R. D'Agata, M. Pinti, E. Bianchini, M.D. Lauro, S. Casalini, A. Cossarizza, M. Berggren, D. Simon, G. Spoto, F. Biscarini, and C.A. Bortolotti, "EGOFET peptide aptasensor for label-free detection of inflammatory cytokines in complex fluids", *Adv. Biosyst,* vol. 2, no. 2, p. 1700072, 2018.

http://dx.doi.org/10.1002/adbi.201700072

[55] M.H. Park, D. Han, R. Chand, D.H. Lee, and Y.S. Kim, "Mechanism of label-free DNA detection using the floating electrode on pentacene thin film transistor", *J. Phys. Chem. C,* vol. 120, no. 9, pp. 4854-4859, 2016.

[56] F.A. Viola, B. Brigante, P. Colpani, G. Dell'Erba, V. Mattoli, D. Natali, and M. Caironi, "A 13.56 MHz rectifier based on fully inkjet-printed organic diodes", *Adv. Mater.,* vol. 32, no. 33, p. 2002329, 2020.

   http://dx.doi.org/10.1002/adma.202002329 PMID: 32648300

[57] F.A. Viola, B. Brigante, P. Colpani, G. Dell'Erba, V. Mattoli, D. Natali, and M. Caironi, "A 13.56 MHz rectifier based on fully inkjet printed organic diodes", *Adv. Mater.,* vol. 32, no. 33, p. e2002329, 2020.

   PMID: 32648300

[58] K. S. Novoselov, A. K. Geim, S. V. Morozov, D. E. Jiang, Y. Zhang, S. V. Dubonos, and A. A. Firsov, "Electric field effect in atomically thin carbon films", *Science,* vol. 306, no. 5696, pp. 666-669, 2004.

[59] Y. Zhang, Y. W. Tan, H. L. Stormer, and P. Kim, "Experimental observation of the quantum Hall effect and Berry's phase in graphene", *Nature,* vol. 438, no. 7065, pp. 201-204, 2005.

[60] H. Dai, "Carbon nanotubes: synthesis, integration, and properties", *Acc Chem. Res.,* vol. 35, no. 12, pp. 1035-1044, 2002.

   PMID: 12484791

[61] F. Schedin, A.K. Geim, S.V. Morozov, E.W. Hill, P. Blake, M.I. Katsnelson, and K.S. Novoselov, "Detection of individual gas molecules adsorbed on graphene", *Nat. Mater.,* vol. 6, no. 9, pp. 652-655, 2007.

   PMID: 17660825

[62] Q. Wang, X. Guo, L. Cai, Y. Cao, L. Gan, S. Liu, Z. Wang, H. Zhang, and L. Li, "TiO_2-decorated graphenes as efficient photoswitches with high oxygen sensitivity," *Chemical Science*, vol. 2, no. 9, 2011.

[63] S. Liu, and X. Guo, "Carbon nanomaterials field-effect-transistor-based biosensors. NPG Asia Materials, 4(8), e23-e23. Chen, R. J., Choi, H. C., Bangsaruntip, S., Yenilmez, E., Tang, X., Wang, Q., & Dai, H. (2004). An investigation of the mechanisms of electronic sensing of protein adsorption on carbon nanotube devices", *J. Am Chem. Soc.,* vol. 126, no. 5, pp. 1563-1568, 2012.

[64] X. Tang, S. Bansaruntip, N. Nakayama, E. Yenilmez, Y.L. Chang, and Q. Wang, "Carbon nanotube DNA sensor and sensing mechanism", *Nano Lett,* vol. 6, no. 8, pp. 1632-1636, 2006.

   PMID: 16895348

[65] E.L. Gui, L.J. Li, K. Zhang, Y. Xu, X. Dong, X. Ho, P.S. Lee, J. Kasim, Z.X. Shen, J.A. Rogers, and S.G. Mhaisalkar, "DNA sensing by field-effect transistors based on networks of carbon nanotubes", *J. Am Chem. Soc.,* vol. 129, no. 46, pp. 14427-14432, 2007.

   PMID: 17973383

[66] H.R. Byon, and H.C. Choi, "Network single-walled carbon nanotube-field effect transistors (SWNT-FETs) with increased Schottky contact area for highly sensitive biosensor applications", *J. Am Chem. Soc.,* vol. 128, no. 7, pp. 2188-2189, 2006.

   PMID: 16478153

[67] Periyasami, A., & Kumar, P. (2025). Label-free biosensor with multidimensional aspects of a Molybdenum di-sulfide dual-gate Schottky tunnel field-effect transistor (DG-STFET). Micro and Nanostructures, 208178.

[68] B.R. Goldsmith, J.G. Coroneus, V.R. Khalap, A.A. Kane, G.A. Weiss, and P.G. Collins, "Conductance-controlled point functionalization of single-walled carbon nanotubes", *Science,* vol. 315, no. 5808, pp. 77-81, 2007.

PMID: 17204645

[69] B.R. Goldsmith, J.G. Coroneus, A.A. Kane, G.A. Weiss, and P.G. Collins, "Monitoring single-molecule reactivity on a carbon nanotube", *Nano Lett,* vol. 8, no. 1, pp. 189-194, 2008.

PMID: 18088152

[70] Q. Mu, W. Liu, Y. Xing, H. Zhou, Z. Li, and Y. Zhang, "Protein binding by functionalized multiwalled carbon nanotubes is governed by the surface chemistry of both parties and the nanotube diameter", *J. Phys. Chem. C,* vol. 112, no. 9, pp. 3300-3307, 2008.

[71] A. Anam, S. I. Amin, D. Prasad, N. Kumar, and S. Anand, "Analysis of III-V material-based dual source T-channel junction-less TFET with metal implant for improved DC and RF performance", *Micro and Nanostructures,* p. 207629, 2023.

[72] W. Huang, S. Taylor, K. Fu, Y. Lin, D. Zhang, and T.W. Hanks, "Attaching proteins to carbon nanotubes via diimide-activated amidation", *Nano Lett,* vol. 2, no. 4, pp. 311-314, 2002.

[73] K. Jiang, L.S. Schadler, R.W. Siegel, X. Zhang, H. Zhang, and M. Terrones, "Protein immobilization on carbon nanotubes via a two-step process of diimide-activated amidation", *J. Mater. Chem.,* vol. 14, no. 1, pp. 37-39, 2004.

[74] D. Pantarotto, C.D. Partidos, R. Graff, J. Hoebeke, J.P. Briand, M. Prato, and A. Bianco, "Synthesis, structural characterization, and immunological properties of carbon nanotubes functionalized with peptides", *J. Am Chem. Soc.,* vol. 125, no. 20, pp. 6160-6164, 2003.

http://dx.doi.org/10.1021/ja034342r PMID: 12785847

[75] P. Ji, H. Tan, X. Xu, and W. Feng, "Lipase covalently attached to multiwalled carbon nanotubes as an efficient catalyst in organic solvent", *AIChE J.,* vol. 56, no. 11, pp. 3005-3011, 2010.

http://dx.doi.org/10.1002/aic.12180

[76] B. Zhang, Y. Xing, Z. Li, H. Zhou, Q. Mu, and B. Yan, "Functionalized carbon nanotubes specifically bind to α-chymotrypsin's catalytic site and regulate its enzymatic function", *Nano Lett,* vol. 9, no. 6, pp. 2280-2284, 2009.

http://dx.doi.org/10.1021/nl900437n PMID: 19408924

[77] B.Y. Kim, and H.J. Shaw, "Gated phase-modulation feedback approach to fiber-optic gyroscopes", *Opt. Lett,* vol. 9, no. 6, pp. 263-265, 1984.

http://dx.doi.org/10.1364/OL.9.000263 PMID: 19721565

[78] H.L. Pang, J. Liu, D. Hu, X.H. Zhang, and J.H. Chen, "Immobilization of laccase onto 1-aminopyrene functionalized carbon nanotubes and their electrocatalytic activity for oxygen reduction", *Electrochim Acta.,* vol. 55, no. 22, pp. 6611-6616, 2010.

http://dx.doi.org/10.1016/j.electacta.2010.06.013

[79] M.A. Alonso-Lomillo, O. Rüdiger, A. Maroto-Valiente, M. Velez, I. Rodríguez-Ramos, F.J. Muñoz, V.M. Fernández, and A.L. De Lacey, "Hydrogenase-coated carbon nanotubes for efficient H2 oxidation", *Nano Lett,* vol. 7, no. 6, pp. 1603-1608, 2007.

http://dx.doi.org/10.1021/nl070519u PMID: 17489639

[80] K. Besteman, J.O. Lee, F.G.M. Wiertz, H.A. Heering, and C. Dekker, "Enzyme-coated carbon nanotubes as single-molecule biosensors", *Nano Lett,* vol. 3, no. 6, pp. 727-730, 2003.

http://dx.doi.org/10.1021/nl034139u

[81] C.Z. Dinu, G. Zhu, S.S. Bale, G. Anand, P.J. Reeder, K. Sanford, G. Whited, R.S. Kane, and J.S. Dordick, "Enzyme-based nanoscale composites for use as active decontamination surfaces", *Adv. Funct. Mater.,* vol. 20, no. 3, pp. 392-398, 2010.

http://dx.doi.org/10.1002/adfm.200901388

[82] K. Matsuura, T. Saito, T. Okazaki, S. Ohshima, M. Yumura, and S. Iijima, "Selectivity of water-soluble proteins in single-walled carbon nanotube dispersions", *Chem. Phys. Lett,* vol. 429, no. 4-6, pp. 497-502, 2006.

http://dx.doi.org/10.1016/j.cplett.2006.08.044

[83] D. Nepal, and K.E. Geckeler, "pH-sensitive dispersion and debundling of single-walled carbon nanotubes: lysozyme as a tool", *Small,* vol. 2, no. 3, pp. 406-412, 2006.

http://dx.doi.org/10.1002/smll.200500351 PMID: 17193060

[84] P. Asuri, S.S. Karajanagi, H. Yang, T.J. Yim, R.S. Kane, and J.S. Dordick, "Increasing protein stability through control of the nanoscale environment", *Langmuir,* vol. 22, no. 13, pp. 5833-5836, 2006.

http://dx.doi.org/10.1021/la0528450 PMID: 16768515

[85] P. Asuri, S.S. Karajanagi, E. Sellitto, D.Y. Kim, R.S. Kane, and J.S. Dordick, "Water-soluble carbon nanotube-enzyme conjugates as functional biocatalytic formulations", *Biotechnol. Bioeng.,* vol. 95, no. 5, pp. 804-811, 2006.

http://dx.doi.org/10.1002/bit.21016 PMID: 16933322

[86] W. Feng, and P. Ji, "Enzymes immobilized on carbon nanotubes", *Biotechnol. Adv.,* vol. 29, no. 6, pp. 889-895, 2011.

PMID: 21820044

[87] R.J. Chen, S. Bangsaruntip, K.A. Drouvalakis, N. Wong Shi Kam, M. Shim, Y. Li, W. Kim, P.J. Utz, and H. Dai, "Noncovalent functionalization of carbon nanotubes for highly specific electronic biosensors", *Proc Natl Acad. Sci. USA,* vol. 100, no. 9, pp. 4984-4989, 2003.

http://dx.doi.org/10.1073/pnas.0837064100 PMID: 12697899

[88] H. Gao, M. Sun, C. Lin, and S. Wang, "Electrochemical DNA biosensor based on graphene and TiO2 nanorods composite film for the detection of transgenic soybean gene sequence of MON89788", *Electroanalysis,* vol. 24, no. 12, pp. 2283-2290, 2012.

http://dx.doi.org/10.1002/elan.201200403

[89] K.S. Mun, S.D. Alvarez, W.Y. Choi, and M.J. Sailor, "A stable, label-free optical interferometric biosensor based on TiO2 nanotube arrays", *ACS Nano,* vol. 4, no. 4, pp. 2070-2076, 2010.

http://dx.doi.org/10.1021/nn901312f PMID: 20356100

[90] A.L. Eckermann, D.J. Feld, J.A. Shaw, and T.J. Meade, "Electrochemistry of redox-active self-assembled monolayers", *Coord Chem. Rev.,* vol. 254, no. 15-16, pp. 1769-1802, 2010.

PMID: 20563297

[91] M. Curreli, C. Li, Y. Sun, B. Lei, M.A. Gundersen, M.E. Thompson, and C. Zhou, "Selective functionalization of In2O3 nanowire mat devices for biosensing applications", *J. Am Chem. Soc.,* vol. 127, no. 19, pp. 6922-6923, 2005.

PMID: 15884914

[92] J. Li, C. Chen, S. Liu, J. Lu, W.P. Goh, and H. Fang, "Ultrafast electrochemical expansion of black phosphorus toward high-yield synthesis of few-layer phosphorene", *Chem. Mater.,* vol. 30, no. 8, pp. 2742-2749, 2018.

[93] D.J. Guo, A.I. Abdulagatov, D.M. Rourke, K.A. Bertness, S.M. George, Y.C. Lee, and W. Tan, "GaN nanowire functionalized with atomic layer deposition techniques for enhanced immobilization of biomolecules", *Langmuir,* vol. 26, no. 23, pp. 18382-18391, 2010.

PMID: 21033757

[94] E.H. Williams, A.V. Davydov, V.P. Oleshko, K.L. Steffens, I. Levin, and N.J. Lin, "Solution-based functionalization of gallium nitride nanowires for protein sensor development", *Surf Sci.,* vol. 627, pp. 23-28, 2014.

[95] H.K. Chang, F.N. Ishikawa, R. Zhang, R. Datar, R.J. Cote, M.E. Thompson, and C. Zhou, "Rapid, label-free, electrical whole blood bioassay based on nanobiosensor systems", *ACS Nano,* vol. 5, no. 12, pp. 9883-9891, 2011.

PMID: 22066492

[96] Y. Huang, H.G. Sudibya, D. Fu, R. Xue, X. Dong, L.J. Li, and P. Chen, "Label-free detection of ATP release from living astrocytes with high temporal resolution using carbon nanotube network", *Biosens Bioelectron,* vol. 24, no. 8, pp. 2716-2720, 2009.

http://dx.doi.org/10.1016/j.bios.2008.12.006 PMID: 19135355

[97] I. Heller, W.T.T. Smaal, S.G. Lemay, and C. Dekker, "Probing macrophage activity with carbon-nanotube sensors", *Small,* vol. 5, no. 22, pp. 2528-2532, 2009.

http://dx.doi.org/10.1002/smll.200900823 PMID: 19697305

[98] H.G. Sudibya, J. Ma, X. Dong, S. Ng, L.J. Li, X.W. Liu, and P. Chen, "Interfacing glycosylated carbon-nanotube-network devices with living cells to detect dynamic secretion of biomolecules", *Angew Chem. Int. Ed,* vol. 48, no. 15, pp. 2723-2726, 2009.

http://dx.doi.org/10.1002/anie.200805514 PMID: 19263455

[99] C. W. Wang, C. Y. Pan, H. C. Wu, P. Y. Shih, C. C. Tsai, K. T. Liao, and Y. T. Chen, "In situ detection of chromogranin A released from living neurons with a single-walled carbon-nanotube field-effect transistor", *Small,* vol. 3, pp. 1350-1355, 2007.

[100] A. Star, E. Tu, J. Niemann, J.C.P. Gabriel, C.S. Joiner, and C. Valcke, "Label-free detection of DNA hybridization using carbon nanotube network field-effect transistors", *Proc Natl Acad. Sci. USA,* vol. 103, no. 4, pp. 921-926, 2006.

http://dx.doi.org/10.1073/pnas.0504146103 PMID: 16418278

[101] M.T. Martínez, Y.C. Tseng, N. Ormategui, I. Loinaz, R. Eritja, and J. Bokor, "Label-free DNA biosensors based on functionalized carbon nanotube field effect transistors", *Nano Lett,* vol. 9, no. 2, pp. 530-536, 2009.

http://dx.doi.org/10.1021/nl8025604 PMID: 19125575

[102] P. Singh, and D.S. Yadav, "Impact of work function variation for enhanced electrostatic control with suppressed ambipolar behavior for dual gate L-TFET", *Curr. Appl. Phys.,* vol. 44, pp. 90-101, 2022.

http://dx.doi.org/10.1016/j.cap.2022.09.014

[103] L.J. Edgar, U.S. Patent No. 1,745,175.

[104] T.H. Lee, *The design of CMOS radio-frequency integrated circuits.* Cambridge university press, 2003.

http://dx.doi.org/10.1017/CBO9780511817281

[105] R. Puers, L. Baldi, M. Van de Voorde, S.E. Van Nooten, Eds., *Nanoelectronics: Materials, Devices, Applications.* , 2017.

[106] M. Grundmann, and M. Grundmann, "Kramers–Kronig relations", *The Physics of Semiconductors: An Introduction including Nanophysics and Applications,* pp. 775-776, 2010.

[107] J.I. Nishizawa, "Junction field-effect devices", In: *Semiconductor Devices for Power Conditioning.* Springer US: Boston, MA, 1982, pp. 241-272.

http://dx.doi.org/10.1007/978-1-4684-7263-9_11

[108] A.S. Dylan, *Metal Oxide Semiconductor (MOS).* Transistor Demonstrated, The Silicon Engine, Computer History Museum, 1960.

[109] B. Duncan, *High Performance Audio Power Amplifiers.* Elsevier, 1996.

[110] L.M. Moretto, K. Kalcher, Eds., *Environmental analysis by electrochemical sensors and biosensors,* vol. Vol. 1. Springer: New York, 2014.

http://dx.doi.org/10.1007/978-1-4939-0676-5

[111] S.S. Alabsi, A.Y. Ahmed, J.O. Dennis, M.M. Khir, and A.S. Algamili, "A review of carbon nanotubes field effect-based biosensors. IEEE Access, 8, 69509-69521. Veetil, J. V., & Ye, K. (2007). Development of immunosensors using carbon nanotubes", *Biotechnol. Prog,* vol. 23, no. 3, pp. 517-531, 2020.

[112] A. Weis, F. Liang, J. Gao, R. T. Barnard, and S. Corrie, "RNA and DNA diagnostics on microspheres: current and emerging methods", *RNA and DNA Diagnostics,* pp. 205-224, 2015.

[113] R.K. Thines, N.M. Mubarak, S. Nizamuddin, J.N. Sahu, E.C. Abdullah, and P. Ganesan, "Application potential of carbon nanomaterials in water and wastewater treatment: A review", *J. Taiwan Inst Chem. Eng.,* vol. 72, pp. 116-133, 2017.

http://dx.doi.org/10.1016/j.jtice.2017.01.018

[114] P. Sudarsanam, Y. Yamauchi, P. Bharali, Eds., *Heterogeneous Nanocatalysis for Energy. and Environmental Sustainability, Volume 1: Energy. Applications.* , 2022.

http://dx.doi.org/10.1002/9781119772057

[115] K.I. Kabel, A.A. Farag, E.M. Elnaggar, and A.G. Al-Gamal, "Removal of oxidation fragments from multi-walled carbon nanotubes oxide using high and low concentrations of sodium hydroxide", *Arab J. Sci. Eng.,* vol. 41, no. 6, pp. 2211-2220, 2016.

http://dx.doi.org/10.1007/s13369-015-1897-1

[116] P. Pantano, "Nanomaterials for Biosensors", In: *Nanotechnologies for the Life. Sciences,* vol. 8. , 2007.

[117] M. Willander, K. Khun, and Z. Ibupoto, "Metal oxide nanosensors using polymeric membranes, enzymes and antibody receptors as ion and molecular recognition elements", *Sensors,* vol. 14, no. 5, pp. 8605-8632, 2014.

http://dx.doi.org/10.3390/s140508605 PMID: 24841244

[118] M. Lynn Crismon, D.S. West-Strum, K. Dowling-McClay, I. Drame, T.J. Hastings, P. Jumbo-Lucioni, K.K. Marwitz, A. Spence, D. Farrell, and R. Walker, "The report of the 2021-2022 AACP research and graduate affairs committee", *Am J. Pharm. Educ.,* vol. 87, no. 1, p. ajpe9454, 2023.

http://dx.doi.org/10.5688/ajpe9454 PMID: 36781185

[119] A.C. Power, B. Gorey, S. Chandra, and J. Chapman, "Carbon nanomaterials and their application to electrochemical sensors: a review", *Nanotechnol. Rev.,* vol. 7, no. 1, pp. 19-41, 2018.

http://dx.doi.org/10.1515/ntrev-2017-0160

[120] H. Zhang, F. Osawa, H. Okamoto, Y. Qiu, Z. Liu, N. Ohshima, ... and H. Sone, "Ultrasensitive specific detection of anti-influenza A H1N1 hemagglutinin monoclonal antibody using silicon nanowire field effect biosensors," *ACS Appl. Bio Mater.,* 2025.

[121] A. Molazemhosseini, F.A. Viola, F.J. Berger, N.F. Zorn, J. Zaumseil, and M. Caironi, "A rapidly stabilizing water-gated field-effect transistor based on printed single-walled carbon nanotubes for biosensing applications", *ACS Appl. Electron. Mater.,* vol. 3, no. 7, pp. 3106-3113, 2021.

http://dx.doi.org/10.1021/acsaelm.1c00332 PMID: 34485915

[122] M. Sedki, Y. Shen, and A. Mulchandani, "Nano-FET-enabled biosensors: Materials perspective and recent advances in North America", *Biosens Bioelectron,* vol. 176, p. 112941, 2021.

http://dx.doi.org/10.1016/j.bios.2020.112941 PMID: 33422922

[123] J.J. Davis, R.J. Coles, H. Allen, and O. Hill, "Protein electrochemistry at carbon nanotube electrodes", *J. Electroanal Chem,* vol. 440, no. 1-2, pp. 279-282, 1997.

http://dx.doi.org/10.1016/S0022-0728(97)80067-8

[124] R.J. Chen, Y. Zhang, D. Wang, and H. Dai, "Noncovalent sidewall functionalization of single-walled carbon nanotubes for protein immobilization", *J. Am Chem. Soc.,* vol. 123, no. 16, pp. 3838-3839, 2001.

http://dx.doi.org/10.1021/ja010172b PMID: 11457124

[125] A. Thess, R. Lee, P. Nikolaev, H. Dai, P. Petit, J. Robert, and R. E. Smalley, "Crystalline ropes of metallic carbon nanotubes", *Science,* vol. 273, no. 5274, pp. 483-487, 1996.

[126] J. Wang, M. Musameh, and Y. Lin, "Solubilization of carbon nanotubes by Nafion toward the preparation of amperometric biosensors", *J. Am Chem. Soc.,* vol. 125, no. 9, pp. 2408-2409, 2003.

http://dx.doi.org/10.1021/ja028951v PMID: 12603125

[127] G. Che, B.B. Lakshmi, E.R. Fisher, and C.R. Martin, "Carbon nanotubule membranes for electrochemical energy storage and production", *Nature,* vol. 393, no. 6683, pp. 346-349, 1998.

http://dx.doi.org/10.1038/30694

[128] Y. Yu, A. Cimeno, Y.C. Lan, J. Rybczynski, D.Z. Wang, T. Paudel, Z.F. Ren, D.J. Wagner, M.Q. Qiu, T.C. Chiles, and D. Cai, "Assembly of multi-functional nanocomponents on periodic nanotube array for biosensors", , vol. 4, no. 1, pp. 27-33, 2009.

PMID: 19829755

[129] N. Gupta, S.M. Gupta, and S.K. Sharma, "Carbon nanotubes: Synthesis, properties and engineering applications", *Carbon Letters,* vol. 29, pp. 419-447, 2019.

[130] N. Fogh-Andersen, B.M. Altura, B.T. Altura, and O. Siggaard-Andersen, "Composition of interstitial fluid", *Clin. Chem.,* vol. 41, no. 10, pp. 1522-1525, 1995.

PMID: 7586528

[131] P. Humbert, F. Fanian, H. Maibach, Eds., *Agache's measuring the skin.* Springer International Publishing: New York, 2019, pp. 477-486.

[132] J. Heikenfeld, A. Jajack, B. Feldman, S.W. Granger, S. Gaitonde, G. Begtrup, and B.A. Katchman, "Accessing analytes in biofluids for peripheral biochemical monitoring", *Nat. Biotechnol.,* vol. 37, no. 4, pp. 407-419, 2019.

PMID: 30804536

[133] B. Ciui, M. Tertiş, A. Cernat, R. Săndulescu, J. Wang, and C. Cristea, "Finger-based printed sensors integrated on a glove for on-site screening of Pseudomonas aeruginosa virulence factors", *Anal. Chem.,* vol. 90, no. 12, pp. 7761-7768, 2018.

PMID: 29851349

[134] K.Y. Goud, C. Moonla, R.K. Mishra, C. Yu, R. Narayan, I. Litvan, and J. Wang, "Wearable electrochemical microneedle sensor for continuous monitoring of levodopa: toward Parkinson management", *ACS Sens.,* vol. 4, no. 8, pp. 2196-2204, 2019.

PMID: 31403773

[135] S. Sharma, Z. Huang, M. Rogers, M. Boutelle, and A.E. Cass, "Evaluation of a minimally invasive glucose biosensor for continuous tissue monitoring", *Anal. Bioanal Chem.,* vol. 408, no. 29, pp. 8427-8435, 2016.

PMID: 27744480

CHAPTER 6

Scope and Challenges of Nano-FET for Digital Circuit Design

Jyoti Kandpal[1,*] and **Swagata Devi**[2]

[1]*Depertment of Electronics and Communication Engineering, Graphic Era Hill University Dehradun, Uttarkhand, India*

[2]*Faculty of Engineering, Assam downtown University, Sankar Madhab Path, Gandhi Nagar, Panikhaiti, Guwahati, Assam, India*

Abstract: Over the previous thirty years, the scaling of complementary metal-oxide-semiconductor (CMOS) technology has stood crucial to the continued advancement of the silicon-based semiconductor industry. However, when technological scaling reaches the nanoscale zone, CMOS devices face several significant challenges, including higher leakage currents, difficulty increasing on-current, massive parameter changes, low yield and reliability, increased manufacturing costs, etc. In order to sustain previous advances, numerous developments in CMOS technologies and device topologies have been developed and put into practice. Simultaneously with these investigations, some innovative nanoelectronic devices, labelled as "Beyond CMOS Devices," are currently intensively investigated and developed as potential replacements or supplements for eventually scaled classic CMOS devices. Despite offering system integration at extremely high densities, these nanoelectronic devices continue to be in their infancy and confront numerous challenges, including high variations and low dependability. The actual implementation of these promising technologies necessitates substantial study at the device and system architectural levels.

Keywords: Fin-FET technology, Nanosheet FET (NS-FET), Semiconductor design, Sub-5-nm technology, Transistors.

INTRODUCTION

Three broad categories are used by the European Nanoelectronics Initiative Advisory Council (ENIAC) (its prediction shown in Fig. **1**) to analyze the silicon-based micro- or nanoelectronics industry:

---

* Corresponding author Jyoti Kandpal: Depertment of Electronics and Communication Engineering, Graphic Era Hill University Dehradun, Uttarkhand, India; E-mail: jayakandpal27@gmail.com

Dharmendra Singh Yadav & Prabhat Singh (Eds.)

1. **Advanced CMOS (More Moore)**: To continue downsizing transistors, particularly in the improved use of metal gates with suitable work functions, high-k oxides, and high-k oxides as insulators, to ensure an effective throughput while decreasing the leakage via gate stack.

2. **More than Moore**: Fig. (**2**) shows the concept of more law. Modern CMOS technology has demonstrated itself as inherently constrained. Radiofrequency, analogue circuits, switches with high-voltage actuators, and motion sensors are non-digital functions that call for a combination of technologies customized to a particular need. To overcome these obstacles and implement new features like mechanics, optics, acoustics, ferroelectrics, etc., "more than Moore" is needed.

3. **Beyond CMOS.** New materials, whether inorganic or organic, new operating principles, such as those that replace electrons with magnetic excitation or spin, and new architectural designs, are covered. Examples of alternatives beyond CMOS embrace innovative materials to fabricate interconnects and transistors such as nanowires and carbon nanotubes, switches working with resistive change polymers for memories, the electronic characteristics of organic compounds, memory, and computing architectures to fully utilize the capabilities of these new devices (as shown in Fig. **3**).

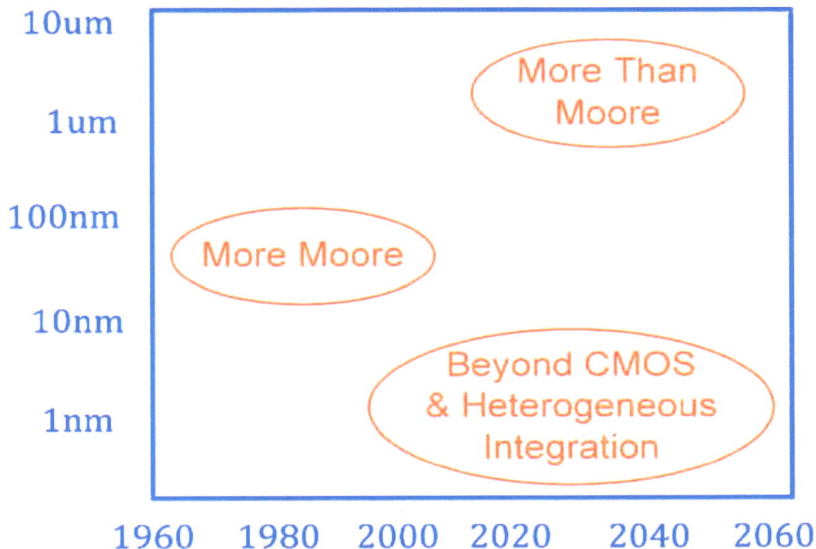

Fig. (1). ENIAC predication for microelectronics future [1].

The designing of electronic systems on a single chip requires optimum exploitation of "system-on-chip" devices, and it will be a vital component of the future of nanoelectronics, integrating "More Moore" as well as "More than Moore" with innovative "heterogeneous integration" technological advances. Another advancement is "system-in-package", which uses different optimized process technologies for combining multiple distinct sub-systems in a single package.

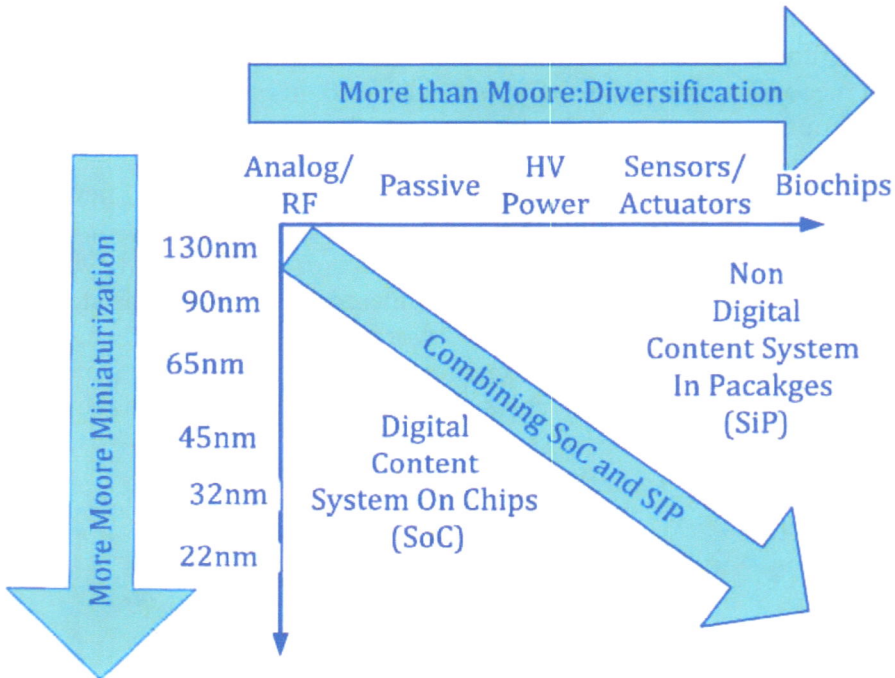

Fig. (2). Moore's law and more [1].

Today's MOSFET is developed at the nanoscale. A tiny chip might contain millions of MOSFETs. When the MOSFET was first designed, it was rather massive and had less functionality than it does now. For decades, engineers have worked to improve MOSFETs' size, power consumption, and other characteristics. As a result, current MOSFET technology has achieved considerable advances. High-performance MOSFETs now have better power handling capabilities and produce less heat. New materials and designs allow quicker switching rates, increasing efficiency in high-frequency circuits. Advanced MOSFETs have reduced on-resistance, allowing for more current flow, and advancements such as silicon carbide (SiC) and gallium nitride (GaN) MOSFETs provide improved performance in high-voltage and high-power applications. Despite advances in current MOSFET technology, numerous limits remain. One key source of worry is the growing demand for power density and miniaturization, which presents issues in heat dissipation and dependability. As MOSFETs reduce size, concerns like leakage current and quantum effects become more noticeable, impacting overall performance and energy efficiency. To solve these restrictions, future research should concentrate on creating novel materials, such as 2D materials or wide-bandgap semiconductors, to increase power management and minimize leakage. Furthermore, optimizing MOSFET design and architecture, researching novel Gate dielectric materials, and improving heat management techniques will be critical for realizing their full potential in next-generation electronics.

A major factor in the continuous expansion of communications, integrated circuit technology, and computing methods has been the progress made in silicon-based CMOS. This continuous enhancement has been mainly driven by its multidimensional scaling, leading to exponential development in performance and density of devices. Economic productivity has consistently increased as a result of the cost-per-function decline with every new technology generation. Modern integrated circuits mostly consist of CMOS transistors because of their scalability as well as special device attributes including isolation from others, significant input resistance, reduced static power dissipation, as well as ease of design and manufacturing. CMOS chips are employed in broadcasting and portable electronics among other applications nowadays [2].

As device scaling approaches physical size limits, the technological cycle slows due to increased process variance, power consumption, and fabrication costs. Several studies have been done to address ever-increasing problems to sustain historical development over the next decade. Table **1** shows the advancement of conventional MOSFETs in a variety of ways. [3-9].

NANODEVICES BEYOND CMOS

Many investigations are being conducted to find a different strategy to keep Moore's law going forward as CMOS transistor dimensional scaling approaches its basic physical limitations. Many substitute logic and memory devices, known as "Beyond CMOS Devices," have formerly been suggested as part of these efforts. These ultra-high densities are integrated electronic computers. At lower supply voltages, these nanodevices use ballistic transport properties and quantum mechanical processes, resulting in relegated power utilization. Because of their tiny size, these devices will likely be used in computers with integrated electronics and ultra-high density. The general overview of promising emerging logic nanodevices, including graphene nanoribbon (GNR) transistors, quantum-dot cellular automata (QCA), single electron transistors (SETs), nanowire (NW), nanosheet transistors and carbon nanotube field-effect transistors (CNTFETs) will be covered in the following sections. These include information on the devices' current development status, basic operating principles, and commercialization challenges. The general overview of promising emerging logic nanodevices, including QCA, GNR transistors, SETs, CNTFETs, NW and nanosheet transistors, will be covered in the following sections. These include information on the devices' current development status, basic operating principles, and commercialization challenges.

Table 1. Shows the advanced CMOS technology with its advantages and disadvantages.

Planar Technology	Advantages	Disadvantages	Diagram
Complementary Metal Oxide Semiconductor (CMOS)	Low static Power Dissipation Robust	Requires More Transistor Latch-Up	
SOI-Based MOSFETs	A layer of oxide is developed before the silicon channel layer. Junction capacitance is absent. Reduced parasitic properties latch-up problem is absent.	Overshooting Drain current Kink effect	

(Table 1) cont.....

	Leakage current is reduced.		
Double Gate (DG) MOSFET Technology	Reduce the short channel effect like lower DIBL. Better Scalability High Drive Current Subthreshold slope near ideal Lower subthreshold current and lower Gate leakage Vt variation elimination resulting from fluctuating random dopants.	For the fabrication, a standard fabrication method is required. An enormous issue is maintaining a thin, uniform channel thickness.	

(Table 1) cont.....

Fin Field Effect Transistor (FinFET)	For the fabrication, a standard fabrication method is required.	FinFET is challenging to manufacture.	
	An enormous issue is maintaining a thin, uniform channel thickness— Quasi planner structure.		
	High drive current.		
	Appropriate for SRAM implementation.		
	Due to the significant current driving capability, switching times are shorter.		

(a) 3D Structure   (b) Cross-sectional View

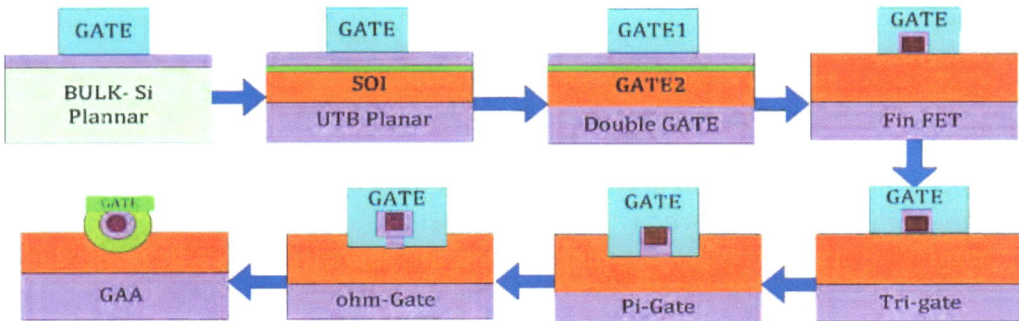

Fig. (3). Advancement of conventional MOSFETs.

Carbon Nanotube Field-effect Transistors (CNTFETs)

Carbon nanotubes also known as CNTs have gained widespread interest from scientists and engineers in a variety of sectors since their discovery in 1991, due to their unusual material characteristics [10-11]. Graphene sheets are rolled into cylinders to form CNTs. Chirality, or the direction in which CNTs are rolled up, determines whether they have metallic or semiconductive properties. The threshold voltage is readily adjusted because the bandgaps of CNTs are negatively correlated with their diameters [12-14]. All-metal nanotubes are appealing as forthcoming interconnects because of their superior material features, including substantial, outstanding mechanical, current carrying capacity, high thermal conductivity, and thermal stability [15].

Semiconductor nanotubes offer major advantages as a channel substrate for efficient (HP) FETs in addition to these characteristics. Because of the absence of dangling bonds, CNTFETs can include high-k dielectrics more easily. Furthermore, because both PMOS and NMOS transistors have nearly equal I-V properties, it greatly benefits the circuit design of CMOS. Moreover, they are highly appealing towards the Silicon-based semiconductor sector because CNTFETs significantly increase device performance parameters such as high speed and low power.

CNT-based electronics show enormous potential, but numerous significant difficulties must be overcome. No extant nanotube synthesis or growth procedures are capable of producing tubes with comparable chiralities and diameters. There have been several reports of purifying advancements. These encouraging preliminary results, however, fall well short of the high level of positioning accuracy that would be necessary. In addition, device fabrication needs more precise gate control via highly abrupt doping profiles and ultrathin high-k gate dielectrics [16].

Graphene Nanoribbon (GNR) Transistor

Graphene, also known as an unrolled carbon nanotube (CNT), is one atomic sheet, comprised of atoms of carbon organized in a two-dimensional (2D) hexagonal structure that has shown enormous promise for nanoelectronics devices [17]. It shares many benefits of metallic CNTs, including high carrier velocity for quick switching, elevated carrier mobility in ballistic transportation, superior heat-transmission efficiency, and more. The capability to manufacture wafer-sized graphene films with entire planar device processing offers robust incorporation possibility with existing CMOS manufacturing techniques, extending a notable benefit over carbon nanotubes [18].

With nearly negligible noise margin and a leakage energy increase of more than five times, dense memory—the greatest potential application for devices based on graphene— is especially prone to differences and defects, even though the GNR material provides faster latency as well as more compact dimensions, and low-energy FETs. These impacts must be carefully considered in future graphene-based electronics technologies' performance evaluation and design optimization.

Single-electron Transistors (SETs)

Since SET devices are small and consume less power at higher latency, they hold great promise for large-scale integration in the future. Three terminals make up the SET structure: drain, source, and gate. On the other hand, the second gate is optional. As can be seen in Fig. (**4**), the SET schematic and the MOSFET schematic are comparable. SET has a small conductive isle that is connected to the gate probe through the capacitance of the gate Cg. The amount of particles on the barrier is set because the growing gate bias only attracts electrons that pass through the source or drain tunnel barriers. When there is a small voltage applied between the drain and source electrodes, the increasing gate bias causes electrons to pass one by one, resulting in the "Coulomb blockade" phenomenon [19-21].

The problem of this device is the limited fan-out; however, this problem is resolved by ground-breaking circuit layouts for example the diagram of binary decisions [22].

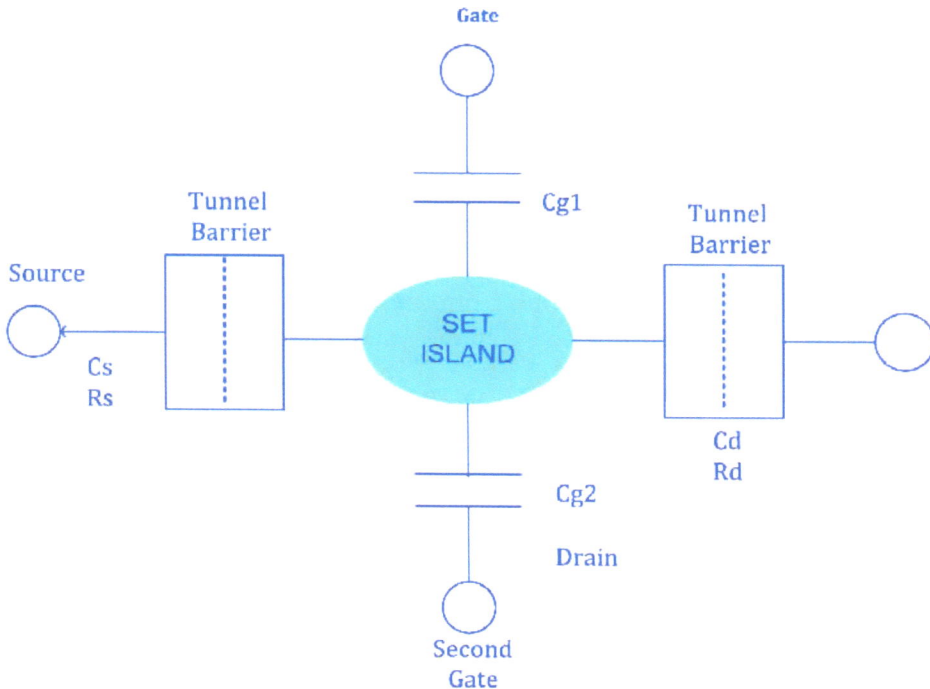

Fig. (4). Schematic of Single electron transistor.

Quantum-dot Cellular Automata (QCA)

Operating speed may be severely limited when the amount of transistors on one chip rises because the heat produced during an alternate cycle is unable to be scattered. There is an innovative concept that acquires smaller feature sizes. Lent et al. [22] introduced QCA, a nanostructure model that uses ranges of related quantum dots for the execution of operations with Boolean logic. The advantage of QCA is in the exceptionally higher densities of packing available given the dots are of minuscule size, the more straightforward connections, and the inadequate product of power and delay [23-24].

Since QCA devices are single-electron devices prone to background charges, they present significant challenges. As of this moment, single electron systems' background charge immunity has no practical remedies. The inability of solid-state QCA systems to function at room temperature is another major drawback of QCA devices [25-26].

Nanowire (NW) Transistors

NWFETs or Nanowire field-effect transistors, additionally referred to as surround gate FETs or FETs with a narrow nanowire channel that is gate-all-around, have received a lot of interest. Its nonplanar design offers enhanced control over electrostatic fields more so than with conventional planar devices over the channel. The ability to develop radial along with axial nanowire heterostructures, their respective positions flat surfaces, their affordable "bottom-up" fabrication design that prevents some manufacture difficulties, and greater carrier movement because of the decrease in dispersion resulting from the crystalline structure are just a few of the important reasons that nanowire studies are gaining traction[27].

The inversion charge, however, moves from surfaces to bulk inversion because of quantum confinement because of the smaller nanowire widths. Defects in the fabrication process that result in variations in the size of nanowires might thereby reduce the properties of charge transport by causing carrier potential and dispersal perturbations. The FET voltage at the threshold may also vary depending on variations in nanowire diameters. One of the main challenges in turning nanowire FETs into a useful technology is minimizing unpredictability. Nanowire transistor modeling is further complicated by quantum confinement phenomena[28].

Nano Sheet

In the complex field of digital circuit development, the Nanosheet FET (NS-FET) has become a revolutionary force that is changing the semiconductor technology landscape. This investigation explores the various uses of NS-FET in digital circuits and clarifies its significant impact on power efficacy, performance, and the overall advancement of digital systems. This ground-breaking technique promises to revolutionize semiconductor capabilities while simultaneously addressing the difficulties associated with circuit design. The impact of NS-FET goes beyond little steps toward progress; it opens the door to a future in which digital systems function with unprecedented performance and increased efficiency. Through its horizontal gate layout, NS-FET technology introduces unmatched channel control, revolutionizing digital circuit processing and enabling quick and effective switching. A comparative study reveals that it performs better than its traditional Fin-FET and NW-FET equivalents, exhibiting improved gate control essential to digital applications [29]. This technological breakthrough represents a significant development in semiconductor design, improving digital circuits' accuracy and speed. The unique qualities of NS-FET improve existing procedures and potentially revolutionize the field of digital applications by raising the bar for performance and efficiency in the rapidly advancing field of semiconductor technology. NS-FET paves the way in reducing power consumption, pioneering a change in digital circuit architecture. Because of its unique characteristics, power requirements are actively reduced, which promotes the creation of energy-efficient systems. NS-FET exemplifies a critical step into sustainable digital technology by reducing power consumption without sacrificing performance through appealing case studies [30]. This ground-breaking development not only solves the escalating issues around power efficiency but also establishes NS-FET as a driving force behind more commercially and environmentally sustainable digital circuitry in the future. The technology sets a standard for a harmonious trade-off involving performance and power efficiency in the digital sphere. NS-FET is at the pinnacle of digital semiconductor innovation, leading to a significant decrease in power consumption. Its distinctive characteristics actively reduce power needs, promoting the creation of systems that use less energy. Enormous case studies demonstrate how NS-FETs can cut power consumption without sacrificing performance, which opens up new possibilities for sustainable digital innovation [31]. This innovation meets the urgent demand for energy efficiency and establishes NS-FET as a leading innovator in developing high-performing, environmentally friendly digital solutions. The technology's ability to balance performance and efficiency heralds a revolutionary leap into a more sophisticated and environmentally friendly era of digital circuit design. Incorporating NS-FET into digital circuits improves signal integrity, reduces heat generation, and strengthens system reliability. These developments work in harmony to address the needs of the modern world and establish a solid basis for future breakthroughs. This integration ensures excellent dependability and effective operation by addressing significant problems in digital circuitry. By minimizing heat problems and improving signal integrity, NS-FET satisfies present needs and paves the way for a highly developed future in which digital systems perform with increased dependability, flexibility, and efficiency. Another application examines NS-FET integration difficulties and investigates design complications, including fabrication subtleties. Researchers are encouraged to look for novel solutions through this inquiry, which offers insightful information about the constantly changing field of digital circuits. By tackling the obstacles related to NS-FET integration, scientists can create space for innovative strategies and techniques to promote progress in semiconductor technology. In addition to helping to overcome present difficulties, this proactive analysis is essential in influencing the direction of electronic circuit development in the future, guaranteeing that NS-FET operates effectively and efficiently when included in complicated systems [32-34].

Issues for nanoscale MOSFETs

Due to small-geometry effects, the following issues arise in channel, Gate, source /Drain, and Bulk shown in Fig. (**5**).

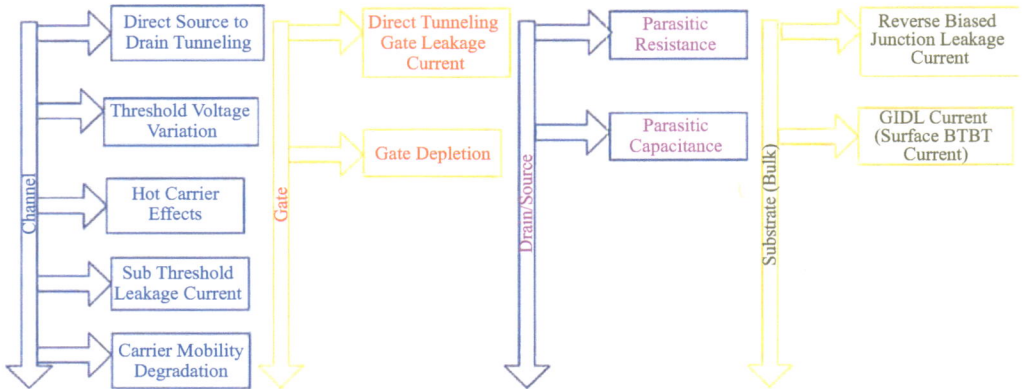

Fig. (5). Issues For Nanoscale.

Solution of Issues

To minimize oxide thickness, short-channel effect and non-uniform and higher doping will be introduced when the devices are entered into a nanometer structure. Fig. (**6**) shows the solution which arises due to the nanostructure.

Nano Device Application

Beyond the silicon CMOS device scaling roadmap: logic devices. The project's scope includes innovative nanoelectronics materials, such as graphene and carbon nanotubes, novel concepts of nanoelectromechanical (NEM) relays, as well as device physics, modelling, circuit design, and device fabrication. Fig. (**7**), presents different types of nanoelectronics devices [35-40].

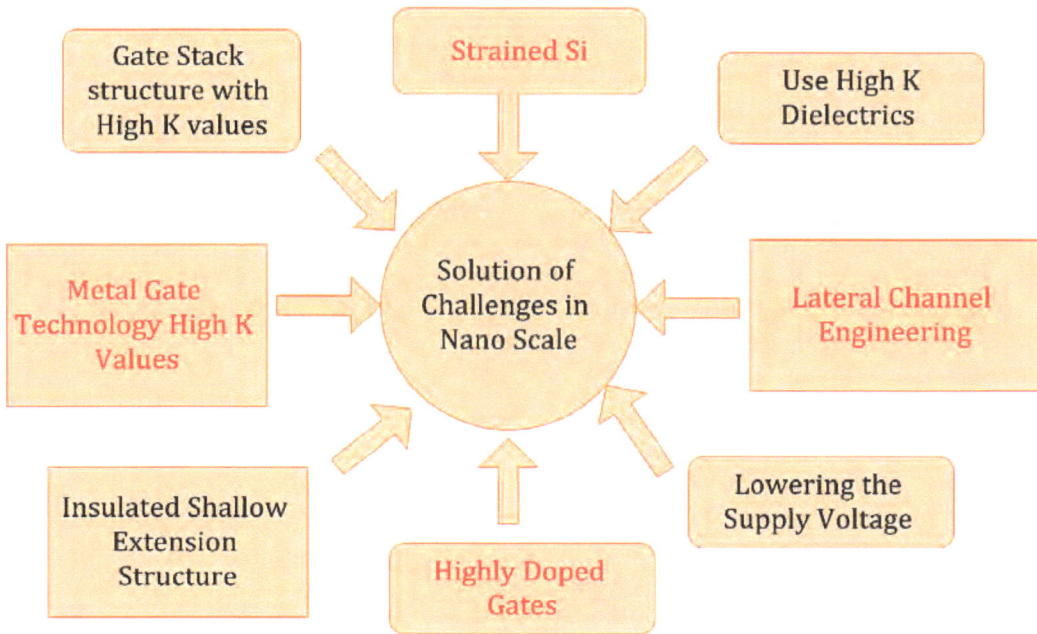

Fig. (6). Solutions of challenges in Nano Scale.

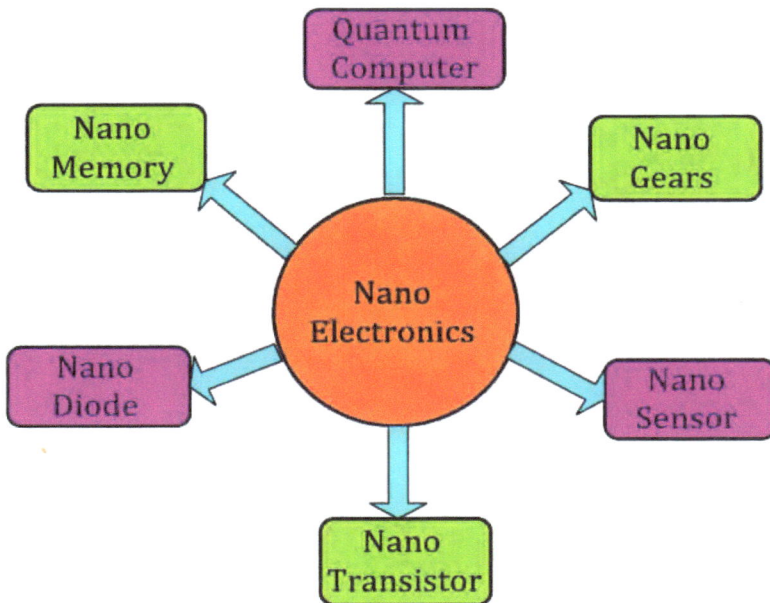

Fig. (7). Nano Device Examples.

Nanoelectronics Transistors

While nanotechnology has already been subtly integrated into current high-technology production processes employing nanoelectronic transistors, these processes are still dependent on conventional top-down approaches. Computer processors could potentially become more potent with nanoelectronics than is currently conceivable with traditional semiconductor production methods. Numerous strategies are being investigated right now, such as novel types of nanolithography and the substitution of conventional CMOS components with nanomaterials like nanowires or tiny molecules.

Both heterostructured nanowires and semiconducting carbon nanotubes have created field-effect transistors. The MOSFET transistor with an 18 nm diameter was tested in 1999 by a CMOS transistor designed at the Grenoble, a France-based Laboratory for Electronics and Information Technology. In this design, about 70 atoms were arranged side by side. This was nearly ten times smaller than the smallest industrial transistor in 2003 (130 nm in 2003, next 90 nm in 2004, also 65 nm in 2005, and 45 nm in 2007). Theoretically, to fit seven billion connections on a coin of 1 Euro was made feasible. In 1999, CMOS transistor was invented. It demonstrated the function of CMOS technology when operated on a molecular scale. Executing the same on an industrial scale and controlling the coordinated assembly of these transistors on a circuit have been some challenging tasks. Fig. (**8**) shows a transistor and a nano transistor.

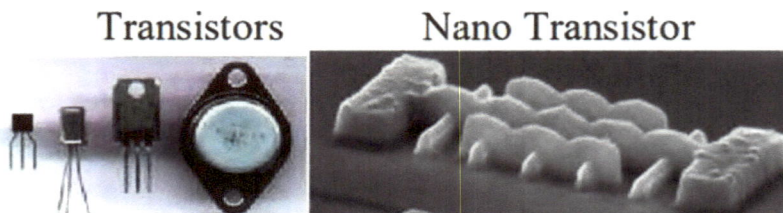

Fig. (8). Example of Transistor and Nano Transistor.

Nano Memory (Spintronic)

The study and application of magnetic moment and electric charge with electron spin in solid-state devices is known as spintronics.

Spintronics are electronic components that execute logic operations based on a carrier's spin and electrical charge. For example, information could be transferred or stored by electrons in their spin-up or spin-down states. The injection, transport, and detection of spin-polarized carriers are among the problems in this recent field of study. Currently under investigation are the effects of spin injection on the electrical framework and nanoscale order of the ferromagnetic-semiconductor interface, the controlled synthesis of new ferromagnetic semiconductors, and the potential use of nanostructured features for spin manipulation.

Nano-Sensors

Numerous products, exciting disciplines, and uses for photonic sensors are made possible by nanotechnology. Existing applications can be improved, such as digital cameras view accommodation of more pixels on a sensor compared to what is currently possible. Additionally, sensors can be created at the nanoscale, improving their quality and possibly eliminating defects. Ultimately, this would lead to larger, more precise pictures. A communication network will use photonic sensors to transform optical data into electricity (*i.e.* photons to electrons transformation). Photonic sensors made at the nanoscale will be more effective and benefit from the same advantages as other nanoscale materials [41].

The progress in Nano-FETs faces significant obstacles due to the precision required for miniaturization and the challenges in fabricating reliable devices. Carbon Nanotube FETs (CNTFETs) are promising due to their superior properties but struggle with variability in nanotube chirality and width, leading to inconsistencies in device performance and manufacturing challenges despite advancements in purification processes.

1. Graphene Nanoribbon (GNR) silicon transistors offer high carrier mobility and fast switching speeds but face challenges with higher leakage currents and noise sensitivity due to fabrication variability, complicating their integration into existing CMOS processes and necessitating novel strategies to manage these issues while maintaining performance benefits.

2. Single-Electron Transistors (SETs) offer low power consumption and potential for large-scale integration but face challenges such as limited fan-out, sensitivity to background charge effects, and difficulties in controlling the Coulomb blockade effect at the nanoscale.

3. Quantum-dot Cellular Automata (QCA) promise higher density and lower power consumption but face challenges such as susceptibility to background charge interference and limited effectiveness at room temperature, underscoring the need for ongoing research to enhance their robustness and temperature tolerance.

4. Nanowire transistors (NWFETs) face challenges such as quantum confinement effects and dimensional variability, which impact charge transport and device reliability, necessitating advancements in fabrication techniques for uniformity and precise control.

5. Nanosheet FETs (NS-FETs) promise enhanced power efficiency and performance with their improved channel control, integrating them into existing technologies presents challenges in fabrication, material compatibility, and system design.

Therefore, advancing Nano-FET technologies demands overcoming fabrication, material, and integration challenges through multidisciplinary advancements to fully realize their potential in future electronic systems.

CONCLUSION

The development of an alternative technique to uphold Moore's law as the geometrical scalability of CMOS transistors reaches its fundamental limits has been the focus of numerous investigations. As part of these attempts, several "beyond CMOS" products, such as logic devices and alternative memory, have been proposed. These devices use ballistic transport properties and quantum mechanical processes at reduced supply voltages, resulting in low power consumption. It is also expected that these tiny devices will find application in ultra-density combined computing devices.

Even though nanoelectronics may combine hundreds of millions of electronics into a single system, it additionally leads to imperfections and variations during fabrication and chip operations. Thus, in addition to the traditional high-latency, low-power, and high-density design considerations, there has to be another limitation called "reliability." In order for the industry to forecast, maximize, and development of nanoelectronic reliability and performance in advance, this additional constraint requires the creation of a novel structure for understanding and dependability in nanoelectronics. A broad field including devices, connects, electronic design, computer-aided design, and manufacturing issues should be the focus of the new paradigm.

Learning Objectives/Key Points

(1) Focuses on shrinking transistor sizes by integrating new materials like high-k oxides, improving performance, and reducing leakage to sustain Moore's Law.

(2) Expands CMOS capabilities by adding functionalities (analog, RF, sensors) through diverse technologies, enabling complex systems beyond digital logic for broader applications.

(3) Explores new materials and principles (nanowires, spintronics) to overcome CMOS scaling limits, aiming for breakthroughs in power efficiency and computational performance at the nanoscale.

(4) NS-FETs revolutionize semiconductor technology with superior gate control and efficiency, significantly enhancing digital circuits' performance and energy efficiency.

(5) Spintronics introduces quantum mechanical properties for denser, faster memory solutions, representing a significant shift from traditional electronic memory devices.

(6) Nanosensors, leveraging nanoscale material properties, promise revolutionary improvements in sensitivity and efficiency for applications like environmental monitoring and medical diagnostics.

Multiple Choice Questions (MCQs)

1. What property allows carbon nanotubes (CNTs) to be considered for use as channel material in high-performance field-effect transistors (FETs)?

a) Their ability to exhibit either metallic or semiconductive characteristics based on chirality.

b) Their uniform diameter across all types.

c) Their inability to integrate with high-k dielectrics.

d) Their high power consumption.

Answer: a) Their ability to exhibit either metallic or semiconductive characteristics based on chirality.

2. Which of the following is NOT a challenge faced by graphene nanoribbon (GNR) transistors?

a) Integration with current CMOS manufacturing techniques.

b) Variations and flaws leading to near-zero noise margins.

c) Leakage power increase by more than 5 times.

d) Low carrier mobility.

Answer: d) Low carrier mobility.

3. What phenomenon characterizes the operation of a Single-electron transistor (SET)?

a) Coulomb blockade.

b) Quantum tunneling without restrictions.

c) High fan-out capability.

d) High power consumption.

Answer: (a) Coulomb blockade.

4. Quantum-dot cellular automata (QCA) are known for their:

a) High operating temperatures.

b) Immunity to background charges.

c) Exceptionally high packing densities.

d) Complex connections.

Answer: (c) Exceptionally high packing densities.

5. Which feature gives nanowire transistors (NWFETs) superior electrostatic control over the channel?

a) Their planar design.

b) The gate-all-around (or surround gate) design.

c) Their high power consumption.

d) Large variations in nanowire sizes.

Answer: (b) The gate-all-around (or surround gate) design.

6. What is a significant benefit of using carbon nanotubes (CNTs) as future interconnects?

a) Their low current carrying capacity.

b) The requirement for uniform chirality and diameters.

c) Their high thermal conductivity and stability.

d) Their compatibility with low-k dielectrics.

Answer: (c) Their high thermal conductivity and stability.

7. The main challenge in commercializing graphene nanoribbon (GNR) transistors is:

a) Their high cost of production.

b) Difficulty in achieving wafer-scale uniformity.

c) The impact of variations and defects on device performance.

d) Their inherently high static power dissipation.

Answer: (c) The impact of variations and defects on device performance.

8. What is the primary focus of Nanosheet FET (NS-FET)?

a) Heat reduction

b) Increased channel width

c) Enhanced gate control

d) Lower signal integrity

Answer: (c) Enhanced gate control

9. What distinguishes NS-FET from traditional Fin-FET and NW-FET?

a) Reduced gate control

b) Inferior performance

c) Horizontal gate layout

d) Limited channel control

Answer: (c) Horizontal gate layout

10. How does NS-FET contribute to power efficiency?

a) By increasing heat generation

b) By reducing energy consumption

c) By decreasing signal integrity

d) By amplifying system reliability

Answer: (b) By reducing energy consumption

11. What does NS-FET integration aim to improve?

a) Signal distortion

b) Heat generation

c) Fabrication complexities

d) System reliability

Answer: (D) System reliability

12. Which field does spintronics primarily focus on?

a) Electron charge

b) Electron spin

c) Semiconductor interfaces

d) Magnetic moment

Answer: (b) Electron spin

13. What is the potential application of nanoscale photonic sensors?

a) Increasing defects

b) Reducing pixel count

c) Improving picture quality

d) Decreasing sensor size

Answer: (c) Improving picture quality

14. What is the main challenge addressed by "beyond CMOS" devices?

a) Increased power consumption

b) Reduced supply voltages

c) Overcoming manufacturing flaws

d) Approaching fundamental scaling limits

Answer: (d) Approaching fundamental scaling limits

15. What additional constraint is introduced in nanoelectronics besides high-latency and low-power design considerations?

a) Performance optimization

b) Chip manufacturing

c) Reliability

d) System integration

Answer: (c) Reliability

REFERENCES

[1] K. Salah, "More than Moore and beyond CMOS: New interconnect schemes and new circuits architectures", *2017 IEEE 19th Electronics Packaging Technology Conference (EPTC)*, 2017, pp. 1-6

[2] H. Iwai, "CMOS technology after reaching the scale limit", *Extended Abstracts-2008 8th International Workshop on Junction Technology (IWJT'08)*, 2008pp. 1-2

     http://dx.doi.org/10.1016/S0039- 6028(01)01558-8

[3] D.S. Yadav, P. Singh, and P. Roat, "Assessing the Impact of Source Pocket Length Variation to Examine DC/RF to Linearity Performance of DG-TFET", *Nano,* vol. 18, no. 04, p. 2350027, 2023.

[4] A. Anam, S. I. Amin, D. Prasad, N. Kumar, and S. Anand, "Analysis of III-V material-based dual source T-channel junction-less TFET with metal implant for improved DC and RF performance", *Micro and Nanostructures,* p. 207629, 2023.

[5] P. Singh, and D.S. Yadav, "Impact of work function variation for enhanced electrostatic control with suppressed ambipolar behavior for dual gate L-TFET", *Curr. Appl. Phys.,* vol. 44, pp. 90-101, 2022.

[6] A. Raman, K.J. Kumar, D. Kakkar, R. Ranjan, and N. Kumar, "Performance investigation of source delta-doped vertical nanowire TFET", *J. Electron. Mater.,* vol. 51, no. 10, pp. 5655-5663, 2022.

[7] P. Singh, and D. S. Yadav, "Design and investigation of F-shaped tunnel FET with enhanced analog/RF parameters", *Silicon,* pp. 1-16, 2021.

[8] P. Singh, and D.S. Yadav, "Assessing the Impact of Drain Underlap Perspective Approach to Investigate DC/RF to Linearity Behavior of L-Shaped TFET", *Silicon,* vol. 14, no. 17, pp. 11471-11481, 2022.

[9] N. Kumar, and A. Raman, "Design and analog performance analysis of charge-plasma based cylindrical GAA silicon nanowire tunnel field effect transistor", *Silicon,* vol. 12, no. 11, pp. 2627-2634, 2020.

[10] S. Iijima, "Helical microtubules of graphitic carbon," *Nature*, vol. 354, pp. 56–58, 1991.

      http://dx.doi.org/10.1038/354056a0

[11] M.S. Dresselhaus, and P. Avouris, "Introduction to carbon materials research", In: *Carbon nanotubes: synthesis, structure, properties, and applications.* Springer Berlin Heidelberg: Berlin, Heidelberg, 2001, pp. 1-9.

[12] P. Avouris, J. Appenzeller, R. Martel, and S.J. Wind, "Carbon nanotube electronics", *Proc IEEE,* vol. 9, no. 11, pp. 1772-1784, 2003.

http://dx.doi.org/10.1109/JPROC.2003.818338

[13] N. Srivastava and K. Banerjee, "Performance analysis of carbon nanotube interconnects for VLSI applications," In: *Proc. 2005 IEEE/ACM Int. Conf. Comput.-Aided Design (ICCAD)*, 2005, pp. 383–390.

[14] V.V. Zhirnov, J.A. Hutchby, G.I. Bourianoff, and J.E. Brewer, "Emerging research logic devices", *IEEE Circuits Devices,* vol. 21, no. 3, pp. 37-46, 2005.

http://dx.doi.org/10.1109/MCD.2005.1438811

[15] J. Appenzeller, "Carbon nanotubes for high-performance electronics—Progress and prospect," *Proc. IEEE*, vol. 96, no. 2, pp. 201–211, 2008.

http://dx.doi.org/10.1109/JPROC.2007.911051

[16] K. S. Novoselov, A. K. Geim, S. V. Morozov, D. E. Jiang, Y. Zhang, S. V. Dubonos, and A. A. Firsov, "Electric field effect in atomically thin carbon films", *Science,* vol. 306, no. 5696, pp. 666-669, 2004.

[17] W.A. De Heer, C. Berger, E. Conrad, P. First, R. Murali, and J. Meindl, "Pionics: The emerging science and technology of graphene-based nanoelectronics", In: *2007 IEEE International Electron. Devices Meeting.* IEEE, 2007, pp. 199-202.

[18] G.V. Angelov, D.N. Nikolov, and M.H. Hristov, "Technology and modeling of nonclassical transistor devices", *J. Electr. Comput. Eng.,* vol. 2019, pp. 1-18, 2019.

http://dx.doi.org/10.1155/2019/4792461

[19] K.K. Likharev, "Single-electron devices and their applications", *Proc IEEE,* vol. 87, no. 4, pp. 606-632, 1999.

http://dx.doi.org/10.1109/5.752518

[20] C. Wasshuber, *Computational Single-Electronics.* Springer: Wien, New York, 2001.

[21] N. Asahi, M. Akazawa, and Y. Amemiya, "Single-electron logic device based on the binary decision diagram", *IEEE Trans Electron. Dev.,* vol. 44, no. 7, pp. 1109-1116, 1997.

http://dx.doi.org/10.1109/16.595938

[22] C.S. Lent, and P.D. Tougaw, "Lines of interacting quantum-dot cells: A binary wire", *J. Appl. Phys.,* vol. 74, no. 10, pp. 6227-6233, 1993.

http://dx.doi.org/10.1063/1.355196

[23] P.D. Tougaw, and C.S. Lent, "Logical devices implemented using quantum cellular automata", *J. Appl. Phys.,* vol. 75, no. 3, pp. 1818-1825, 1994.

[24] C.S. Lent, P.D. Tougaw, W. Porod, and G.H. Bernstein, *Nanotechnology,* vol. 4, p. 49, 1993.

http://dx.doi.org/10.1088/0957-4484/4/1/004

[25] G.L. Snider, A.O. Orlov, V. Joshi, R.A. Joyce, H. Qi, and K.K. Yadavalli, "Electronic quantum-dot cellular automata", *2008 9th International Conference on Solid-State and Integrated-Circuit Technology,* , 2008pp. 549-552

[26] V.V. Zhirnov, J.A. Hutchby, G.I. Bourianoffls, and J.E. Brewer, "Emerging research logic devices", *IEEE Circuits and Devices Magazine,* vol. 21, no. 3, pp. 37-46, 2005.

[27] J. Kandpal, and G. Rawat, "Role of Nanotechnology in Nanoelectronics", *Nanoelectronics Devices: Design, Materials, and Applications,* no. Part I, p. 1, 2023.

[28] E. Goel, A. Pandey, Eds., *Nanoscale Field Effect Transistors: Emerging Applications.* 2023.

[29] J. Ajayan, D. Nirmal, S. Tayal, S. Bhattacharya, L. Arivazhagan, A.A. Fletcher, P. Murugapandiyan, and D. Ajitha, "Nanosheet field effect transistors-A next generation device to keep Moore's law alive: An intensive study", *Microelectronics J.,* vol. 114, p. 105141, 2021.

[30] V.B. Sreenivasulu, and V. Narendar, "Design insights of nanosheet FET and CMOS circuit applications at 5-nm technology node", *IEEE Trans Electron. Dev.,* vol. 69, no. 8, pp. 4115-4122, 2022.

[31] A. Rahimifar, and Z. Ramezani, "Scope and challenges with nanosheet FET-based circuit design", In: *Device Circuit Co-Design Issues in FETs.* CRC Press, pp. 161-180.

[32] F.I. Sakib, M.A. Hasan, and M. Hossain, "Performance analysis of nanowire and nanosheet NCFETs for future technology nodes", *Engineering Research Express,* vol. 3, no. 4, p. 045044, 2021.

[33] S. Valasa, S. Tayal, L. R. Thoutam, J. Ajayan, and S. Bhattacharya, "A critical review on performance, reliability, and fabrication challenges in nanosheet FET for future analog/digital IC applications", *Micro and Nanostructures,* p. 207374, 2022.

[34] A. Bisht, Y. P. Pundir, and P. K. Pal, "Nanosheet Transistor with Inter-bridge Channels for Superior Delay Performance: A Comparative Study", *Silicon,* pp. 1-11, 2023.

[35] H. Han, C.H. Kim, and S. Jung, "Vertical integration: a key concept for future flexible and printed electronics", *Flexible and Printed Electronics,* vol. 7, no. 2, p. 023003, 2022.

[36] R.W. Whatmore, "Nanotechnology--what is it? Should we be worried?", *Occup Med. (Lond),* vol. 56, no. 5, pp. 295-299, 2006.

PMID: 16868126

[37] M.H. Fulekar, *Nanotechnology: importance and applications.* I.K. International Pvt Ltd., 2010.

[38] T.T. Tran, and A. Mulchandani, "Carbon nanotubes and graphene nano field-effect transistor-based biosensors", *Trends Analyt Chem.,* vol. 79, pp. 222-232, 2016.

[39] J. Grollier, D. Querlioz, and M.D. Stiles, "Spintronic nanodevices for bioinspired computing", *Proc IEEE,* vol. 104, no. 10, pp. 2024-2039, 2016.

PMID: 27881881

[40] B.A. Parviz, D. Ryan, and G.M. Whitesides, "Using self-assembly for the Fabrication of nanoscale electronic and photonic devices", *IEEE Trans Adv. Packag,* vol. 26, no. 3, pp. 233-241, 2003.

[41] Y. Zhou, M. Zhang, Z. Guo, L. Miao, S.T. Han, and Z. Wang, "Recent advances in black phosphorus-based photonics, electronics, sensors and energy devices", *Mater. Horiz,* vol. 4, no. 6, pp. 997-1019, 2017.

CHAPTER 7

Analysis and Device Physics of HTFET-based 14T SRAM for Next-Generation Memory Excellence

B.V.V Satyanarayana[1*], M. Parvathi[2], G. Prasanna Kumar[1], A.K.C Varma[1], T.S.S Phani[3] and T. Saran Kumar[3]

[1]*Department of ECE, Vishnu Institute of Technology, Bhimavaram, Andhra Pradesh, India*

[2]*Department of Electronics and Communication Engineering, BVRIT Hyderabad College of Engineering for Women, Hyderabad, Telangana, India*

[3]*Department of ECE, Bonam Venkata Chalamayya Engineering College (A), Odalarevu, Andhra Pradesh, India*

Abstract: In this chapter, we address the limitations of device scaling imposed by the subthreshold value restriction of 60mV/decade in the CMOS VLSI design. The focus of current research primarily revolves around effective power methods for cutting-edge electronic devices with additional attributes. Instead of conventional homo-junction MOS devices, our investigation explores the utilization of heterojunctions with SiGe and Ge as these materials have a lower bandgap. By employing a Heterojunction Tunneling Field Effect Transistor (HTFET), we demonstrate a reduction in the subthreshold swing value and achieve low leakage current. We present a revolutionary HTFET design with Gate Oxide Overlap onto Source (GOS) to improve the futuristic features of low-power devices for ultra-low-power memory applications. We implement both n-type and p-type GOS HTFETs, contributing to energy-efficient SRAM cells, by combining low bandgap materials such as SiGe or Ge with high-k dielectrics. The suggested devices show large improvements in Miller capacitance together with a noteworthy decrease in subthreshold swing, high current ratios from ON to OFF, and an increased drive current proportion in the ON state. Expanding the application scope, the proposed device is integrated into a radiation-hardened 14T SRAM cell, showcasing superior performance compared to traditional designs. Memory activities are accelerated, and the chapter concludes with a comparative power and delay analysis of HTFETs-based SRAM cells.

Keywords: BTBT, heterojunctions, TFETs, subthreshold swing, SiGe/Ge HTFETs, 14T SRAM.

---

* Corresponding author B.V.V Satyanarayana: Department of ECE, Vishnu Institute of Technology, Bhimavaram, Andhra Pradesh, India; Tel: +91 9908891456; E-mail: vvsatya.b@gmail.com

Dharmendra Singh Yadav & Prabhat Singh (Eds.)

INTRODUCTION

The design and implementation of a low-power SRAM cell specifically suited for ultra-low-power semiconductor memory applications are the focus of this chapter. The main goal is to investigate heterojunction tunnel field-effect transistor (HTFET) topologies at the device level, paying particular attention to important variables including Miller capacitance, ON-state driving current, subthreshold swing, and ON-OFF current ratio. We hope to provide a low-power, low-voltage device that satisfies the requirements of modern semiconductor memory applications by carefully examining these factors [1].To evaluate the efficacy of the proposed low-power device in practical scenarios, various SRAM cells are essential for circuit-level testing. Consequently, an exploration of different SRAM cell structures suitable for efficient memory devices becomes imperative [2-3]. This chapter elucidates the physical features of the device necessary for optimizing energy-efficient devices and provides insights into various SRAM cell configurations.

Additionally, this chapter emphasizes the significance of investigating HTFET structures as a pivotal component of the low-power SRAM cell design. The choice of materials, particularly SiGe and Ge, is explored for their low bandgap properties, offering a promising avenue for subthreshold swing reduction below the conventional 60mV/decade limit [4-5]. The incorporation of Gate Oxide Overlap onto the Source (GOS) further enhances device performance, contributing to the achievement of sub-60mV/decade subthreshold swing values and lower leakage currents. Through detailed simulations and validation using TCAD Silvaco tools, the proposed HTFET designs are thoroughly evaluated at the device level, ensuring their suitability for energy-efficient semiconductor memory applications [6-7].

Furthermore, the practical application of the designed low-power SRAM cell is demonstrated through the implementation of a radiation-hardened 14T SRAM cell. Comparative analyses with traditional designs, including 6T, 7T, and 8T memory cells, highlight the superior performance of the proposed device. The read-and-write delays of these memory cells are measured, showcasing accelerated memory activities [8]. Finally, a comprehensive comparison with other state-of-the-art devices, such as conventional HTFET and others is presented. This comprehensive exploration not only advances the understanding of low-power semiconductor memory but also propels the discourse on the practical application of innovative HTFET designs in memory technology [9].

The subsequent sections delve into the comprehensive research findings on Heterojunction Tunneling Field Effect Transistors (HTFETs) and their integration with SRAM cells. Through detailed discussions and analyses, we unveil the intricacies of our research work, shedding light on the promising advancements in the realm of low-power semiconductor memory technologies.

Analysis of Traditional HTFETs

Heterojunctions emerge as a highly suitable choice for low-leakage memory applications, particularly when integrated into SRAM cells [10]. Substituting homojunctions with heterojunctions in SRAM cells results in a noteworthy reduction in power consumption and delay during read and write operations. Various techniques involving heterojunctions have been proposed by researchers to enhance the performance of these memory cells [11-13]. The structure of traditional HTFET is shown in Fig. (1). The concept of HT Fin FETs, employing a double-gate configuration was introduced in a study [14]. This innovative design optimizes the surface potential, considering the region of overlap. Utilizing the principle of superposition by solving 2D surface potential Poisson's equations, the model is validated with the introduction of HfO_2 as high-k dielectric. Operating with a fixed output voltage of 0.7V and input voltages ranging from 0V to 0.6V, this approach showcases improved power efficiency.

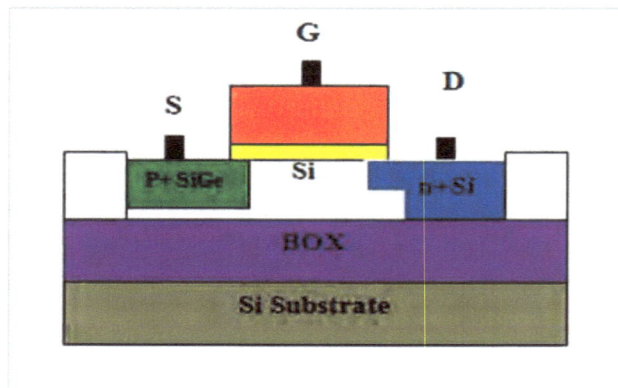

Fig. (1). 2D Structure of traditional HTFET.

The Gate-on-Source-Channel SOI TFETs (GOSC TFET), addressing the reduction in Miller capacitance for enhanced switching speed is proposed in a study [15]. This device, with an input voltage of 0.4V, provides a comprehensive analysis of parameters including output voltage, thickness of oxide, carrier concentration level, and input voltage, validated through TCAD. The SiGe/Si heterojunction n-TFET was introduced, employing HfO_2/SiO_2 as the gate oxide material and silicon-germanium as the source [16]. The proposed system's analysis covers parameters such as OFF-state current, power dissipation, ON-state current, subthreshold slope, and switching delay, showcasing promising results with an input voltage range from 0V to 2.5V.

The SiGe Ferroelectric Schottky Barrier FET (SiGe Fe SBFET), employing an LBG material was proposed [17]. The ferroelectric SBFET enhances the performance of germanium and silicon materials, providing lower SS and a high current ratio. Simulation using Silvaco Atlas Simulator validates the design, with an input voltage of 0.6V. These diverse heterojunction-based approaches contribute significantly to advancing the efficiency of low-power semiconductor memory technologies. An innovative approach with the Double Gate Isosceles Trapezoid Tunnel FET, aiming to mitigate leakage currents by enhancing the drive current was introduced [18]. Employing a SOI substrate, this device is optimized for improved characteristics intransient and DC analysis. The proposed system, analyzed using TCAD simulation, addresses various currents, and output capacitance with less delay of 1nsec. The incorporation of a Si-Ge heterojunction enhances performance at the source-channel junction.

A study investigated Si/SiGe materials-based HTFETs circuit designs [19]. They highlighted the key distinctions between conventional MOSFETs and HTFETs, and proposed a 7T SRAM cell using HTFETs to overcome the limitations of traditional devices, resulting in a substantial reduction in leakage power. TCAD tools facilitated the development of device characteristics, further validated through Verilog-A simulation. Limitations such as increased Miller capacitance and asymmetric current flow were acknowledged. The HTFET is presented by introducing a pocket region for vertical tunneling featuring steep SS with good drive current [20] Utilizing wide bandgap materials in the channel, the Vertical Tunneling Field-Effect Transistor (VTFET) exhibited better current ratio, steeper SS, and best scaling possibility at an operating voltage of 0.4V to evaluate performance, emphasizing optimization opportunities for heterojunction band offsets, pocket dimensions, length, and doping profiles.

The physics of tunneling phenomena and the device physics of TFETs are presented in [21]. Comprehensive explanations of TFET characteristics, coupled with TCAD tool-based modeling and simulation, form the foundation of their work. The authors tackle the intricate task of solving Poisson's equation based on TFET boundary conditions, discussing various heterostructures and highlighting design challenges in their implementation. This collective body of research underscores the diverse avenues explored in the realm of heterojunction TFETs, showcasing their potential to revolutionize semiconductor technology.

In the pursuit of advancing semiconductor technology, numerous innovative models of Tunnel Field Effect Transistors (TFETs) have been developed, each offering unique advantages for specific applications. Biswas et al. [22] introduced a novel TFET model with a source overlapping the gate stack, tailored for nano-FETs with less operating voltages, particularly designed in non-volatile memories. The use of heterostructures involving InAs-Si nanowires further enhances TFET development. An alternative approach with an L-shaped TFET, focusing on improving BTBT through overlapping the source (S), gate (G), and channel regions is explored in a study [23]. The geometrical Quantum Confinement Effect (QCE) plays a crucial role in the thin layer of channel overlapping, creating distinct conduction and valence bands in the overlapped region. However, the distinct behavior of the conduction band impacts BTBT, leading to a reduction in the ON-state drive current.

An innovative heterojunction TFET employing nitride material, demonstrating the significance of proper Well size selection in engineering the nitride has also been presented [24]. The Alloy Engineered Nitride TFET (AEN-TFET) provides the reduced well size, resulting in improved behavior and enhanced current drive capability. This design offers high speed and low energy consumption, making it suitable for low-power systems. A T-shaped SOI TFET where the asymmetric structure of the source (S) and drain (D) regions suppresses amplifier switching [25]. This design achieves a high switching ON to OFF current ratio, and the integration of TFET with MOSFET allows for the implementation of an efficient controlling device. Furthermore, the device generates a pair of sub-threshold work areas, rendering it suitable for use with voltage sensors.

The concept of Gate-All-Around was introduced in DMATFETs using Ferroelectrics [26] as FE-DMGAA-TFETs are used for better drain current in the design of energy-efficient semiconductor memories. The negative capacitance in these devices enhances drain current and strengthens surface potential, offering a unique solution to address the limitations of other models. Recognizing the restriction of MOSFETs in semiconductor memory applications, TFETs have emerged as alternative devices [27]. However, challenges such as a good current ratio with an excellent improvement in Miller capacitance persist. To address these challenges, researchers focus on developing heterojunction structures using low bandgap materials, offering higher ON current and lower leakage current [28].

These heterostructures present themselves as viable alternatives to MOSFETs, particularly the best choice for semiconductor memory applications. The remaining sections delve into the specifics of various heterostructures, exploring their device physics and their characteristics.

Heterojunction Vertical TFETs (HVTFETs)

Heterojunction Vertical Tunnel Field-Effect Transistors (HVTFETs) exhibit a significantly lower subthreshold swing of 16mV/decade with an operating voltage of 0.4V in comparison to traditional TFETs, which typically yields a value of 60mV/decade [29]. This substantial reduction in SS is instrumental for scaling w.r.t. the voltage and power of the device. Leveraging wide bandgap materials, VTFETs, and specifically HVTFETs, enhance the switching characteristics. The incorporation of wide bandgap materials results in improved switching characteristics for VTFETs. The expansion in the area of tunneling contributes to heightened drive current and diminished leakage current in standby mode. This increase in the area of tunneling is facilitated by the introduction of a halo or pocket region, which is heavily doped and strategically positioned in TFETs. The cross-sectional view of the HVTFET is illustrated in Fig. (2). Optimizing device parameters further enhances the performance characteristics of HVTFETs. Table **1** provides a comprehensive overview of the device parameters associated with heterojunction VTFETs, shedding light on the nuanced features that contribute to their improved performance.

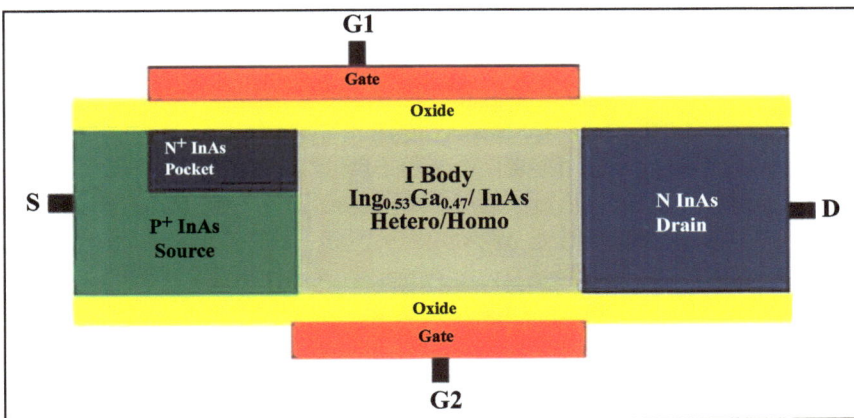

Fig. (2). 2D structure of HVTFET.

Table 1. HVTFET Device Physics.

Parameter	Value
Source	InAs(P$^+$)
Drain	InAs(N)
Channel	InGaAs
Pocket material	InAs(N$^+$)
Source doping	$5 \times 10^{19} \text{cm}^{-3}$
Pocket doping	$5 \times 10^{19} \text{cm}^{-3}$
Drain doping	$5 \times 10^{18} \text{cm}^{-3}$
Channel doping	$1 \times 10^{15} \text{cm}^{-3}$
Dielectric k value	15.4
Body Thickness	10nm
Bandgap of InAs	0.5437eV
Bandgap of In$_{0.53}$Ga$_{0.47}$	0.9130eV
EOT	1.2nm
Channel length	20nm
Pocket Area	10nm X 3.6nm
Supply Voltage/Operating Voltage(V$_D$)	0.4 V

Heterojunction Vertical BTBT transistors, particularly Heterojunction Vertical Tunnel Field-Effect Transistors (HVTFETs), exhibit a remarkably steep subthreshold swing of 26mV/decade at a voltage of 0.4V. This SS value is notably smaller than the typical value observed in conventional devices at 27^0C. The design intricacies of the TFET include a gate-to-source overlap area featuring a pocket region, characterized by a heavily doped halo. This additional pocket region plays a pivotal role in extending the tunneling area of the TFET. Consequently, the device achieves a higher ON-state current while concurrently reducing the leakage current in the off-state. The incorporation of the heavily doped halo amplifies the tunneling characteristics, allowing HVTFETs to surpass traditional MOSFETs in terms of subthreshold swing, offering a promising avenue for advanced semiconductor applications.

Heterojunction Intra-band TFETs (HIBTFETs)

Heterojunction intra-band tunnel (HIBT) Field-Effect Transistors (FETs) emerge as a robust solution for applications demanding no drift in output and a significant reduction in leakage at low V$_{DD}$, specifically at V$_{DD}$ = 0.4V, making them a preferred choice in Static Random Access Memories (SRAMs) [30]. These HIBTFETs exhibit minimal variations in the current ratio as V_{DD} undergoes variations. The device configuration employs silicon (Si) for the source/drain regions, while the channel incorporates Gallium Phosphate (GaP) with different band offsets. At the interface between the source/drain and the channel, intra-band tunneling occurs, driven by the disparate band offsets. Importantly, this intra-band tunneling exclusively takes place if the device is ON and contributing to the HTFET's bidirectional nature. Modulating the device geometry offers a pathway for assessing the sensitivity of HIBTFETs, ensuring symmetry in drain current behavior. Illustrative examples include Si-GaP, Ge-GaAs, and Si-GaAs HIBTFETs, with Fig. (**3**) showcasing the Si-GaP HIBTFET. A comprehensive overview of device parameters is presented in Table **2**, highlighting the device's superior characteristics, notably lower variations in current ratios across supply voltage (VDD) fluctuations.

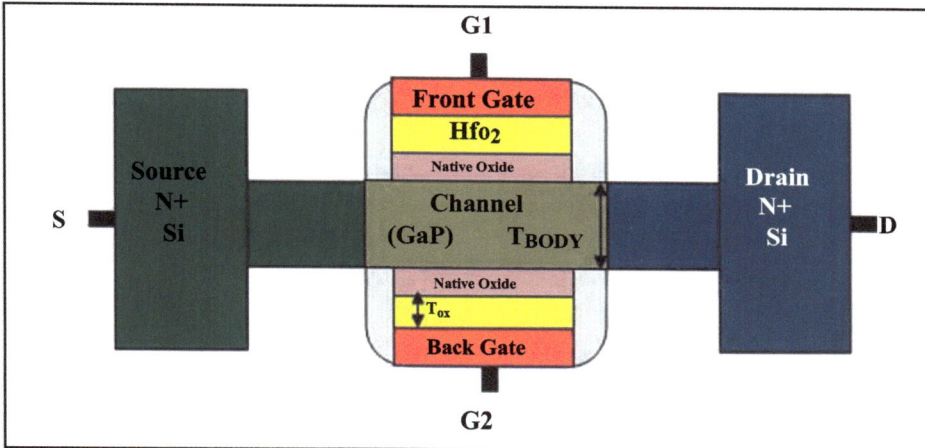

Fig. (3). Cross section of HIBTFET.

Table 2. Device parameters for HIBTFET.

Parameter	Value
Source	$Si(n^+)$
Drain	$Si(n^+)$
Channel	GaP
Source doping	$10^{20}cm^{-3}$
Drain doping	$10^{20}cm^{-3}$
Channel doping	intrinsic

(Table 2) cont.....

Dielectric Constant (k)	26
Body Thickness	4.5nm
Bandgap of Si	1.12eV
Bandgap of GaP	2.26eV
EOT	0.471nm
Channel length	10.8nm
V_D	0.4V

Strained Staggered Heterojunction TFETs (SSHTFETs)

The strained staggered Heterojunction Tunnel Field-Effect Transistors (SSHTFETs) present an innovative asymmetric design aimed at enhancing Band-to-Band Tunneling (BTBT) in heterojunctions. Distinguished by their asymmetry, these SSHTFETs exhibit a remarkable combination of a steeper SS, better current ratio [31], and peak performance at voltages below 0.5V. Despite facing challenges related to tunneling by the large bandgap of Silicon, these devices capitalize on the resulting increased BTBT and reduced band gap. The confluence in characteristics that include high ON current and low OFF current, positions SSHTFETs as ideal candidates for low-power applications requiring a steep SS. Moreover, they prove effective in mitigating the short-channel effects commonly encountered in Metal-Oxide-Semiconductor Field-Effect Transistors (MOSFETs). The structural representation of SSHTFET is illustrated in Fig. (**4**), while Table **3** provides details on the crucial device physics for strained-staggered HTFETs. The notable performance attributes of SSHTFETs, particularly their suitability for low-power applications with a steep SS, ON current, and low OFF current, underscore their potential in advancing semiconductor technologies.

Heterojunction SOI-TFETs (HSOI TFET)

The heterojunction Silicon-On-Insulator Tunnel Field-Effect Transistor (HSOI TFET) stands out as a novel device, demonstrating superior characteristics by elevating the Miller capacitance and achieving a better subthreshold swing of 22mV/decade [32]. This unique device configuration incorporates gate oxide overlapping on the source, a key design feature that significantly enhances its performance metrics.

The overlap of the gate oxide onto the source results in an augmented ON-OFF current ratio, an increased area of Band-to-Band Tunneling (BTBT), and an elevated ON current, accompanied by a heightened ON-state drive current. These improvements in characteristics stem from the expanded BTBT area facilitated by the gate oxide overlap on the source, setting the HSOI TFET apart from other devices. The heightened gate-to-source capacitance (C_{gs}) associated with the increased BTBT area concurrently leads to a decrease in Miller capacitance (C_{gd}). The reduction in Miller capacitance addresses the challenges of transients, eliminating over/undershoots. In the HSOI TFET, Germanium (Ge) is employed as the source material, while Silicon (Si) serves as both the drain and channel material. Leveraging Ge as LBG material, the overlapping with the channel enhances device performance in terms of switching and current ratio, contributing to a reduction in Miller capacitance.

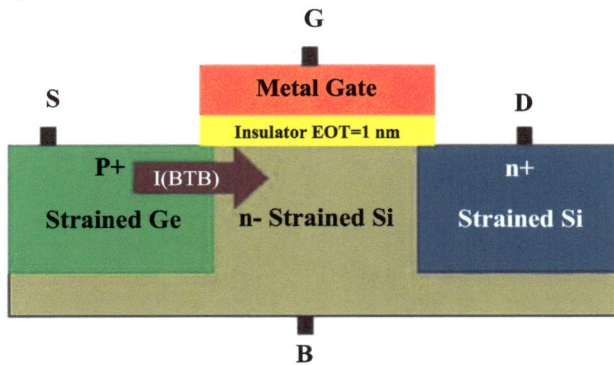

Fig. (4). Cross section of SSHTFET.

Table 3. Device parameters for SSHTFET.

Parameter	Value
Source	Strained-Ge(p^+)
Drain	Strained-Si(n^+)
Channel	Strained-Si(n)
N_d	10^{15}cm^{-3}
N_a	10^{20}cm^{-3}
Dielectric Constant(k)	9
Strained-Si BG	0.89eV
Strained-Ge BG	0.58eV
EOT	1nm
Channel length	35nm
Effective bandgap	0.25eV
V_D	0.4V

With $V_{DS} = 0.4V$ as the operating voltage, the HSOI TFET achieves an impressive ON current (I_{ON}) of $18\mu A/\mu m$ and an I_{ON}/I_{OFF} ratio of 1010. Simulation results indicate the exponential increase in output current, demonstrating the elimination of Miller capacitance. Figs. (**5a** and **5b**) visually depict SOI-TFET with the overlap technique. Comprehensive device parameters are presented in Table **4**, showcasing the excellence of oxide and gate-overlapped heterojunction SOI-TFETs in delivering a remarkable SS of 22mV/decade, coupled with an enhanced current ratio at V_{DS}=0.4V. The HSOI TFET emerges as a promising solution in the realm of heterojunction tunnel FETs, offering a superior SS, increased ON current, improved ON-OFF current ratio, and reduced Miller capacitance.

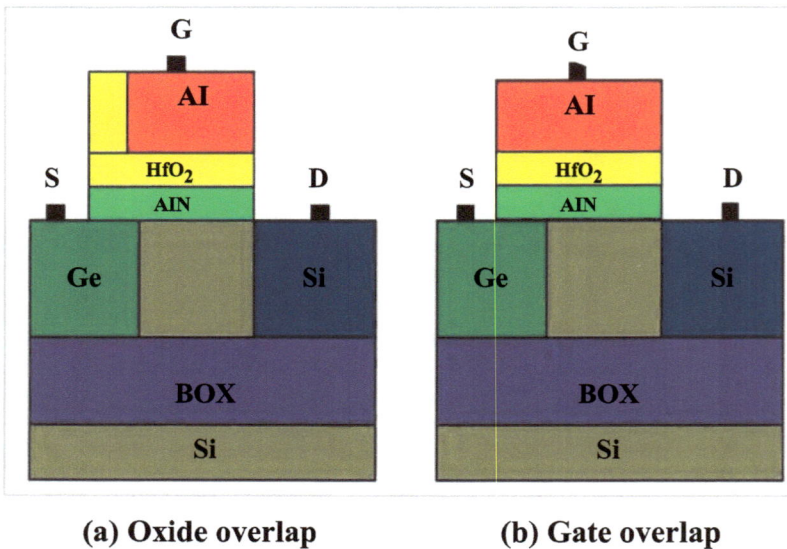

(a) Oxide overlap   **(b) Gate overlap**

Fig. (5). Cross sections of HSOI TFET.

Table 4. Device parameters for HSOI-FET.

Parameter	Value
Source	$Ge(P^+)$
Drain	$Si(N^+)$
Channel	Si
Source doping	$10^{19}cm^{-3}$
Drain doping	$10^{19}cm^{-3}$
Channel doping	$10^{16}cm^{-3}$

(Table 4) cont.....

Dielectric Constant(k)	$29(HfO_2)$ & $9(AlN)$
Si BG	1.12eV
Ge BG	0.67eV
EOT	0.68nm
Channel length	30nm
V_D	0.4V
Overlapped area	3nm
Gate	Al
Insulation layer	HfO_2+AlN

Staggered heterojunction TFETs (SHTFETs)

The SHTFETs represent a distinct category of TFETs designed to surmount the limitations inherent in Metal-Oxide-Semiconductor Field-Effect Transistors (MOSFETs). The structural configuration of SHTFETs is depicted in Fig. (**6**), showcasing their innovative design [33]. These transistor structures are optimized for operation with 0.3V, striking a nuanced balance between performance and power, albeit with a trade-off due to the utilization of III-V materials. SHTFETs emerge as highly relevant for low-power applications, where they exhibit exceptional characteristics. Operating at an impressive 0.3V, these devices deliver a robust drive current of 0.4mA/μm in the ON state, coupled with a minimal leakage current of 50nA/μm in the OFF state. The resulting I_{ON}/I_{OFF} ratio stands at an exemplary 10^4, affirming the superior balance achieved between high current ratio. This performance is notably advantageous compared to homojunction devices, as SHTFETs enhance tunneling efficiency and reduce leakage current.

A comprehensive presentation of device parameters is provided in Table **5**, encapsulating the remarkable attributes of SHTFETs, particularly their ability to concurrently achieve high current ratio. This balance positions SHTFETs as promising contenders for applications demanding optimal performance with minimized power consumption.

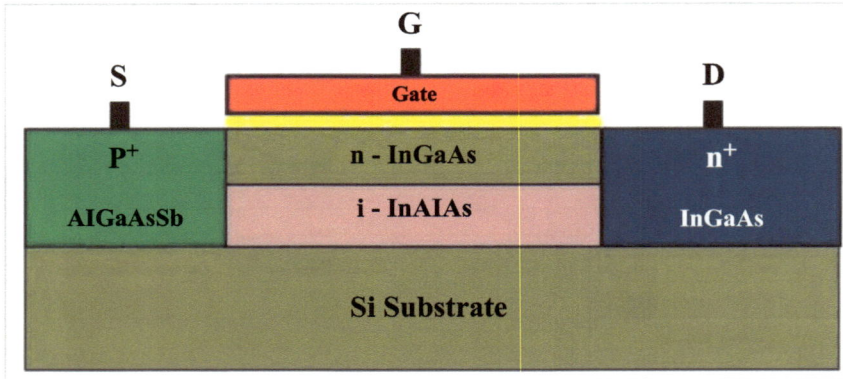

Fig. (6). Cross section of SHTFET.

Table 5. Device parameters for SHTFETs.

Parameter	Value
Source	AlGaAsSb(p$^+$)
Drain	InGaAs(n$^+$)
Channel	InGaAs
Source doping	2×10^{19}cm^{-3}
Drain doping	5×10^{19}cm^{-3}
Channel doping	5×10^{16}cm^{-3}
Dielectric Constant(k)	9
Oxide Thickness	20nm
AlGaAsSb BG	1.36eV
InGaAs BG	0.50eV
EOT	0.6nm
Channel length	100nm
V_D	0.3V

COMPARATIVE ANALYSES OF HETEROJUNCTION TFETs

Table **6** presents the drain current (Ids) and gate-to-source voltage (V$_{gs}$) data for diverse heterojunction Tunnel Field-Effect Transistors (TFETs), facilitating the depiction of their transfer characteristics. The analysis of these characteristics allows for the calculation of essential specifications such as SS and current ratio. Fig. (**7**) further visualizes the comparative evaluation of transfer characteristics among these heterojunction TFETs, providing valuable insights into their performance distinctions.

Table 6. I_{ds} and V_{gs} of HTFETs.

V_{gs} (V)	Drain Current (I_{ds})				
	HVTFETs I_d(mA/μm)	HIBTFETs I_d(A/μm)	SSHTFETs I_d(A/μm)	HSOI-TFETs I_d(A/μm)	SHTFETs I_d(mA/μm)
0.0	$10^{-6.0}$	$10^{-7.5}$	$10^{-9.0}$	10^{-15}	$10^{-4.2}$
0.1	$10^{-8.0}$	$10^{-6.0}$	$10^{-5.5}$	10^{-14}	$10^{-2.0}$
0.2	$10^{-9.0}$	$10^{-5.0}$	$10^{-4.2}$	$10^{-8.0}$	$10^{-1.0}$
0.3	$10^{-9.5}$	$10^{-4.0}$	$10^{-4.0}$	$10^{-6.5}$	$10^{-0.5}$
0.4	$10^{-9.0}$	$10^{-3.5}$	$10^{-3.8}$	$10^{-6.0}$	$10^{-0.5}$
0.5	$10^{-8.0}$	$10^{-3.0}$	$10^{-3.8}$	$10^{-5.5}$	$10^{-0.4}$
0.6	$10^{-3.0}$	$10^{-2.5}$	$10^{-3.7}$	$10^{-5.2}$	$10^{-0.4}$
0.7	$10^{-2.0}$	$10^{-2.3}$	$10^{-3.7}$	$10^{-5.1}$	$10^{-0.4}$
0.8	$10^{-1.5}$	$10^{-2.2}$	$10^{-3.6}$	$10^{-5.0}$	$10^{-0.3}$
0.9	$10^{-1.5}$	$10^{-2.2}$	$10^{-3.6}$	$10^{-4.9}$	$10^{-0.3}$
1.0	$10^{-1.0}$	$10^{-2.1}$	$10^{-3.5}$	$10^{-4.8}$	$10^{-0.3}$

In the realm of heterojunction Tunnel Field-Effect Transistors (TFETs), diverse structures exhibit unique trade-offs in key performance metrics. Heterojunction Intra-Band Tunnel (HIBT) FETs, while boasting high ON-state drive current, face challenges with elevated subthreshold swing, diminishing circuit speed, and increasing power consumption due to constraints on reducing supply voltage and threshold voltage. Conversely, Heterojunction Vertical BTBT Transistors (HVTFETs) showcase a low SS of 16mV/decade but suffer from a reduced current ratio, leading to slower operational speeds. Staggered Strained Heterojunction TFETs (SSHTFETs) and Staggered Heterojunction TFETs (SHTFETs) demonstrate comparable drive currents, with SSHTFETs excelling in ON current at reduced power supply, outperforming other configurations. However, SSHTFETs exhibit a higher subthreshold swing compared to SHTFETs.

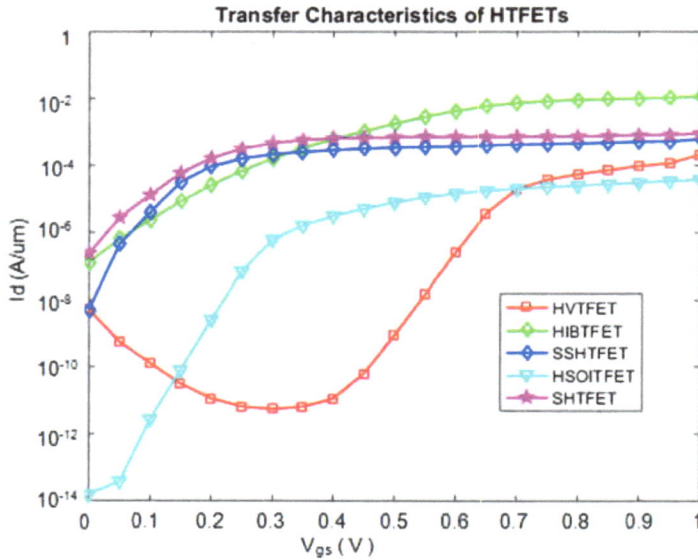

Fig. (7). Transfer characteristics of various HTFETs.

Heterojunction Silicon-On-Insulator TFETs (HSOI-TFETs) present an advantageous scenario with improved subthreshold swing and enhanced Miller capacitance. Nevertheless, these devices suffer from a lower ON-state drive current. In conclusion, each heterostructure offers specific advantages: HVTFETs are preferable for achieving lower subthreshold swing, HIBTFETs excel in high current ratio, SHTFETs are adept for LV and LP operations, SSHTFETs provide improved Miller capacitance, and HSOI-TFETs offer reduced OFF-state current. The choice among these heterojunction structures depends on the specific requirements and priorities of the intended application.

SRAM Circuits

The novel device under consideration is subjected to rigorous testing involving various Static Random Access Memory (SRAM) cell configurations, namely 14T, to assess its power efficiency and delay characteristics. In contrast, several conventional SRAM cells proposed by researchers are also evaluated for comparative analysis. An innovative low-power 6T SRAM cell, incorporating a global bit line technique, is introduced in [34]. This design aims to diminish power and enhance storing capability. The reduction in read and write time, along with improvements in area and power metrics, is achieved. Two SRAM cells are specifically devised for a 4 Kb memory core, operating at a supply voltage of 1.8V.

A read assist-based 7T SRAM cell tailored to operate in subthreshold regions is presented [35]. This configuration mitigates the read-write conflicts encountered by the basic circuit at voltages below 0.4V. The dynamic write ability is augmented for energy-efficient semiconductor memory applications, and performance metrics of VLSI design are meticulously addressed. The proposed design exhibits lower data retention voltage and enhancements in the ON-OFF current ratio, leakage current reduction, and overall power efficiency. A read/write assist with negative bias-based 8T SRAM cell designed for LV and LP is proposed [36]. Leveraging a negative bias technique for V_{SS} and bit line (BL), the design not only achieves LP and high speed but demonstrates substantial improvements in stability in read and write operations. The fabricated SRAM design operating at 0.6V showcases a remarkable 66% reduction in power consumption compared to conventional cells.

A 14T SRAM cell with radiation-hardened tailored for LV and LP semiconductor memory applications is presented [37]. This design exhibits a 65% improvement in write speed and a 50% reduction in power consumption. The design not only withstands single-event upsets (SEU) but also demonstrates immunity to partial single-event–multiple-node upsets (SEMNUs).

Findings

The literature survey reveals significant findings and insights that guide the design and implementation of GOSHTFET-based SRAM cells such as the following.

1. Limitation of MOSFET Subthreshold Swing: The SS of MOSFETs is constrained to 60mV/dec, rendering them unsuitable for ultra-low power memory applications.

2. TFET Advantages and Limitations: TFETs offer a steeper subthreshold swing and reduced leakage current, addressing some issues of MOSFETs. However, the ON-state drive current of TFETs is comparatively lower than that of MOSFETs.

3. Advantages of Heterojunction Tunnel Transistors: Heterojunction tunnel transistors emerge as a favorable alternative, providing advantages such as a reduction in leakage, a steeper SS, and a high current ratio.

4. Impact of Low Bandgap Materials: The use of low bandgap materials like Ge/SiGe demonstrates potential improvements in BTBT and ON-OFF state current ratios, enhancing overall transistor performance.

5. Subthreshold Operation with Lower Supply Voltage: Heterojunction tunnel transistors exhibit the capability to operate with lower voltages, contributing to energy-efficient memory applications.

6. Preference for Heterojunctions in Low Voltage, Low Power: In the context of low voltage, low power memory applications, heterojunctions outperform homojunctions, making them the preferred choice.

7. Suitability of Heterojunction TFETs for SRAM Cells: Heterojunction TFETs prove to be well-suited for designing low leakage current SRAM cells, offering advantages such as LP and reduced delays.

Summary of Traditional Approaches

This section comprehensively reviews the literature pertinent to various HTFETs, encompassing an exploration from device considerations to circuit analyses in the context of low-power SRAM design. The focus of the review centers on various heterojunction TFET architectures, delving into critical device parameters including SS, current ratio, etc. Through an in-depth analysis of transistor characteristics, the suitability of these devices for energy-efficient semiconductor memory applications is scrutinized. In particular, the chapter delves into the examination of heterojunction TFET-based SRAM cells with configuration 14T. This detailed exploration extends to the intricacies of 14T SRAM cells, shedding light on their potential applications in the realm beyond CMOS technologies. The synthesis of findings derived from the literature survey, spanning the spectrum from individual device attributes to broader circuit-level implications, is presented conclusively at the conclusion of the chapter.

Design of GOS-HTFET

Because of lower SS characteristics, the GOS is a promising technique in HTFETs with better BTBT for memory-based digital systems. The typical model generates improved SS with a significant effect on the drive current in The ON-state, however, there are yet unidentified hidden physics laws governing SS, which could lead to structural complexity. Therefore, increased device scaling with structural alteration is the focus of the GOS HTFET design. However, the cause of the SCEs is the extreme scaling process. However, the novel transistor designs made of low band gap materials boost the overall performance while mitigating such short channel effects. Similarly, appropriate geometric structural alignment significantly increases the tunneling rate of BTBT. A GOS HTFET with an improved SS idea is present in the proposed HTFET paradigm. The selection of materials, such as Si, InGaAs, and Ge has the greatest potential to achieve reduced SS, is mostly responsible for achieving lower SS. The N and P-type GOS-HTFET structures' proposed designs are depicted in Figs. (**8** and **9**), respectively.

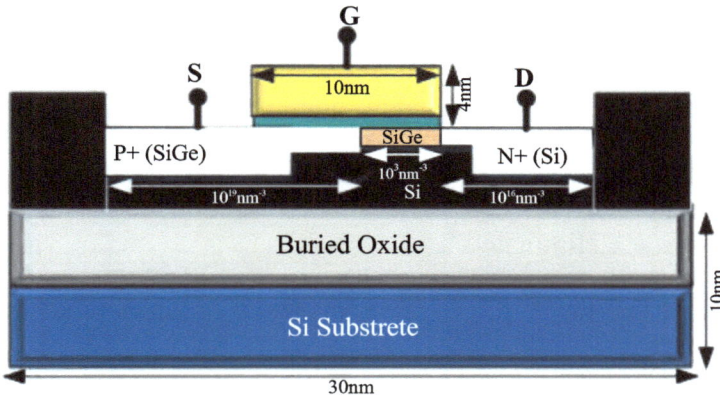

Fig. (8). The Structure of n-type GOS-HEFET design.

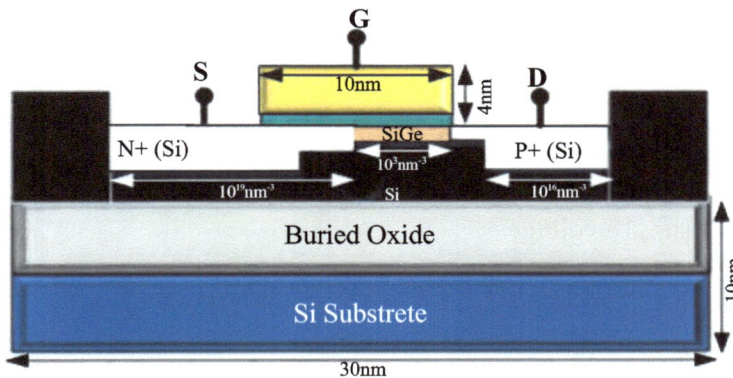

Fig. (9). The structure of the p-type GOS-HEFET design.

Features of GOS HTFET

Applications involving ultra-low power memory are proposed for the GOS HTFET. Listed below are key features of the GOS HTFET that help it meet the specifications needed to be designed and implemented as a low-power, low-voltage device.

- Higher drive current and current ratio by increasing the BTBT area and current through Gate Oxide Overlap on to Source (GOS).

- The device uses low bandgap materials, like SiGe and Ge, to boost BTBT and drive current.

- Because of the overlapping technique, a higher BTBT is obtained in comparison to other heterojunctions.

- High drive current in the ON state when compared to TFETs.

- High-k dielectric materials and heterojunctions result in low leakage current.

- By increasing the area at the source-to-gate interface, Miller capacitance is improved and the effects of overshoot and undershoot in the switching characteristics are eliminated.

- Higher BTBT and low bandgap materials result in lower subthreshold swing.

- Reduced energy usage.

- Fast operation speed.

- It is feasible for the device to be further scaled.

- Obtaining low-voltage operation is simple and easy.

- Enhanced ratio of ON to OFF current.

- Decrease in supply voltage.

APPLICATION OF GOS HTFET IN SRAM DESIGN

GOS HTFET-based 14T SRAM Cell

Fig. (10) illustrates a radiation-hardened 14T SRAM cell incorporating GOS HTFETs for the forthcoming book chapter. The innovative design of this SRAM cell, rooted in GOS HTFETs, significantly mitigates leakage current, thereby enhancing both power efficiency and operational delay. While maintaining the fundamental attributes of conventional SRAM cells, the proposed circuit introduces additional advantages. In the realm of space applications necessitating a radiation-hardened 14T SRAM cell, this novel circuit finds widespread utilization. Despite the increased transistor count inherent in the 14T configuration, the performance of the proposed circuit exhibits remarkable stability, rendering GOS HTFETs adaptable to various SRAM cell topologies.

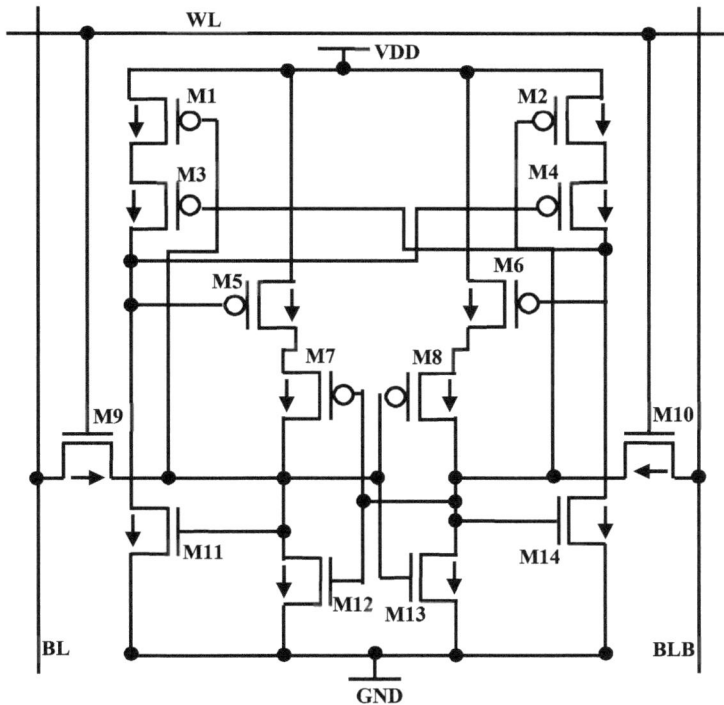

Fig. (10). 14T SRAM using GOS HTFETs.

Power Analysis of 14T SRAM Cell

Furthermore, the novel GOSHTFET can be seamlessly integrated into any configuration of the SRAM cell tailored to user-defined specifications and application-specific requirements. The GOSHTFET finds application in 14T SRAM with the radiation-hardened technique, particularly where delay and optimal power characteristics are paramount. This novel GOS HTFET-based device exhibits commendable performance in specific application domains, akin to its traditional SRAM counterparts. Both the conventional and proposed designs are subjected to a comprehensive analysis, considering average power, maximum power, and minimum power.

Average Power Assessment

Fig. (**11**) illustrates the average power consumption of the 14T SRAM cell, employing GOSHTFETs across temperatures ranging from 0°C to 25°C. A comparative analysis with a standard MOSFET-based design reveals that the incorporation of the low-power GOSHTFET significantly diminishes average power consumption. This reduction holds true across various temperatures, highlighting the efficacy of the proposed device in achieving superior power efficiency for both higher transistor counts and upgraded SRAM memory units.

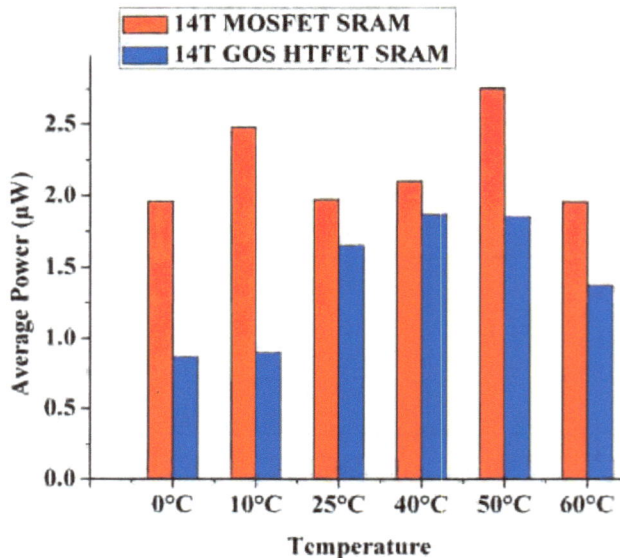

Fig. (11). Average power analysis of 14T SRAM.

Maximum Power Assessment

Fig. (**12**) demonstrates the maximum power characteristics of the 14T SRAM cell utilizing GOS HTFETs at different temperatures, presenting a noteworthy reduction when compared to conventional approaches. The proposed GOS HTFET contributes to a substantial decrease in maximum power consumption across diverse temperature ranges, affirming its efficacy in optimizing power efficiency for the 14T SRAM configuration.

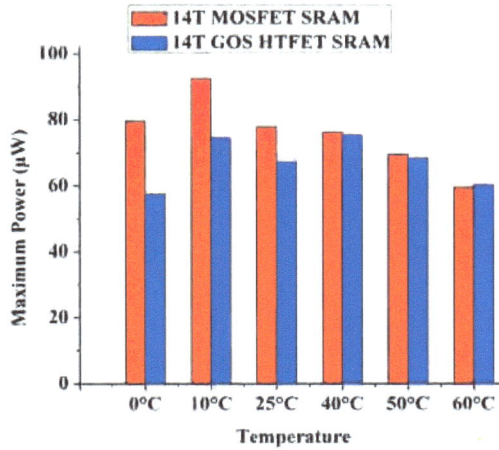

Fig. (12). Maximum power analysis of 14T SRAM.

Minimum Power Assessment

Fig. (**13**) showcases the minimum power analysis of the 14T SRAM cell incorporating GOS HTFETs, offering a comparison with conventional architectures at varying temperatures. Notably, the GOS HTFET outperforms MOSFET designs, leading to a significant reduction in minimum power consumption across different temperature scenarios.

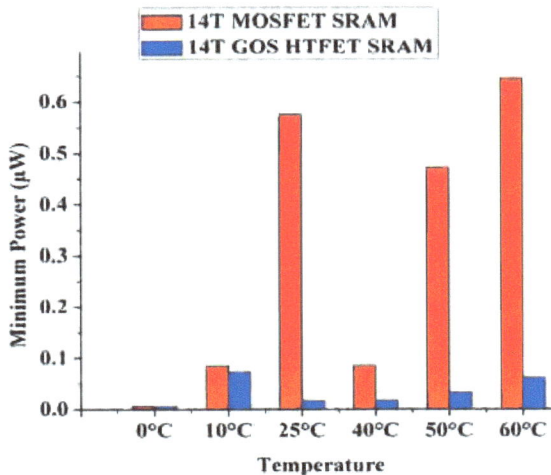

Fig. (13). Minimum power analysis of 14T SRAM.

Delay Analysis of SRAM Cells

In contemporary digital systems, memory plays a pivotal role, being a fundamental component that directly or indirectly influences nearly every system function [38]. The speed of read and write operations in memory is crucial, in determining the overall system speed and power consumption [39]. Speed, an intricate but essential design metric, is inversely linked to delay, making it a critical factor in the design of advanced electronic systems [40]. SRAMs, preferred over DRAMs in practical applications due to reduced leakage current, fast access time, and superior performance, are particularly significant for meeting the demands of portable and feature-rich devices [41-42]. The ongoing focus on CMOS integrated technology development is the latency of SRAM memory units, where read and write operations govern the operational speed of digital systems [43].

Write Delay Assessment

Utilizing the Tanner tool, the delay of a 14T SRAM cell is analyzed, comparing conventional MOSFETs with GOS HTFETs. The proposed GOS HTFET, characterized by a smaller SS and high current ratio with LV and HP, significantly reduces write operation delay and enhances overall operational speed. Fig. (**14**) presents the write delay comparison of SRAM cells at 27°C, employing both MOSFET and GOSHTFET at a voltage of 1.2V. The comparison reveals that, across different SRAM cell configurations, the GOS HTFET-based SRAM consistently outperforms traditional systems, resulting in accelerated system speed and enhanced processor efficiency, particularly in multitasking scenarios.

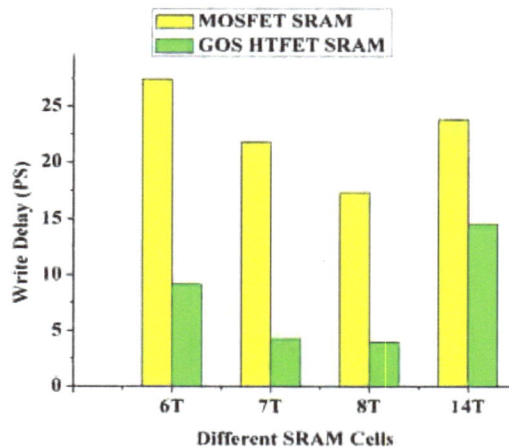

Fig. (14). SRAM Write delay Analysis at 27°C.

Read Delay Assessment

Analogous to the write delay analysis, various GOS HTFET-based SRAM memory cells exhibit significantly lower read delays compared to conventional systems. Fig. (15) illustrates the read delay study of diverse SRAMs at 27°C, employing a voltage of 1.2V. Both conventional and proposed methods are employed to assess and compare the read delay of multiple SRAM cells at room temperature. Across all configurations, the GOS HTFET-based SRAM consistently outperforms conventional systems, showcasing markedly reduced read delays.

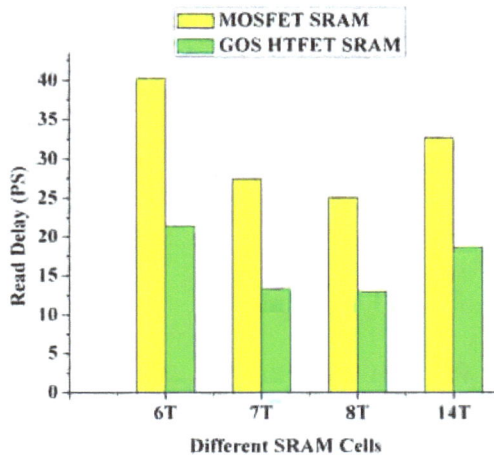

Fig. (15). SRAM Read delay analysis at 27°C.

Performance Trade-offs of 14 SRAM in Different Applications

When comparing the 14T SRAM based on HTFET with other advanced memory technologies such as FinFET-based SRAM, MRAM, and RRAM, itis essential to highlight the specific benefits and trade-offs in various application contexts.

Low-power IoT Devices: The 14T HTFET-based SRAM excels in ultra-low-power consumption due to its reduced subthreshold swing and leakage currents, making it ideal for IoT applications where power efficiency is critical. Compared to FinFET-based SRAM, the power dissipation of HTFET SRAM is significantly lower, though the access speed may be slightly slower.

High-Performance Computing (HPC): In scenarios where speed is more critical, FinFET-based SRAM has a performance advantage due to its higher electron mobility and faster switching times. However, the 14T SRAM shows greater resilience to radiation, which may be more beneficial in radiation-heavy environments like aerospace or nuclear computing.

Portable Embedded Systems: HTFET SRAM provides a better balance between power and performance in embedded systems. While MRAM and RRAM offer non-volatility, they typically suffer from slower write speeds and endurance limitations, whereas the 14T HTFET SRAM provides faster access times and greater durability under extensive read/write cycles.

Mobile and Wearable Devices: The demand for ultra-low power consumption is critical in mobile and wearable technologies, where battery life is a top priority. The 14T HTFET SRAM, with its low subthreshold swing and minimal leakage current, is highly advantageous in this context. Compared to FinFET-based SRAM, HTFET-based SRAM offers superior energy efficiency, extending battery life. However, mobile devices often require fast data access, where the FinFET SRAM might outperform HTFET SRAM due to its higher switching speed. Thus, while HTFET SRAM provides low power consumption, it might be slightly slower for some latency-sensitive mobile applications.

Data Centers: In large-scale data centers, the balance between power efficiency and performance is crucial. HTFET-based SRAM cells offer significant power savings, which can help reduce the energy footprint of large memory arrays. However, due to the complexity of the 14T architecture, its implementation may require careful optimization to achieve competitive performance metrics in terms of latency and bandwidth. In contrast, FinFET-based SRAMs and DRAMs might offer higher access speeds, though at the cost of higher power consumption. Therefore, HTFET SRAM is more suited to data centers where power efficiency is more important than the absolute fastest access times.

Table **7** below is a comparative table summarizing the trade-offs in different application scenarios.

Challenges and Solutions in the Manufacturing of Mass-Production of 14T SRAM

Mass production of 14T SRAM based on HTFET technology involves several challenges, especially given the use of non-traditional materials like SiGe or Ge for heterojunctions. The following challenges and potential solutions can be discussed:

Table 7. Summary of the trade-offs in different applications.

Parameter	14T HTFET SRAM	FinFET SRAM	Magneto resistive RAM	Resistive RAM	3D NAND Flash	Dynamic RAM
Power Consumption	Very Low	Moderate	Low	Very Low	High	High
Speed	Moderate	High	Moderate	Moderate	Low	High
Resistance	High	Moderate	High	High	Low	Low
Read/Write Cycles	High	High	Moderate	Moderate	Low	Moderate
Volatility	Volatile	Volatile	Non-volatile	Non-volatile	Non-volatile	Volatile
Memory Density	Moderate	High	Low	High	High	Moderate
Scalability	High	High	Moderate	High	Very High	High
Temperature Stability	High	High	Moderate	Low	Low	High
Fabrication Complexity	High	Moderate	High	Moderate	Moderate	Low
Cost	Moderate	Low	High	Moderate	Low	Low

Material Integration: SiGe and Ge, which are essential for lowering the bandgap in HTFETs, are not as mature as pure silicon in terms of manufacturing processes. The challenge lies in ensuring compatibility with existing CMOS fabs. The solution involves adopting advanced epitaxial growth techniques and improving material quality control during deposition to reduce defects and interface states.

Gate Oxide Overlap (GOS) Manufacturing: The fabrication of GOS HTFETs requires precise alignment to ensure that the gate overlaps the source region correctly. This increases the complexity of the lithography steps and may affect yield. Advanced lithography techniques like EUV (Extreme Ultraviolet Lithography) and improved self-aligned processes can mitigate this challenge.

Cost: The additional steps required for heterojunction formation, such as selective area doping and deposition of high-k dielectrics, increase the cost of fabrication. However, these costs can be offset by scaling production and improving yield rates through refined process control.

Scalability: While HTFET-based SRAM provides low power and improved performance, there is a need for significant refinement in process variability to ensure uniform performance across a large number of chips. This can be achieved by incorporating more robust design-for-manufacturability (DFM) techniques.

Power Delivery Network Complexity: As the density of SRAM cells increases, ensuring a reliable and efficient power delivery network (PDN) becomes more challenging. In traditional SRAM designs, the PDN complexity is relatively straightforward, but with the 14T HTFET SRAM's low-power nature and its demand for ultra-low-voltage operation, delivering stable power without introducing significant voltage fluctuations becomes critical. To overcome this challenge, advanced power gating techniques can be used, along with optimizing the layout of the supply network to reduce fluctuations.

Thermal Management: As devices scale down and memory density increases, heat dissipation becomes a significant concern, particularly for low-power devices like HTFET-based SRAM. While HTFET technology itself reduces power consumption, the additional complexity of the 14T SRAM could result in heat generation in high-density memories. Managing this heat is critical to ensure reliability. Solutions to this challenge include incorporating thermal-aware design strategies during the early stages of chip layout, optimizing power consumption patterns, and using advanced cooling techniques like heat sinks, thermal Vias, or cooling systems in large-scale applications.

Table **8** below is a comparative table summarizing the challenges and solutions to the manufacturing of mass-production of 14T SRAM:

Table 8. Summary of the challenges and solutions the manufacturing of 14T SRAM.

Challenge	Description	Potential Solution
Material Integration	Integration of SiGe/Ge into standard CMOS processes is complex.	Advanced epitaxial growth techniques, and improved material quality control.
GOS Manufacturing	Achieving precise gate-source overlap increases lithography complexity.	EUV lithography improved self-aligned processes.
Yield Rate	Variability in the process can result in lower chip yields.	Process optimization, statistical process control, and better design rules.
Increased Manufacturing Cost	Additional steps for heterojunction and high-k dielectrics increase cost.	Scale production, yield improvement, and process optimization.
Process Variability	Ensuring uniform performance across chips is challenging.	Design-for-manufacturability (DFM) techniques, improved process control.
Tool Requirements	New tools may be needed for the deposition and etching of heterojunction materials.	Investment in advanced toolsets like atomic layer deposition (ALD).
Thermal Budget	HTFET fabrication requires precise temperature control.	Use low-temperature processes and optimize thermal budgets across steps.
Compatibility with Existing Infrastructure	Compatibility with existing CMOS lines is limited due to different material properties.	Hybrid CMOS-HTFET integration or process retrofitting to support new materials.
Power Delivery Network Complexity	High-density SRAM cells require robust power delivery for consistent performance.	Implement advanced power gating techniques and optimize supply network layout.
Thermal Management	Heat dissipation during high-density SRAM operation can lead to reliability issues.	Incorporate thermal-aware design strategies and use heat sinks or other cooling technologies.

CONCLUSION

The limitations posed by MOSFETs, constrained by a minimum subthreshold swing of 60mV/decade, hinder their scalability, rendering them unsuitable for ultra-low power systems. The escalating demand for energy-efficient alternatives in portable embedded digital systems necessitates innovative solutions, particularly in memory design, which significantly impacts overall power consumption. In response to this demand, this study introduces and constructs a heterojunction tunnel transistor with oxide overlap (GOSHTFET), showcasing a high ON-OFF current ratio, low leakage current, high drive current, and a steeper subthreshold swing at low voltage. Leveraging the Tanner tool for modeling, simulation, and validation, key parameters of the proposed device are presented. Integration of the GOSHTFET into various SRAM topologies demonstrates superior performance. Future integration of CMOS and HTFET technologies holds promise for digital logic implementation, where CMOS offers overall superior performance and HTFET technology excels in leakage reduction. A systematic assessment of the proposed GOSHTFET in diverse SRAM cells reveals its superiority, as evidenced by power studies and delay analyses, highlighting its potential for advancing low-power SRAM designs.

Learning Objectives/Key Points

(1) Practical Insights for Designers: This chapter aims to provide practical insights for designers aiming to develop low-power SRAM cells for ultra-low-power semiconductor memory applications. By examining crucial variables such as Miller capacitance, ON-state driving current, subthreshold swing, and ON-OFF current ratio, designers can better understand the factors influencing the performance of these memory devices and make informed design decisions.

(2) Comprehensive Evaluation Methodology: Through the use of detailed simulations and validation using TCAD Silvaco tools, the efficacy of the proposed HTFET designs is thoroughly evaluated at the device level. This comprehensive evaluation methodology ensures that the designed low-power SRAM cells meet the requirements of modern semiconductor memory applications, paving the way for their practical implementation.

(3) Exploration of SRAM Cell Structures: The exploration of various SRAM cell structures suitable for efficient memory devices is essential for circuit-level testing and practical application. By elucidating the device physics necessary for optimizing energy-efficient devices and providing insights into different SRAM cell configurations, this chapter equips readers with the knowledge needed to design and implement low-power SRAM cells effectively.

(4) Advancement in Semiconductor Memory Technologies: By delving into the comprehensive research findings on Heterojunction Tunneling Field Effect Transistors (HTFETs) and their integration with SRAM cells, this chapter advances the understanding of low-power semiconductor memory technologies. The detailed discussions and analyses contribute to propelling the discourse on the practical application of innovative HTFET designs in memory technology, fostering further advancements in the field.

(5) Criteria for Selecting HTFETs: HVTFETs showcase a low SS of 16mV/decade but suffer from a reduced current ratio, leading to slower operational speeds. SSHTFETs and SHTFETs demonstrate comparable drive currents, with SSHTFETs excelling in ON current at reduced power supply. However, SSHTFETs exhibit a higher subthreshold swing compared to SHTFETs. HSOI-TFETs present an advantageous scenario with improved subthreshold swing and enhanced Miller capacitance. The choice of these heterojunction structures depends on the specific requirements and priorities of the intended application.

(6) Enhanced Performance Characteristics: GOS HTFET design offers a higher drive current and current ratio by increasing the BTBT area and utilizing low band gap materials like SiGe and Ge. The incorporation of high-k dielectric materials and heterojunctions results in low leakage current, ensuring efficient power utilization.

(7) Improved Device Efficiency: By increasing the area at the source-to-gate interface, GOSHTFET mitigates Miller's capacitance effects, eliminating overshoot and undershoot in switching characteristics. The device achieves lower subthreshold swing, reduced energy usage, fast operation speed, and enhanced ON-to-OFF current ratio, making it suitable for ultra-low power memory applications.

(8) Enhanced Performance in 14T SRAM Design: GOS HTFET-based radiation-hardened 14 T SRAM cell exhibits superior power efficiency and operational delay, mitigating leakage current and maintaining stability. Despite increased transistor count, GOSHTFETs demonstrate remarkable stability, making them adaptable to different SRAM cell topologies.

(9) Power Consumption Analysis: GOS HTFETs significantly reduce average power consumption across temperatures, outperforming standard MOSFET-based designs. Noteworthy reduction in maximum power consumption across temperature ranges, affirms the efficacy of GOS HTFETs in optimizing power efficiency. GOS HTFETs lead to a significant decrease in minimum power consumption compared to conventional architectures, demonstrating superior efficiency across different temperature scenarios.

(10) Improved Operational Speed: GOS HTFET-based SRAM cells consistently exhibit reduced write operation delay, resulting in accelerated system speed and enhanced processor efficiency. GOS HTFET-based SRAMs demonstrate significantly lower read delays compared to conventional systems, ensuring faster data access and improved overall system performance.

Multiple Choice Questions (MCQs)

1. What is the main goal of investigating heterojunction tunnel field-effect transistor (HTFET) topologies?

a) Maximizing input voltage

b) Minimizing Miller capacitance

c) Increasing subthreshold swing

d) Enhancing ON-OFF current ratio

Answer: d) Enhancing ON-OFF current ratio

2. What is the significance of Gate Oxide Overlap onto Source (GOS) in HTFET designs?

a) Increases leakage current

b) Reduces Miller capacitance

c) Enhances ON-state driving current

d) Improves subthreshold swing

Answer: b) Reduces Miller capacitance

3. What advantage does a radiation-hardened 14T SRAM cell incorporating GOS HTFETs offer?

a) Lower power efficiency

b) Decreased operational delay

c) Increased leakage current

d) Reduced stability

Answer: b) Decreased operational delay

4. What characteristic of GOS HTFETs makes them adaptable to various SRAM cell topologies?

a) Higher transistor count

b) Remarkable stability

c) Increased leakage current

d) Reduced operational delay

Answer: b) Remarkable stability

5. What is the primary advantage of GOS HTFETs over conventional MOSFET-based designs in terms of power consumption?

a) Increased average power consumption

b) Lower maximum power consumption

c) Higher minimum power consumption

d) Reduced average power consumption

Answer: d) Reduced average power consumption

6. What aspect of GOS HTFET-based SRAM cells leads to accelerated system speed?

a) Reduced leakage current

b) Increased transistor count

c) Reduced write operation delay

d) Higher average power consumption

Answer: c) Reduced write operation delay

7. What is the primary application domain of GOS HTFET-based SRAM cells?

a) High-temperature environments

b) Low-power memory systems

c) Radiofrequency (RF) communication

d) Quantum computing

Answer: b) Low-power memory systems

8. What is the primary advantage of GOS HTFETs over conventional devices in terms of subthreshold swing?

a) Increased subthreshold swing

b) Reduced subthreshold swing

c) No effect on subthreshold swing

d) Unpredictable subthreshold swing

Answer: b) Reduced subthreshold swing

9. Which type of materials is used in GOS HTFETs to optimize Band-to-Band Tunneling (BTBT)?

a) Wide bandgap materials

b) High bandgap materials

c) Low bandgap materials

d) Conventional materials

Answer: c) Low bandgap materials

10. What characteristic of GOS HTFETs contributes to their suitability for ultra-low power memory applications?

a) High leakage current **b)** High transistor count

c) Low subthreshold swing **d)** High operational delay

Answer: c) Low subthreshold swing

11. How does the gate oxide overlap with the source impact the performance of GOS HTFETs?

a) Reduces drive current **b)** Increases Miller capacitance

c) Decreases tunneling rate **d)** Enhances Band-to-Band Tunneling (BTBT)

Answer: d) Enhances Band-to-Band Tunneling (BTBT)

12. Which technique is emphasized in the chapter to mitigate short-channel effects in SRAM cells?

a) Gate Oxide Overlap onto Source (GOS)

b) Decreasing transistor count

c) Increasing Miller capacitance

d) Using high bandgap materials

Answer: a) Gate Oxide Overlap onto Source (GOS)

13. How does GOS HTFET-based SRAM compare to conventional systems in terms of average power consumption?

a) Higher average power consumption

b) Lower average power consumption

c) Similar average power consumption

d) Depends on transistor count

Answer: b) Lower average power consumption

14. What aspect of GOS HTFET-based SRAM cells contributes to enhanced power efficiency?

a) Increased transistor count

b) Reduced operational delay

c) Higher leakage current

d) Lower subthreshold swing

Answer: d) Lower subthreshold swing

15. How does the gate oxide overlap with the source impact the switching characteristics of GOS HTFETs?

a) Increases overshoot and undershoot

b) Reduces drive current

c) Improves Band-to-Band Tunneling (BTBT)

d) Decreases tunneling rate

Answer: c) Improves Band-to-Band Tunneling (BTBT)

REFERENCES

[1]   B.V.V. Satyanarayana and M. Durga Prakash, "Device and circuit level design, characterization and implementation of low power 7T SRAM cell using heterojunction tunneling transistors with oxide overlap," *Microprocessors and Microsystems*, vol. 77, p. 103164, 2020.

[2]   M. Parvathi, "Architectural Designs and Performance Analysis of Adiabatic-based 6T, 9T, and 12T SRAM Cells", *International Conference on Automation, Computing and Renewable Systems (ICACRS)*, , 2022pp. 88-92

[3]   T.V. Reddy and K.M. Rao, "Performance & functionality of novel Subthreshold SRAM's using low power techniques for SOC designs," In: *Proc. 3rd Int. Conf. Communication and Electronics Systems (ICCES)*, 2018, pp. 259–263.

[4]   B.V.V. Satyanarayana, and M.D. Prakash, "Design analysis of GOS-HEFET on lower Subthreshold Swing SOI", *Analog Integr Circ. Sig Process.,* vol. 109, pp. 683-694, 2021.

[5]   B.V.V. Satyanarayana, and M. Durga Prakash, "Lower subthreshold swing and improved miller capacitance heterojunction tunneling transistor with overlapping gate", *Mater. Today Proc,* vol. 45, pp. 1997-2001, 2021.

[6]   D.S. Yadav, P. Singh, and P. Roat, "Assessing the Impact of Source Pocket Length Variation to Examine DC/RF to Linearity Performance of DG-TFET", *Nano,* vol. 18, no. 4, p. 2350027, 2023.

http://dx.doi.org/10.1142/S1793292023500273

[7]   A. Anam, S. I. Amin, D. Prasad, N. Kumar, and S. Anand, "Analysis of III-V material-based dual source T-channel junction-less TFET with metal implant for improved DC and RF performance", *Micro and Nanostructures,* p. 207629, 2023.

[8]   C. Meriga, R.T. Ponnuri, B.V.V. Satyanarayana, A.A.K. Gudivada, A.K. Panigrahy, and M.D. Prakash, "A Novel Teeth Junction Less Gate All Around FET for Improving Electrical Characteristics", *Silicon,* vol. 14, no. 5, pp. 1979-1984, 2022.

http://dx.doi.org/10.1007/s12633-021-00983-y

[9]   M.D. Prakash, B.V. Krsihna, B.V.V. Satyanarayana, N.A. Vignesh, A.K. Panigrahy, and S. Ahmadsaidulu, "A Study of an Ultrasensitive Label Free Silicon Nanowire FET Biosensor for Cardiac Troponin I Detection", *Silicon,* vol. 14, no. 10, pp. 5683-5690, 2022.

http://dx.doi.org/10.1007/s12633-021-01352-5

[10] M.V. Ganeswara Rao, N. Ramanjaneyulu, and S. Madugula, "Exploring High-Temperature Reliability of 4H-SiC MOSFETs: A Comparative Study of High-K Gate Dielectric Materials", *Trans Electr. Electron. Mater.,* 2023.

[11] P. Singh, and D.S. Yadav, "Impact of work function variation for enhanced electrostatic control with suppressed ambipolar behavior for dual gate L-TFET", *Curr. Appl. Phys.,* vol. 44, pp. 90-101, 2022.

http://dx.doi.org/10.1016/j.cap.2022.09.014

[12] A. Raman, K.J. Kumar, D. Kakkar, R. Ranjan, and N. Kumar, "Performance investigation of source delta-doped vertical nanowire TFET", *J. Electron. Mater.,* vol. 51, no. 10, pp. 5655-5663, 2022.

http://dx.doi.org/10.1007/s11664-022-09840-z

[13] P. Singh, and D. S. Yadav, "Design and investigation of F-shaped tunnel FET with enhanced analog/RF parameters", *Silicon,* pp. 1-16, 2021.

[14] B. Dixit, "Investigation of surface potential for double gate hetero junction tunnel FinFET: Application to high- material HfO2", *2018 International Symposium on Devices, Circuits and Systems (ISDCS),* 2018pp. 1-5

http://dx.doi.org/10.1109/ISDCS.2018.8379663

[15] S.K. Mitra, and B. Bhowmick, "Physics-based capacitance model of Gate-on-Source/Channel SOI TFET", , vol. 13, no. 12, pp. 1672-1676, 2018.

http://dx.doi.org/10.1049/mnl.2018.5214

[16] S.M. Turkane, and A.H. Ansari, "Performance Analysis of TFET Using Si0.35Ge0.65/Si Hetero-junction Hetero-dielectric with Buried Oxide Layer", *2018 IEEE Global Conference on Wireless Computing and Networking (GCWCN),* 2018pp. 91-97

http://dx.doi.org/10.1109/GCWCN.2018.8668644

[17] A. Vinod, P. Kumar, and B. Bhowmick, ""Impact of ferroelectric on the electrical characteristics of silicon--germanium based heterojunction Schottky barrier FET," AEU-International J.", *Electron. Commun,* vol. 107, pp. 257-263, 2019.

[18] H.Y. Gu, and S. Kim, "Design Optimization of Double-Gate Isosceles Trapezoid Tunnel Field-Effect Transistor (DGIT-TFET)", *Micromachines (Basel),* vol. 10, no. 4, p. 229, 2019.

http://dx.doi.org/10.3390/mi10040229 PMID: 30935007

[19] Y. Lee, D. Kim, J. Cai, I. Lauer, L. Chang, S.J. Koester, D. Blaauw, and D. Sylvester, "Low-Power Circuit Analysis and Design Based on Heterojunction Tunneling Transistors (HETTs)", *IEEE Trans Very Large Scale Integr (VLSI) Syst.,* vol. 21, no. 9, pp. 1632-1643, 2013.

http://dx.doi.org/10.1109/TVLSI.2012.2213103

[20] K. Ganapathi, and S. Salahuddin, "Heterojunction vertical band-to-band tunneling transistors for steep subthreshold swing and high on current", *IEEE Electron. Device Lett,* vol. 32, no. 5, pp. 689-691, 2011.

http://dx.doi.org/10.1109/LED.2011.2112753

[21] W. Hu, Q. Zhang, C. Dai, C. Peng, W. Lu, and X. Wu, "Low-power 12T TFET-MOSFET hybrid SRAM bitcell and hybrid 8T SRAM array based on multiplexing strategy," *Microelectronics Journal*, vol. 157, p. 106569, Mar. 2025.

[22] J.K. Mamidala, R. Vishnoi, and P. Pandey, *Tunnel field-effect transistors (TFET): modelling and simulation. ,* 2016.

http://dx.doi.org/10.1002/9781119246312

[23] G. Vijayakumari, U. Rajasekaran, R. Praveenkumar, S. D. Vijayakumar, and V. Kumar, "Breaking Barriers: Junctionless Metal-Oxide-Semiconductor Transistors Reinventing Semiconductor Technology," In: *Field Effect Transistors*, 2025, pp. 125–144.

[24] F. Najam, and Y.S. Yu, "Impact of quantum confinement on band-to-band tunneling of line-tunneling type l-shaped tunnel field-effect transistor", *IEEE Trans Electron. Dev.,* vol. 66, no. 4, pp. 2010-2016, 2019.

http://dx.doi.org/10.1109/TED.2019.2898403

[25] T.A. Ameen, H. Ilatikhameneh, P. Fay, A. Seabaugh, R. Rahman, and G. Klimeck, "Alloy engineered nitride tunneling field-effect transistor: A solution for the challenge of heterojunction tfets", *IEEE Trans Electron. Dev.,* vol. 66, no. 1, pp. 736-742, 2018.

[26] C. Liu, "A T-Shaped SOI Tunneling Field-Effect Transistor With Novel Operation Modes", *IEEE J. Electron. Devices Soc.,* vol. 7, pp. 1114-1118, 2019.

[27] P. Singh, and D.S. Yadav, "Assessing the Impact of Drain under lap Perspective Approach to Investigate DC/RF to Linearity Behavior of L-Shaped TFET", *Silicon,* vol. 14, no. 17, pp. 11471-11481, 2022.

[28] N. Kumar, and A. Raman, "Design and analog performance analysis of charge-plasma based cylindrical GAA silicon nanowire tunnel field effect transistor", *Silicon,* vol. 12, no. 11, pp. 2627-2634, 2020.

[29] S. Ahmad, S.A. Ahmad, M. Muqeem, N. Alam, and M. Hasan, "TFET-based robust 7T SRAM cell for low power application", *IEEE Trans Electron. Dev.,* vol. 66, no. 9, pp. 3834-3840, 2019.

[30] V. Mishra, Y.K. Verma, P.K. Verma, and S.K. Gupta, "EMA-based modeling of the surface potential and drain current of dual-material gate-all-around TFETs", *J. Comput. Electron.,* vol. 17, no. 4, pp. 1596-1602, 2018.
http://dx.doi.org/10.1007/s10825-018-1250-5

[31] V. Mishra, Y.K. Verma, and S.K. Gupta, "Surface potential–based analysis of ferroelectric dual material gate all around (FE-DMGAA) TFETs", *Int. J. Numer Model,* vol. 33, no. 4, p. e2726, 2020.
http://dx.doi.org/10.1002/jnm.2726

[32] T-J.K. Liu, and K. Kuhn, *CMOS and beyond: logic switches for terascale integrated circuits.* Cambridge University Press, 2015.

[33] D.A. Pucknell, and K. Eshraghian, *Basic VLSI design.* Prentice-Hall of India, 1994.

[34] B. Majumdar, and S. Basu, "Low power single bitline 6T SRAM cell with high read stability", *2011 International conference on recent trends in information systems,* 2011pp. 169-174
http://dx.doi.org/10.1109/ReTIS.2011.6146862

[35] S. Gupta, K. Gupta, and N. Pandey, "A 32-nm subthreshold 7T SRAM bit cell with read assist", *IEEE Trans Very Large Scale Integr (VLSI) Syst.,* vol. 25, no. 12, pp. 3473-3483, 2017.
http://dx.doi.org/10.1109/TVLSI.2017.2746683

[36] M. Yabuuchi, K. Nii, Y. Tsukamoto, S. Ohbayashi, Y. Nakase, and H. Shinohara, "A 45nm 0.6V cross-point 8T SRAM with negative biased read/write assist", *2009 Symposium on VLSI Circuits,* 2009pp. 158-159

[37] C. Peng, J. Huang, C. Liu, Q. Zhao, S. Xiao, X. Wu, Z. Lin, J. Chen, and X. Zeng, "Radiation-hardened 14T SRAM bitcell with speed and power optimized for space application", *IEEE Trans Very Large Scale Integr (VLSI) Syst.,* vol. 27, no. 2, pp. 407-415, 2019.
http://dx.doi.org/10.1109/TVLSI.2018.2879341

[38] Y.D. Ykuntam, and B. Penumutchi, Y.D. Ykuntam, B. Penumutchi, and Srilakshmi Gubbala, "Design of Speed and Area Efficient Non Linear Carry Select Adder (NLCSLA) Architecture Using XOR Less Adder Module", In: *Advances in Signal. Processing, Embedded Systems and IoT Lecture Notes in Electrical Engineering.* In: Chakravarthy, V., Bhateja, V., Flores Fuentes, W., Anguera, J., Vasavi, K.P. (eds), Springer, Singapore., 2023.

[39] M.V.G. Rao, P.R. Kumar, and T. Balaji, "A High Performance Dual Stage Face Detection Algorithm Implementation using FPGA Chip and DSP Processor", *Journal of Information Systems and Telecommunication (JIST),* vol. 10, no. 40, pp. 241-248, 2022.

http://dx.doi.org/10.52547/jist.31803.10.40.241

[40] N. Tasneem, M. M. Islam, Z. Wang, H. Chen, J. Hur, D. Triyoso, and A. Khan, "The impacts of ferroelectric and interfacial layer thicknesses on ferroelectric FET design," *IEEE Electron Device Lett.*, vol. 42, no. 8, pp. 1156–1159, 2021.

[41] S.K. Gupta, J.P. Kulkarni, S. Datta, and K. Roy, "Heterojunction intra-band tunnel FETs for low-voltage SRAMs", *IEEE Trans Electron. Dev.,* vol. 59, no. 12, pp. 3533-3542, 2012.

http://dx.doi.org/10.1109/TED.2012.2221127

[42] A. Biswas, S. Tomar, and A. M. Ionescu, *Tunnel FET based non-volatile memory boosted by vertical band-to-band tunneling*, U.S. Patent 9,867,923, 11, 2018.

[43] P. Singh and D. S. Yadav, "Assessment of temperature and ITCs on single gate L-shaped tunnel FET for low power high frequency application," *Engineering Research Express*, vol. 6, no. 1, p. 015319, 2024.

CHAPTER 8

Optoelectronic Characteristics of Long Wave Infrared HgCdTe-based Single- and Dual-Junction Detectors

Shonak Bansal[1],*

[1] *Department of Electronics and Communication Engineering, Chandigarh University, Gharuan, Punjab, India*

Abstract: Mercury Cadmium Telluride ($Hg_{1-x}Cd_xTe$) stands out as the predominant material for developing infrared (IR) detectors. In this chapter, the two-dimensional (2D) p-n (single homojunction and single heterojunction) and p-i-n (dual-heterojunction) architecture models of p+-$Hg_{0.7783}Cd_{0.2217}Te$/n--$Hg_{0.7783}Cd_{0.2217}Te$, p+-$Hg_{0.69}Cd_{0.31}Te$/n--$Hg_{0.7783}Cd_{0.2217}Te$, and n+-$Hg_{0.68}Cd_{0.32}Te$/n--$Hg_{0.7783}Cd_{0.2217}Te$/p+-$Hg_{0.7783}Cd_{0.2217}Te$ are proposed in long-wavelength infrared (LWIR) spectral region. The detectors are designed and analyzed for various optoelectronic characteristic parameters. The outcomes achieved through the Silvaco Atlas TCAD software are compared with those derived from analytical expressions and are found to agree with the analytical results. The proposed detectors are well-suited for their functioning at a wavelength of 10.6 μm under the condition of liquid nitrogen temperature (77 K). The single homojunction-based detector shows an external quantum efficiency (QEext) of 58.29%, a 3-dB cut-off frequency (f3-dB) of 104 GHz with a response time of 3.3 ps, whereas the heterojunction-based detector exhibits a QEext of 67.6%, a f3-dB of 265 GHz with a response time of 1.3 ps, and least dark current density. On the other hand, a dual-junction-based detector exhibits a QEext of 84.92%, a f3-dB of 1.28 THz with a response time of 0.27 ps, further confirming the suitability of the proposed dual-junction detector for low-noise operations.

Keywords: Cut-off frequency, Dark current, Detectivity, HgCdTe, Heterojunction, homojunction, Noise current, Photocurrent, Photodetector, Quantum efficiency, Response time, Responsivity, Spectral response.

INTRODUCTION

Recently, the field of infrared (IR) sensing technology has witnessed significant progress, driven by the need for high-performance detectors with enhanced sensitivity, broader spectral range, and improved operational efficiency. Among the

* **Corresponding author Shonak Bansal:** Department of Electronics and Communication Engineering, Chandigarh University, Gharuan, Punjab, India; E-mail: shonakk@gmail.com

Dharmendra Singh Yadav & Prabhat Singh (Eds.)

various materials explored for infrared photodetection [1-6], mercury cadmium telluride ($Hg_{1-x}Cd_xTe$) has appeared as a promising candidate for enhancing terahertz (THz) and broadband IR detectors, offering improved performance capabilities. This is attributed to its tunable bandgap, low leakage current, relatively high photon absorption coefficient, enhanced stability, low thermal generation rate, moderate dielectric permittivity ensuring small device capacitance, better lattice matching for the growth of high-quality crystals, low thermal expansion coefficient providing device stability, and favorable optoelectronic properties [7-13]. The IR detectors found their major applications in military, optical communications, civilian, thermal and biomedical imaging, remote sensing, missile guidance, fire alarming, gas sensing, satellite remote sensing, motion detection, spectroscopy, surveillance, chemical analysis, telecommunication systems, and night vision. [14-18].

Epitaxial growth of $Hg_{1-x}Cd_xTe$ can be achieved as a chemical compound consisting of both HgTe and CdTe. HgTe exhibits a semi-metallic nature owing to its zero energy bandgap and low resistivity, in contrast to CdTe, which is a semiconductor alloy characterized by an energy bandgap of around 1.6 eV. These chemical compounds possess nearly identical lattice constants, enabling the defect-free growth of $Hg_{1-x}Cd_xTe$ at any composition. The minimal variation in lattice constant with cadmium (Cd) composition facilitates the formation of high-quality layers and heterostructures. The combination of these materials provides the flexibility to attain energy bandgaps from 0 to 1.6 eV. Since HgTe and CdTe have zinc-blende structures, $Hg_{1-x}Cd_xTe$ also has a zinc blende structure for all the Cd compositions [8]. Due to the energy bandgap tunability of $Hg_{1-x}Cd_xTe$ via Cd composition (x), several high-performance IR detectors based on $Hg_{1-x}Cd_xTe$ have been reported. These detectors exhibit various configurations, including p-n [9-18], p-i-n [22, 24], avalanche photodetector [25], barrier detectors [11-13, 26-28], and dual-band IR detectors [10, 29], and have been demonstrated at both cryogenic and room temperatures. The need for cryogenic cooling facilities introduces considerable increases in power consumption, cost, and weight for these IR detectors. The ongoing advancement of $Hg_{1-x}Cd_xTe$ technology primarily serves military-related safety or security applications, particularly in the detection of specific conditions or objects. The primary driving force behind considering alternatives to $Hg_{1-x}Cd_xTe$ is the technological drawbacks associated with this material. One such drawback is the weak bond between Hg and Te, leading to instability in bulk, surface, and interfaces. Concerns persist regarding yield and uniformity, especially in the long-wavelength IR (LWIR: 8-12 μm) spectral region [30]. LWIR spectral region encompassing 10.6 μm is well-suited for challenging weather conditions and serves as a strategic atmospheric window. The primary limitation of $Hg_{1-x}Cd_xTe$-based IR detectors is the elevated dark current restricted by the Auger recombination process and the requirement for working at low temperatures [31, 32]. Therefore it is necessary to create and develop $Hg_{1-x}Cd_xTe$-based IR detectors that exhibit enhanced performance at or near room temperature. Despite these challenges, $Hg_{1-x}Cd_xTe$ continues to be the predominant semiconductor material for IR detectors.

The key challenge in developing any commercially viable detector is integrating it with complementary metal oxide semiconductor (CMOS) technology to deliver advanced detection capabilities. The integration with silicon (Si)-based CMOS technology has garnered extensive interest due to the incorporation of both electronics and optoelectronics on a chip. Si being abundant in nature, is a perfect choice for CMOS technology. It is the primary substrate behind every integrated chip technology. It also facilitates lithography and other fabrication processes in the CMOS technology domain [33-36]. The integration of electronic circuitry follows standard technology, whereas optoelectronic material (such as direct bandgap compound semiconductors) integration with Si (indirect bandgap) is a big challenge due to crystallographic compatibility issues. The real challenge would be to incorporate $Hg_{1-x}Cd_xTe$ as per CMOS process technology due to its incompatibility with Si technology. Despite its Si-based CMOS incompatibility, the lattice-matched $Hg_{1-x}Cd_xTe/CdZnTe$ interface is of great interest [23].

The versatility of HgCdTe makes it a key material for developing single and dual-junction detectors. Therefore, this chapter aims to provide the two-dimensional (2D) architecture models of p^+-$Hg_{0.7783}Cd_{0.2217}Te/n^-$-$Hg_{0.7783}Cd_{0.2217}Te$ (single homojunction), p^+-$Hg_{0.69}Cd_{0.31}Te/n^-$-$Hg_{0.7783}Cd_{0.2217}Te$ (single heterojunction), and n^+-$Hg_{0.68}Cd_{0.32}Te/n^-$-$Hg_{0.7783}Cd_{0.2217}Te/p^+$-$Hg_{0.7783}Cd_{0.2217}Te$ (p-i-n dual-junction) in long wave IR spectral region at liquid cryogenic or liquid nitrogen temperature of 77 K. The detectors are designed and analyzed for various optoelectronic characteristics. The outcomes achieved through the utilization of the Silvaco Atlas TCAD software are compared with those obtained from the analytical model and are found to be in accordance with the analytical expressions. The proposed detectors are well-suited for functioning at a wavelength of 10.6 μm.

MODELING OF HgCdTe

Fundamental Material Properties of $Hg_{1-x}Cd_xTe$

As stated above, $Hg_{1-x}Cd_xTe$ is the semiconductor material alloy, making it a significant material for IR detection. Its placement is determined by the following three primary characteristics [30]:

i.   Tunable energy bandgap over the IR spectral region;

ii.   Larger optical absorption coefficients that permit enhanced quantum efficiency (QE); and

iii.   Intrinsic recombination processes facilitate favorable operations at high temperatures.

Consequently, $Hg_{1-x}Cd_xTe$ can employed for detectors operating in diverse modes, including photodiode, photoconductor, and metal-insulator-semiconductor.

Energy Bandgap

The energy bandgap signifies the gap between the lower boundary of the conduction band energy level and the upper limit of the valence band energy level. The Cd composition, denoted as x, in $Hg_{1-x}Cd_xTe$ adjusts the energy bandgap in the IR spectral region with a cut-off wavelength of λ_c. In $Hg_{1-x}Cd_xTe$, the energy bandgap, represented as $E_g(x,T)$ in eV, is approximated based on both the Cd composition and lattice temperature T, as described by [23]:

$$E_g(x,T) = 0.832x^3 - 0.810x^2 + 1.93x + 5.35 \times 10^{-4} \left(\frac{T^3 - 1822}{T^2 + 255.2} \right) (1 - 2x) - 0.302 \text{ eV} \quad (1)$$

The cut-off wavelength λ_c is determined as the wavelength where the response diminishes to 50% of its maximum value and is estimated from the energy bandgap values by [8]:

$$\lambda_c(x,T) = \frac{1.24}{E_g(eV)} \text{ } \mu m \quad (2)$$

Figs. (**1a** and **b**) illustrate the variation in energy bandgap and cut-off wavelength for $Hg_{1-x}Cd_xTe$ with Cd composition at various temperatures, respectively. It is noticed from Fig. (**1a**) that with an increase in the cadmium

(Cd) composition, the energy bandgap of this composite material progressively rises from HgTe's negative value to CdTe's positive value. At 0 K, the bandgap of $Hg_{1-x}Cd_xTe$ spans from –0.306 eV for the HgTe, crosses 0 eV at nearly $x = 0.17$, and then expands to 1.654 eV for CdTe. Also, the small variation in Cd composition results in a significant fluctuation in the cut-off wavelength of $Hg_{1-x}Cd_xTe$ and considerable non-uniformity over a large area [24-34].

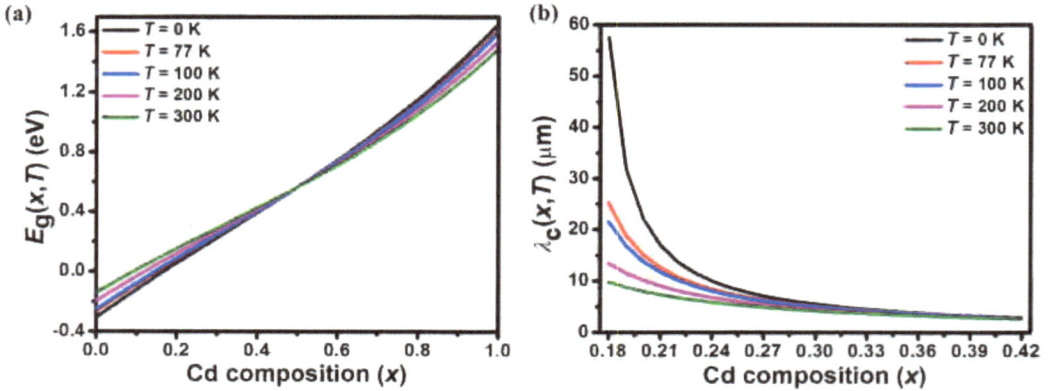

Fig. (1). Variation of the **(a)** energy bandgap and **(b)** cut-off wavelength of $Hg_{1-x}Cd_xTe$ with varying Cd composition at various temperatures.

Carrier Masses

The effective masses of carriers in narrow energy bandgap mercury compounds exhibit similarity and are approximated using the Kane band model. The effective mass of the electron (m_n^*) has been computed from Weiler's expression [35-40]:

$$m_n^* = \frac{m_0}{\left[1+2F+\frac{E_p}{3}\left(\frac{1}{E_g(x,T)+\Delta} + \frac{2}{E_g(x,T)}\right)\right]} \text{ kg} \tag{3}$$

here m_0 denotes the electron effective mass (= 9.1×10^{-31} kg); $F = -0.8$; $E_p = 19$ eV; and $\Delta = 1$eV. The effective mass of the hole $m_p^* = 0.55*m_0$ is commonly employed in the modeling of IR detectors.

Fig. (**2a**) illustrates the change in electron effective mass for $Hg_{1-x}Cd_xTe$ with varying Cd compositions at different temperatures.

(a)

(b)

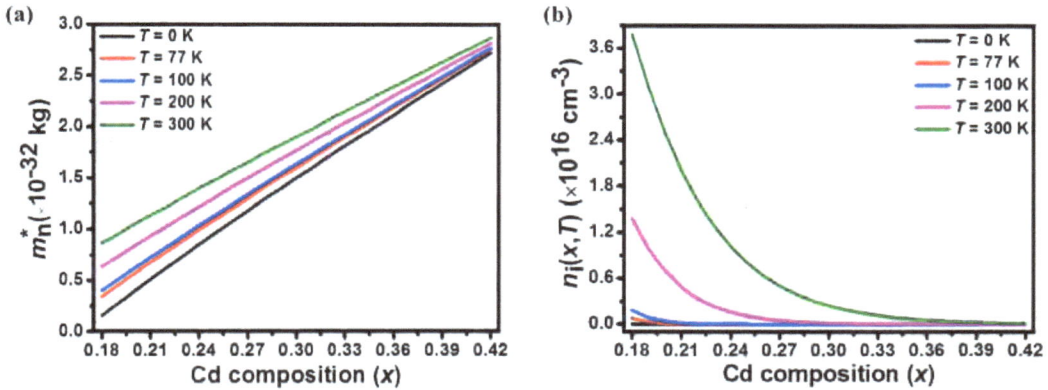

Fig. (2). Variation of the **(a)** electron effective mass and **(b)** intrinsic carrier density for $Hg_{1-x}Cd_xTe$ with varying Cd composition at various temperatures.

Carrier Density

The carrier density plays a crucial role in predicting the electrical characteristics of a semiconductor. Primary dark current resulting from recombination and diffusion processes vary according to the intrinsic carrier density, directly affecting the detector's performance. As the temperature exceeds 0 K, the generation of carriers through the thermal process initiates, causing electrons in the valence band to transition towards the conduction band. The intrinsic carrier density $n_i(x,T)$, a quantity associated with this transition, can be determined by considering both the Cd composition and lattice temperature via [21]:

$$n_i(x,T)=\left(-5.77046x^2 - 3.57290x+1.25942\times10^{-2}xT - 4.74019\times10^{-4}T \right.$$
$$\left. -4.24123\times10^{-6}T^2+5.24256\right)\times10^{14}E_g(x,T)^{0.75}T^{1.5}e^{\left(\frac{-qE_g(x,T)}{2kT}\right)} \text{ cm}^{-3}\textbf{(4)}$$

where q represents the charge of an electron, while k stands for Boltzmann's constant.

Fig. (**2b**) illustrates the variation in intrinsic carrier density for $Hg_{1-x}Cd_xTe$ with Cd composition at various temperatures. A rise in the energy bandgap creates a more substantial energy barrier for thermally generated charge carriers, impeding conduction between the bands. Consequently, as the energy bandgap increases, there is a notable decrease in the $n_i(x,T)$. In simpler terms, the $n_i(x,T)$ decreases with the Cd composition as illustrated in Fig. (**2b**). From Fig. (**2b**), it is evident that the $n_i(x,T)$ rises with the increase in thermal energy as the temperature increases [38-45].

Carrier Mobility

The smallest effective masses in $Hg_{1-x}Cd_xTe$ result in high electron mobilities, whereas the hole mobilities are notably lower by two orders of magnitude. The electron mobility, dependent on both Cd composition and temperature, is determined as [30]:

$$\mu_n(x,T) = \frac{9 \times 10^8 S}{T^{2r}} \text{ cm}^2/\text{Vs}, \quad \text{where } S = \left(\frac{0.2}{x}\right)^{7.5} \text{ and } r = \left(\frac{0.2}{x}\right)^{0.6} \quad (5)$$

The hole mobility is approximated by the following relation:

$$\mu_p = \mu_0\left[1 + \left(\frac{p}{1.8 \times 10^{17}}\right)^2\right]^{-1/4} \text{ cm}^2/\text{Vs}, \quad (6)$$

where $\mu_0 = 440$ and p denotes the hole density in cm^{-3}.

In IR detectors, it is commonly assumed that the electron mobility exceeds the hole mobility to a significant extent, with an estimated value of $\mu_n = 100\mu_p$. The variation of electron mobility for $Hg_{1-x}Cd_xTe$ with Cd composition at various temperatures is shown in Fig. (**3a**). It has been detected that electron mobility decreases with the increase in Cd composition and temperature.

Electron Affinity

The electrical parameters of the detector can be approximated with the energy bandgap diagram, where the performance is evaluated by band bending, discontinuity, and alignment. In heterojunction devices, alterations in the energy bandgap lead to discontinuities or fluctuations in the energy bandgap illustration. Nevertheless, the fluctuations in the energy bandgap do not impact the valence and conduction bands to the same degree. To achieve a precise approximation, a term known as electron affinity, representing the energy difference between the lower edge of the conduction band energy level and the vacuum energy level, is defined. It is estimated based on the Cd composition and temperature using the relation [30, 38]:

$$\chi(x,T) = -0.813 \times (E_g(x,T) - 0.083) + 4.23 \text{ eV} \quad (7)$$

The change in electron affinity for $Hg_{1-x}Cd_xTe$ with varying Cd composition at various temperatures is depicted in Fig. (**3b**). It has been noted that electron affinity decreases with the increase in Cd composition and temperature.

Lattice Constant

$Hg_{1-x}Cd_xTe$ is the narrow bandgap semiconductor material that covers the entire IR spectral region with minimal alterations in the lattice constant and as a function of Cd composition is given by [39]:

$$a(x) = 6.4614 + 0.0084x + 0.01168x^2 - 0.0057x^3 \text{ Å} \qquad (8)$$

The variation of lattice constant with Cd composition is illustrated in Fig. (**3c**). As clear from the figure, there is a small variation in lattice constant which proves advantageous in the production of novel devices utilizing lattice-matched high-quality complex epitaxial layers.

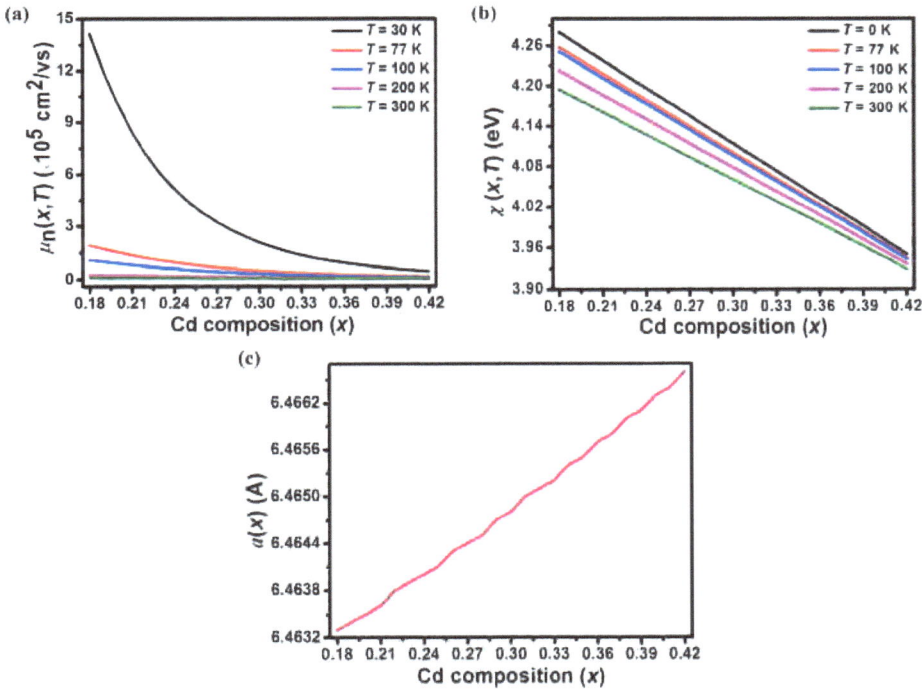

Fig. (3). Variation of the (**a**) electron mobility and (**b**) electron affinity of $Hg_{1-x}Cd_xTe$ with varying Cd composition at various temperatures. (**c**) Variation of the lattice constant of $Hg_{1-x}Cd_xTe$ with varying Cd composition.

Optical Properties of Hg$_{1-x}$Cd$_x$Te

The optical characteristics of Hg$_{1-x}$Cd$_x$Te have mainly been examined in the vicinity of the bandgap energies [8]. The absorption coefficient stands out as a crucial parameter for optoelectronic device optical characterization. Hg$_{1-x}$Cd$_x$Te is a direct energy bandgap material having a strong optical absorption characteristic. The robust optical absorption and the presence of a Hg$_{1-x}$Cd$_x$Te layer with a thickness of ~8-20 μm prove adequate for absorbing a significant percentage of IR radiations, leading to a high quantum efficiency [39]. Chu's empirical relation [40, 41] is utilized to estimate the optical absorption coefficient within the Kane region. In a region where photon energy (E_p) is less than the material's bandgap energy, $E_g(x,T)$ (tail region), and in a region where E_p is greater than $E_g(x,T)$ (Kane region), the optical absorption coefficient $\alpha(\lambda)$ of Hg$_{1-x}$Cd$_x$Te is approximated by:

$$\alpha(\lambda) = \begin{cases} \alpha_0 e^{(\delta/kT)(E_p - E_0)} & E_p < E_g(x,T) \\ \alpha_g(x,T) e^{\beta(x,T)(E_p - Eg(x,T))} & E_p > E_g(x,T) \end{cases} \tag{9}$$

where $\alpha_0 = e^{(45.68x - 18.5)}$; $E_0 = 1.77x - 0.355$. These are the fitting parameters that change with the composition x and

$$\frac{\delta}{kT} = \frac{\ln \alpha_g - \ln \alpha_0}{E_g(x,T) - E_0} \tag{10}$$

here,

$$\alpha_g(x,T) = (8694 - 10.314T)x + 1.88T - 65 \tag{11}$$

$$\beta(x,T) = (21 - 0.13T)x + 0.083T - 1 \tag{12}$$

Fig. (**4a**) demonstrates the variation in the $\alpha(\lambda)$ with wavelength for several Cd compositions at 77 K. It is noted that the absorption intensity typically diminishes with a reduction in the bandgap. This is attributed to both the reduction in the effective mass of the conduction band and the wavelength-dependent nature of the absorption coefficient [8]. Fig. (**4a**) demonstrates a rapid fall in the absorption coefficient beyond wavelength 10.6 μm and designated as the cut-off wavelength for the LWIR detector. The dielectric constant, which is independent of temperature is related to the Cd composition by the empirical relation [8]:

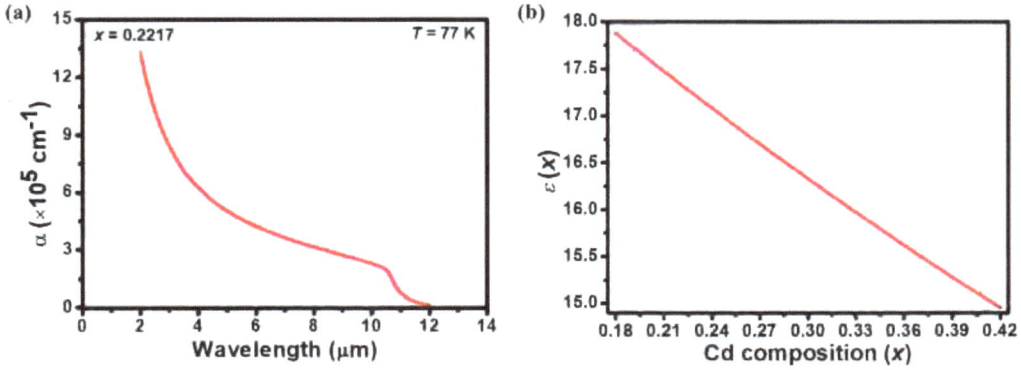

Fig. (4). Variation of the **(a)** optical absorption coefficient and **(b)** dielectric constant for $Hg_{1-x}Cd_xTe$ with varying Cd composition.

$$\varepsilon(x) = 20.5 - 15.6x + 5.7x^2 \tag{13}$$

Fig. **(4b)** demonstrates how the dielectric constant varies with the composition of Cd.

Refractive Index

When analyzing the optical properties, $Hg_{1-x}Cd_xTe$ is characterized by complex refractive indices. The real component of the refractive index for $Hg_{1-x}Cd_xTe$ is expressed as [42-49].

$$n(\lambda, T) = \sqrt{P + \frac{Q}{\left(1-\left(\frac{R}{\lambda}\right)^2\right)} + S\lambda^2} \tag{14}$$

here

$$P = 2.909x^2 - 9.852x + 10^{-3}(300 - T) + 13.173, \tag{15a}$$

$$Q = -0.0961x^2 - 0.246x + 8 \times 10^{-4}(300 - T) + 0.83, \tag{15b}$$

$$R = 8.531x^2 - 14.437x + 7 \times 10^{-4}(300 - T) + 6.706, \tag{15c}$$

$$S = 1.853 \times 10^{-4} x^2 - 0.00128x + 1.953 \times 10^{-4} \qquad \textbf{(15d)}$$

The imaginary component of the refractive index, named, extinction coefficient K for $Hg_{1-x}Cd_xTe$ depends upon both the wavelength and absorption coefficient as [23]:

$$K = \frac{\lambda \alpha(\lambda)}{4\pi} \qquad \textbf{(16)}$$

Fig. (**5**) demonstrates the variation of n and K data with the wavelength in the LWIR region at a cryogenic temperature.

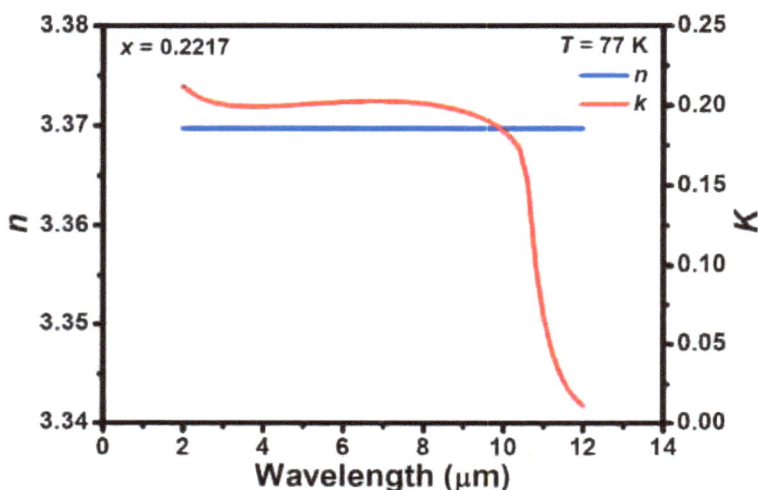

Fig. (5). The plot of n and K data for $Hg_{1-x}Cd_{x=0.2217}Te$ with the wavelength in the LWIR spectral region.

Table **1** provides a summary of the material properties for different Cd compositions of $Hg_{1-x}Cd_xTe$ at cryogenic temperature.

J-V CHARACTERISTIC OF METAL-$Hg_{0.7783}Cd_{0.2217}$Te-METAL DEVICE

Fig. (**6a**) shows an n-type $Hg_{1-x}Cd_{x=0.2217}Te$ with a dopant level at 10^{15} cm^{-3} [17] for LWIR operations at cryogenic temperature. The energy bandgap of the n-$Hg_{0.7783}Cd_{0.2217}Te$ layer is selected to achieve optimal absorption of LWIR radiations at 77 K. The device receives illumination from the top with an illumination intensity (P_{in}) of 1 W/cm^2. The anode and cathode, functioning as electrical ohmic contacts are located at the upper and lower portions of the n-$Hg_{0.7783}Cd_{0.2217}Te$ layer.

Table 1. Material parameters for $Hg_{1-x}Cd_xTe$ at cryogenic temperature.

Parameters	$x = 0.334$	$x = 0.2217$	$x = 0.1867$
E_g (eV)	0.296	0.117	0.060
λ_c (µm)	4.2	10.6	20.6
μ_n (cm^2/Vs)	3.24×10^4	1.18×10^5	1.76×10^5
μ_p (cm^2/Vs)	3.24×10^2	1.18×10^3	1.76×10^3
m_n^* (kg)	1.94×10^{-32}	8.04×10^{-33}	4.22×10^{-33}
m_p^* (kg)	5.01×10^{-31}	5.01×10^{-31}	5.01×10^{-31}
n_i (cm^{-3})	1.84×10^7	8.26×10^{12}	3.82×10^{14}
χ (eV)	4.06	4.2	4.25
ε_r	15.9	17.3	17.8

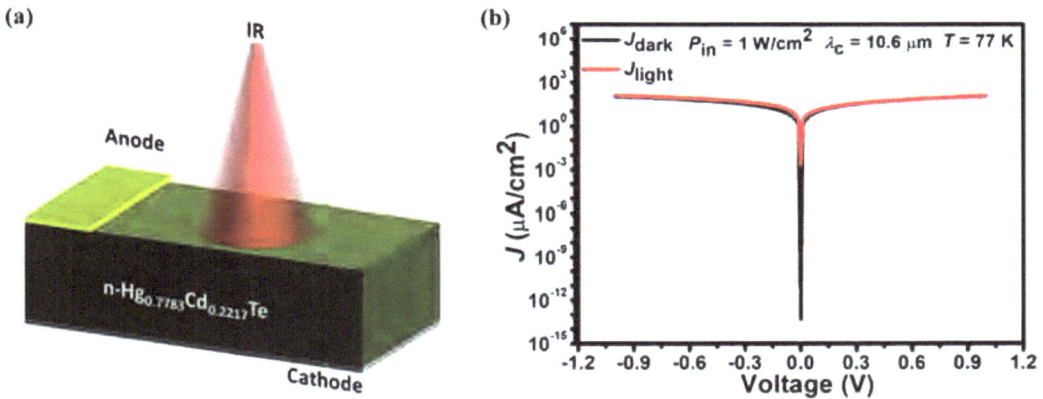

Fig. (6). (a) The schematic of the metal-$Hg_{0.7783}Cd_{0.2217}Te$-metal device under LWIR spectral region. **(b)** The *J-V* characteristics of the metal-$Hg_{0.7783}Cd_{0.2217}Te$-metal device under both the dark and illumination conditions in logarithmic scale at 77 K.

In Fig. (**6b**), the logarithmic-scale representation of the current density-voltage (*J-V*) relationship is shown under both dark and illuminated conditions at 77 K. It is mentioned that the majority of IR detectors based on $Hg_{1-x}Cd_xTe$ are impractical for room temperature operation due to the significant dark current density (J_{dark}) generated by thermal excitation, which can be comparable to or even exceed the photocurrent density (J_{light}) [17]. The J_{dark} of 100.22 $\mu A/cm^2$ and J_{light} of 114.41 $\mu A/cm^2$ is observed at 77 K.

SIMULATION OF $Hg_{1-x}Cd_xTe$ BASED SINGLE- AND DUAL-JUNCTION LWIR DETECTORS

After discussing the material properties and *J-V* characteristics of HgCdTe, the Silvaco TCAD simulation and analytical modeling of $Hg_{1-x}Cd_xTe$-based homojunction and heterojunction detectors are discussed in this section at cryogenic temperature. The LWIR spectral region holds potential applications in free-space optical communication systems, presenting numerous benefits over conventional communication systems. The LWIR spectral region features a strategic atmospheric window at 10.6 μm, making it particularly suitable for challenging weather conditions [17, 23].

The mercury cadmium telluride-based p^+-n^- homojunction, p^+-n^- heterojunction, and n^+-n^--p^+ dual-junction detectors are designed and simulated by using a 2D drift-diffusion approach in the Atlas simulator. All recombination processes are taken into consideration during simulation. The fundamental non-linear equations are resolved to analyze the different optoelectronic characteristics of the detector. The Newton-Richardson method is employed to enhance the efficiency of each iteration, whereas to compute the mobility of the detectors, the concentration-dependent analytic physical model is utilized. The assessment of doping and carrier densities is done by incorporating Fermi-Dirac statistics. The simulation incorporates the consideration of Shockley-Read-Hall (SRH), Auger, tunneling, and optical recombination models to approximate the dark current density [21]. The simulation results are cross-validated with those derived from an analytical model.

p^+-n^- Homojunction Detector

The schematic model of the $Hg_{1-x}Cd_xTe$-based p^+-n^- detector under LWIR condition is shown in Fig. (**7**). The homojunction detector comprises a 4.5 μm thick layer of p^+-$Hg_{0.7783}Cd_{0.2217}Te$ layer on the top of an n^--$Hg_{0.7783}Cd_{0.2217}Te$ layer with a thickness of 10.25 μm. The p^+-$Hg_{0.7783}Cd_{0.2217}Te$ layer is exposed to LWIR radiation with an intensity of 1 W/cm^2 to capture substantial amounts of radiation. The LWIR radiation gets absorbed within the p$^+$- and n$^-$-regions. The concentrations of acceptors and donors are set at 10^{17} and 10^{15} cm^{-3} for the p$^+$- and n$^-$-regions, respectively. The depletion region width x_p^+ and x_n^- near the junction are approximated as [23]:

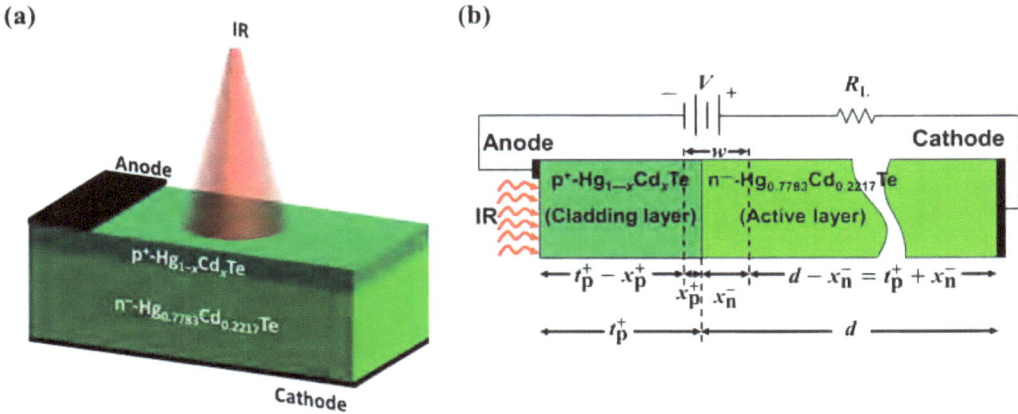

Fig. (7). The schematic representation of p$^+$-n$^-$ HgCdTe-based LWIR detector. **(a)** three-dimensional (3D) view and **(b)** two-dimensional (2D) view. Here R_L is load resistance.

$$x_p^+ = \sqrt{\frac{2\varepsilon_{r1}\varepsilon_{r2}}{q}\left[\frac{N_D}{N_A}\left(\frac{V_{bi}-V}{\varepsilon_{r1}N_A + \varepsilon_{r2}N_D}\right)\right]} \tag{17}$$

$$x_n^- = \sqrt{\frac{2\varepsilon_{r1}\varepsilon_{r2}}{q}\left[\frac{N_A}{N_D}\left(\frac{V_{bi}-V}{\varepsilon_{r1}N_A + \varepsilon_{r2}N_D}\right)\right]} \tag{18}$$

The total depletion region width w is estimated by [23]:

$$w = x_p^+ + x_n^- = \sqrt{\frac{2}{q}\frac{\varepsilon_{r1}\varepsilon_{r2}}{\left(\varepsilon_{r1}N_A + \varepsilon_{r2}N_D\right)}\frac{\left(N_A + N_D\right)^2}{N_A N_D}\left(V_{bi}-V\right)} \tag{19}$$

where $V_{bi} = V_{ip} + V_{in}$ corresponds to the built-in potential. V_{ip} is the barrier potential related to energy band bending in the p$^+$- region whereas V_{in} represents the barrier potential associated with energy band bending in the n$^-$-region.

In Fig. (**8a**), the diagram depicting the energy bandgap of the simulated p$^+$-n$^-$ Hg$_{0.7783}$Cd$_{0.2217}$Te-based homojunction LWIR detector at 0 V is illustrated under no illumination. In simulating the energy bandgap of all the detectors, a ground state degeneracy of 4 is assumed for the valence band, and a degeneracy of 2 is assumed for the conduction band [19, 24].

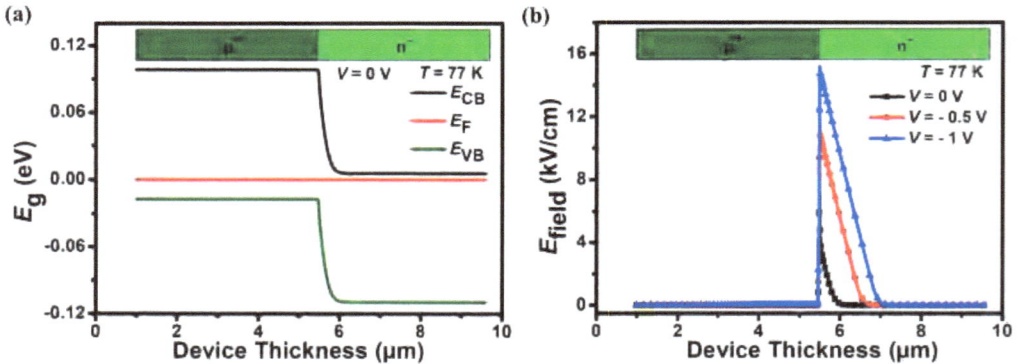

Fig. (8). (**a**) The energy bandgap depiction and (**b**) the electric field profile across the p$^+$-n$^-$ Hg$_{0.7783}$Cd$_{0.2217}$Te-based LWIR homojunction detector. In this context, E_{CB}, E_{VB}, and E_F signify the conduction band, valence band, and Fermi energy levels, respectively.

Results and Discussion

The simulation for the homojunction detector, utilizing Hg$_{1-x}$Cd$_x$Te with x=0.2217, is conducted to operate at a wavelength of 10.6 μm. Electron-hole pairs are produced when the energy of the incident photons exceeds the bandgap of the material. The composition value of Cd in Hg$_{1-x}$Cd$_x$Te is fine-tuned to achieve a wavelength of 10.6 μm.

Fig. (**8b**) depicts the electric field (E_{field}) across the detector's junction without illumination at various biases. The E_{field} exhibits a well-defined triangular shape at the junction, attaining peak values of 7.36 kV/cm at 0 V, 11.65 kV/cm at –0.5 V, and 15.15 kV/cm at –1 V within the detector's junction.

Fig. (**9**) illustrates the dark current density J_{dark} variations with the voltage V. These variations result from the presence of diffusion (diff), generation-recombination (gr), and tunneling (TUN) current components. The tunneling current component encompasses contributions from both band-to-band (BTB) and trap-assisted tunneling (TAT). Accordingly, the voltage and temperature T-dependent overall J_{dark} is expressed as [21]:

$$J_{\text{dark}}(V,T) = J_{\text{diff}} + J_{\text{gr}} + J_{\text{BTB}} + J_{\text{TAT}} \tag{20}$$

where

$$J_{\text{diff}} = \left[(J_{n^-})_{p^+} + (J_{p^+})_{n^-} \right] \left(e^{(qV/kT)} - 1 \right) \tag{21}$$

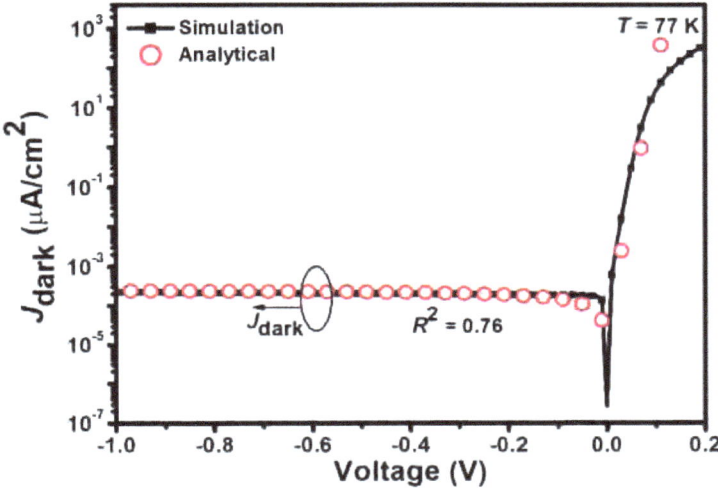

Fig. (9). The J_{dark}-V and $(RA)_{\text{NET}}$-V characteristics of p^+-n^- $Hg_{0.7783}Cd_{0.2217}Te$-based LWIR homojunction detector.

Here, $(J_{n^-})_{p^+}$ and $(J_{p^+})_{n^-}$ signify the diffusion current density components for electrons and holes in the p^+- and n^--regions, respectively, and are calculated by:

$$(J_{n^-})_{p^+} = \frac{q n_{ip}^2}{N_A} \sqrt{\frac{\mu_n kT}{q\tau_n}} \frac{\gamma_n \cosh\left(\frac{t_p^+ - x_p^+}{L_n}\right) + \sinh\left(\frac{t_p^+ - x_p^+}{L_n}\right)}{\cosh\left(\frac{t_p^+ - x_p^+}{L_n}\right) + \gamma_n \sinh\left(\frac{t_p^+ - x_p^+}{L_n}\right)} e^{\left(-\frac{q(V_{bi} + \Delta E_{CB})}{kT}\right)} \tag{22}$$

$$(J_{p^+})_{n^-} = \frac{q n_{in}^2}{N_D} \sqrt{\frac{\mu_p kT}{q\tau_p}} \frac{\gamma_p \cosh\left(\frac{d - x_n^-}{L_p}\right) + \sinh\left(\frac{d - x_n^-}{L_p}\right)}{\cosh\left(\frac{d - x_n^-}{L_p}\right) + \gamma_p \sinh\left(\frac{d - x_n^-}{L_p}\right)} e^{\left(-\frac{q(V_{bi} + \Delta E_{VB})}{kT}\right)} \tag{23}$$

where n_{ip} and n_{in} represent the intrinsic carrier concentrations in p$^+$- and n$^-$-regions, respectively, N_A and N_D are the acceptor and donor concentrations, respectively. μ_n and μ_p denote the electron and hole mobilities, respectively. The duration of the electron's and hole's lifetimes are denoted as τ_n and τ_p, respectively. In the p$^+$- and n$^-$-regions, the surface-to-bulk recombination velocity ratios are indicated via $\gamma_p = L_p S_p / D_p$ and $\gamma_n = L_n S_n / D_n$, respectively.

$L_p = \sqrt{D_p \tau_p}$ and $L_n = \sqrt{D_n \tau_n}$ denote the hole and electron diffusion lengths, respectively, measured in centimeters. Meanwhile, S_p and S_n are the hole's and electron's surface recombination velocities at the heterojunction, respectively. The hole diffusion coefficient D_p and the electron diffusion coefficient D_n are approximated by the following equations [43]:

$$D_p = \frac{\mu_p kT}{q} \quad \text{cm}^2/\text{s} \tag{24}$$

$$D_n = \frac{\mu_n kT}{q} \quad \text{cm}^2/\text{s} \tag{25}$$

Following the creation of the p$^+$-n$^-$ heterojunction, the discontinuities in the valence and conduction band edges are denoted by $\Delta E_{VB} (= \Delta E_g - \Delta E_{CB})$ and $\Delta E_{CB} (= \chi_n - \chi_p)$, respectively. Here, χ_n and χ_p represent the electron affinities in the n$^-$- and p$^+$-regions, respectively with $\chi_n > \chi_p$. The total bandgap discontinuity is represented by $\Delta E_g (= E_{gp} - E_{gn})$, where E_{gp} and E_{gn} are the energy bandgaps in the p$^+$- and n$^-$-regions, respectively. ΔE_{VB} and ΔE_{CB} denote the edge discontinuities in the valence and conduction band, respectively [21].

The migration of charge carriers through the p$^+$-n$^-$ junction is significantly influenced by the trap levels positioned at the junction, especially within the depletion region. The current density components attributed to electrons and holes generated and recombined in the depletion region are expressed as [21]:

$$J_{gr} = \begin{cases} \dfrac{q n_{in} w V}{(V_{bi} - V)\tau_{SRH}} & V < 0 \\[3mm] \dfrac{2 n_{in} w kT}{(V_{bi} - V)\tau_{SRH}} \sinh\left(\dfrac{qV}{2kT}\right) & V > 0 \end{cases} \tag{26}$$

where $\tau_{SRH} = 1/\sigma N_f v_{th}$ denotes the SRH generation-recombination lifetime of carriers. σ signifies the minority carrier's capture cross-section, N_f corresponds to the SRH trap density, and $v_{th} = \sqrt{\dfrac{3kT}{m_n^*}}$ represents the minority carrier's thermal velocity.

The TAT component of current density is analytically assessed by [21]:

$$J_{TAT} = \frac{2\pi^2 q^2 m_n^* W_c^2 N_t \left(V_{bi} - V\right) w}{h^3 \left(\Delta E_g - E_t\right)} e^{\left(-\frac{\sqrt{3} w \Delta E_g^2}{8\sqrt{2} P (V_{bi} - V)} \alpha\left(\frac{E_t}{\Delta E_g}\right)\right)} \tag{27}$$

In this context, $\alpha\left(\dfrac{E_t}{\Delta E_g}\right) = \dfrac{\pi}{2} + \sin^{-1}\left(\pm 1 \pm 2\dfrac{E_t}{\Delta E_g}\right) \pm \left(1 - 2\dfrac{E_t}{\Delta E_g}\right)\sqrt{\dfrac{E_t}{\Delta E_g}\left(1 - \dfrac{E_t}{\Delta E_g}\right)}$ (28)

where, P and W_c correspond to the matrix elements linked to the interband and potential, respectively. N_t characterizes the trap density, distinct from the SRH trap density, and E_t indicates the position of the trap levels within the bandgap.

The current density component due to BTB tunneling mechanism in this work is approximated by [21, 23]:

$$J_{BTB} = \frac{q^3 EV}{4\pi^2 \eta^2}\sqrt{\frac{2m_n^*}{\Delta E_g}} e^{\left(-\frac{4\sqrt{2m_n^* \Delta E_g^3}}{3q\eta E}\right)} \tag{29}$$

where $\eta = h/2\pi$ denotes the reduced Planck's constant and

$$E = \left[\frac{2q}{\varepsilon_0 \varepsilon_{r2}}\left(\frac{\Delta E_g}{q} + V\right)\frac{n_0 p_0}{n_0 + p_0}\right] \tag{30}$$

here, n_0 and p_0 represent the thermal-equilibrium concentration of electrons and holes, respectively.

The resulting net resistance-area product $((RA)_{NET})$ arising from distinct components of current densities is formulated via:

$$\frac{1}{(RA)_{NET}} = \frac{1}{(RA)_{diff}} + \frac{1}{(RA)_{gr}} + \frac{1}{(RA)_{BTB}} + \frac{1}{(RA)_{TAT}} \qquad (31)$$

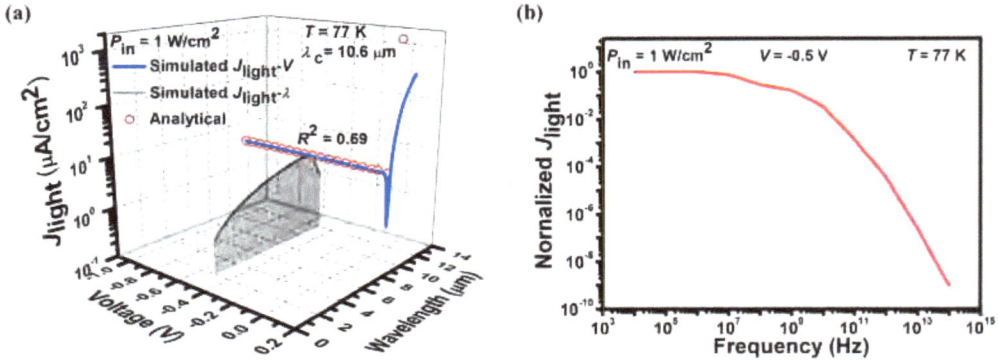

Fig. (10). (a) The J_{light}-V and J_{light}-λ characteristics and (b) normalized frequency response of the p$^+$-n$^-$ Hg$_{0.7783}$Cd$_{0.2217}$Te-based LWIR homojunction detector.

Here $(RA)_x = \left(\dfrac{dJ_x}{dV}\right)^{-1}$. The subscript x represents the distinct components of current density.

The change in J_{light} to applied bias and wavelength (λ) is depicted in Fig. (10a). There is a swift decline in J_{light} following the λ_c of 10.6 µm, affirming the operation of the detector in the LWIR region. Furthermore, as depicted in Fig. (10a), there is a rise in the J_{light} of the device with an escalation in voltage. The analytical approximation for the voltage and temperature-dependent $J_{light}(V,T)$ through the detector is given by [44]:

$$J_{light}(V,T) = J_{dark}(V,T) - \frac{QE_{ext}\,q\lambda_c P_{in}}{hc} \qquad (32)$$

In equation (32), the second term notifies the current density generated due to the incident photons. The simulated J-V characteristics exhibit a satisfactory alignment with those derived from the analytical analysis, with a coefficient of correlation R^2

value varying between 0.69 and 0.76. The detector shows a low value of J_{dark} of 0.26 pA/cm^2, J_{light} of 4.9 µA/cm^2, J_{light}/J_{dark} ratio of 1.9×10^7, and $(R_0A)_{NET}$ of 26.88 Ωcm^2 at zero bias and 77 K. The detector demonstrates a small J_{dark} value attributed to the potential barrier within the detector at 0 V. The detector's SRH carrier lifetime is calculated to be 0.29 µs.

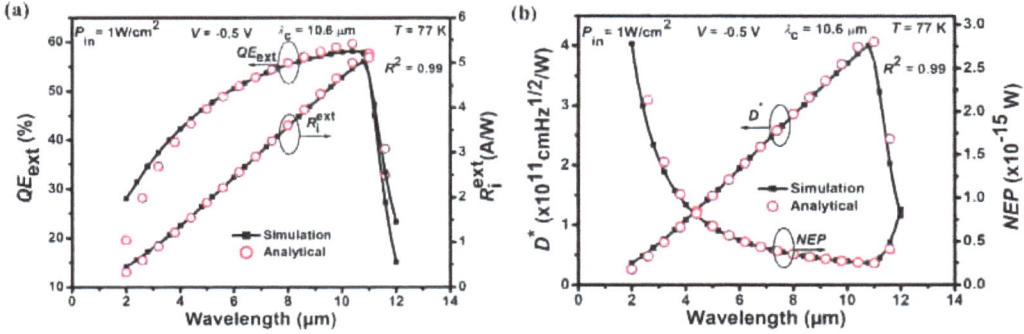

Fig. (11). The simulated and analytical results of **(a)** QE_{ext}, R_i^{ext} , **(b)** D^* and NEP of p$^+$-n$^-$ Hg$_{0.7783}$Cd$_{0.2217}$Te-based LWIR homojunction detector with varying wavelength.

The frequency response of the p$^+$-n$^-$ Hg$_{0.7783}$Cd$_{0.2217}$Te-based LWIR homojunction detector is demonstrated in Fig. **(10b)**. The detector demonstrates a 3-dB cut-off frequency ($f_{3\text{-dB}}$) of 104 GHz with a response time $\tau_r(s) = \dfrac{0.34}{f_{3-dB}(Hz)}$ [45] of 3.3 ps.

Fig. **(11a)** demonstrates the change in external quantum efficiency (QE_{ext}) and external photocurrent responsivity $\left(R_i^{ext}\right)$ with a wavelength of –0.5 V. The variation of specific detectivity (D^*) and noise equivalent power (NEP) with the wavelength is shown in Fig. **(11b)**. In this chapter, NEP for all detectors is calculated at a unity bandwidth.

The overall QE_{ext} of the detector is influenced by three factors present in the p$^+$-, n$^-$-, and depletion regions:

$$QE_{ext} = \left(QE_{ext}\right)_{p^+} + \left(QE_{ext}\right)_{n^-} + \left(QE_{ext}\right)_{dep} \tag{33}$$

where

$$
(QE_{ext})_{p^+} = \frac{(1-R_p)\alpha_p L_n}{\alpha_p^2 L_n^2 - 1} e^{-(\alpha_p t_p^+ + \alpha_n x_n^-)}
$$

$$
\times \left[\frac{(\gamma_n - \alpha_p L_n)e^{-\alpha_n(d-x_n^-)} - \left\{\gamma_n \cosh\left(\dfrac{d-x_n^-}{L_n}\right) + \sinh\left(\dfrac{d-x_n^-}{L_n}\right)\right\}}{\cosh\left(\dfrac{d-x_n^-}{L_n}\right) + \gamma_n \sinh\left(\dfrac{d-x_n^-}{L_n}\right)} + \alpha_p L_n \right] \tag{34}
$$

$$
(QE_{ext})_n = \frac{(1-R_p)(1-R_n)\alpha_n L_p}{\alpha_n^2 L_p^2 - 1}
$$

$$
\times \left[\frac{(\alpha_n L_p + \gamma_p) - e^{-\alpha_n x_p^+}\left\{\gamma_p \cosh\left(\dfrac{x_p^+}{L_p}\right) + \sinh\left(\dfrac{x_p^+}{L_p}\right)\right\}}{\cosh\left(\dfrac{x_p^+}{L_p}\right) + \gamma_p \sinh\left(\dfrac{x_p^+}{L_p}\right)} - \alpha_n L_p e^{-\alpha_n x_p^+} \right] \tag{35}
$$

and

$$
(QE_{ext})_{dep} = (1 - R_p)(1 - R_n)\left[e^{-\alpha_p x_p^+} - e^{-\alpha_n(t_p^+ + x_n^-)}\right] \tag{36}
$$

In this context, R_p and R_n denote the Fresnel reflection coefficients at the entry and p^+-n^- junction, respectively. α_p and α_n are the absorption coefficients in p^+- and n^--regions, respectively [21].

In general, the Fresnel reflection coefficient (R) at the hetero-interfaces is evaluated by:

$$
R = \frac{K^2 + (n_2 - n_1)^2}{K^2 + (n_2 + n_1)^2} \tag{37}
$$

where n_1 and n_2 represent the refractive indices of the entrance and transmitted medium, respectively. In the cases where the extinction coefficient K is negligible, the reflection coefficient can be calculated as [46]:

$$R = \left(\frac{n_2 - n_1}{n_2 + n_1}\right)^2 \qquad (38)$$

The external photocurrent responsivity R_i^{ext} which represents the ratio of photocurrent density to the illuminated power is calculated by using the following relation [1, 23, 47]:

$$R_i^{ext} = QE_{ext}\left(\frac{\lambda}{1.24}\right) \text{ A/W} \qquad (39)$$

where λ is the operating wavelength.

The specific detectivity D^* characterizes the smallest signal detectable by a detector and is computed by:

$$D^* = \frac{R_i^{ext}}{2}\sqrt{\frac{(R_0 A)_{NET}}{kT}} \text{ cmHz}^{1/2}/\text{W} \qquad (40)$$

Where $(R_0 A)_{NET} = \left(\frac{dJ}{dV}\right)^{-1}_{V=0}$ is the net resistance area product at 0 V. Achieving a high specific detectivity is facilitated by a large value of $(R_0 A)_{NET}$.

The noise equivalent power *NEP* represents the smallest signal that a detector can differentiate from noise, and it is estimated by [1, 23, 47, 48]:

$$NEP = \frac{\sqrt{\Delta f A}}{D^*} \text{ W/Hz}^{1/2} \qquad (41)$$

Here, Δf and A represent the bandwidth and active area, respectively.

The results presented in Fig. (**11**) demonstrate a commendable concordance between the findings obtained from the simulation and the analytical model, with an R^2 value of 0.99. At –0.5 V, the detector exhibits QE_{ext} of 58.29%, R_i^{ext} of 4.98 A/W, D^* of 3.96×10^{11} cmHz$^{1/2}$/W, and *NEP* of 2.52×10^{-16} W with an incident wavelength of 10.6 μm. This high value of R_i^{ext} is attributed to the reduction in the drift time of the photoinduced carriers, leading to an electron lifetime longer than the transit time. When the mean free path length of electrons surpasses the thickness of the device, photon interaction becomes negligible. The external terminals collect the photoinduced carriers in the active layer, contributing to a total photocurrent with high QE_{ext}, high R_i^{ext}, high D^*, and least NEP for the detector. Furthermore, by optimizing the device's thickness, its size is reduced, leading to a decrease in the generation of thermal excess carriers and noise in the diffusion-limited operation [49].

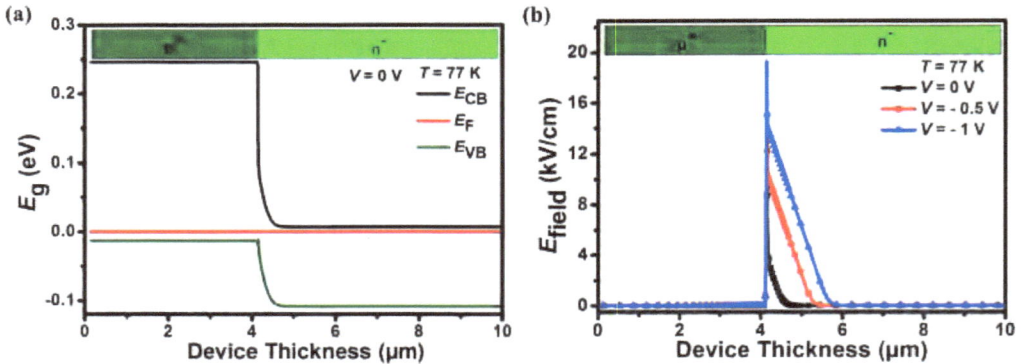

Fig. (12). **(a)** The energy bandgap depiction and **(b)** the electric field profile across the p⁺-Hg$_{0.69}$Cd$_{0.31}$Te/n⁻-Hg$_{0.7783}$Cd$_{0.2217}$Te LWIR heterojunction detector.

p⁺-n⁻ Heterojunction Detector

The heterojunction detector device in Fig. (**7**) comprises a 4 µm thick p⁺-Hg$_{0.69}$Cd$_{0.31}$Te layer over a 10.25 µm thick n⁻-Hg$_{0.7783}$Cd$_{0.2217}$Te layer. In this configuration, the doping level concentrations for the p⁺- and n⁻-regions are set at 2×10^{17} and 9×10^{14} cm⁻³, respectively [23]. The simulated energy bandgap depiction of p⁺-n⁻ Hg$_{1-x}$Cd$_x$Te-based heterojunction LWIR detector at 0 V is depicted in Fig. (**12a**) without any illumination.

Results and Discussion

In Fig. (**12b**), the electric field profile with a triangular shape across the p⁺-Hg$_{0.69}$Cd$_{0.31}$Te/n⁻-Hg$_{0.7783}$Cd$_{0.2217}$Te heterojunction detector is depicted without illumination at various biases. The electric field reaches a maximum value of 18.56, 18.91, and 19.27 kV/cm at 0, −0.5, and −1 V, respectively, at the heterojunction of the detector, which is more than the p⁺-n⁻ homojunction detector. This high electric field quickly drifts the photocarriers toward the external electrodes, contributing to the photocurrent density and enhancing the overall performance.

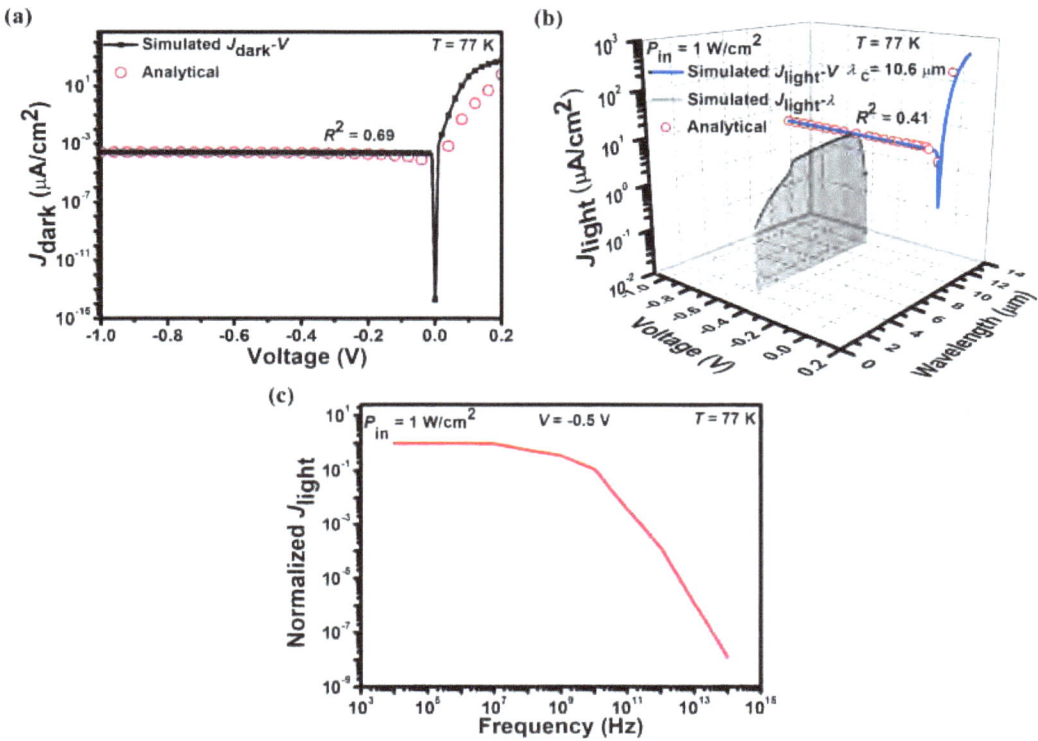

Fig. (13). (a) The J_{dark}-V characteristics of the p$^+$-Hg$_{0.69}$Cd$_{0.31}$Te/n$^-$-Hg$_{0.7783}$Cd$_{0.2217}$Te LWIR heterojunction detector. **(b)** The J_{light}-V and J_{light}-λ characteristics of the p$^+$-Hg$_{0.69}$Cd$_{0.31}$Te/n$^-$-Hg$_{0.7783}$Cd$_{0.2217}$Te LWIR heterojunction detector. **(c)** The normalized frequency response of the p$^+$-Hg$_{0.69}$Cd$_{0.31}$Te/n$^-$-Hg$_{0.7783}$Cd$_{0.2217}$Te LWIR heterojunction detector.

The J_{dark}-V characteristic of the p$^+$-Hg$_{0.69}$Cd$_{0.31}$Te/n$^-$-Hg$_{0.7783}$Cd$_{0.2217}$Te heterojunction detector is shown in Fig. **13(a)**. The J_{dark}-V and J_{light}-λ characteristics are illustrated in Fig. **13(b)**. The simulated J_{dark}-V and J_{light}-V characteristics show a satisfactory agreement with those derived from the analytical analysis with R^2 values of 0.69 and 0.41, respectively. The p$^+$-n$^-$ heterojunction detector shows a low value of J_{dark} of 1.98×10^{-8} pA/cm^2, J_{light} of 5.59 μA/cm^2, J_{light}/J_{dark} ratio of 2.82×10^{14}, and $(R_0A)_{NET}$ of 20.26 MΩcm^2 at zero bias and 77 K. The heterojunction detector's SRH carrier lifetime is determined to be 0.25 μs. The normalized frequency response of the p$^+$-Hg$_{0.69}$Cd$_{0.31}$Te/n$^-$-Hg$_{0.7783}$Cd$_{0.2217}$Te heterojunction detector is depicted in Fig. **13(c)**. The heterojunction detector shows a $f_{3\text{-dB}}$ of 265 GHz with τ_r of 1.3 ps.

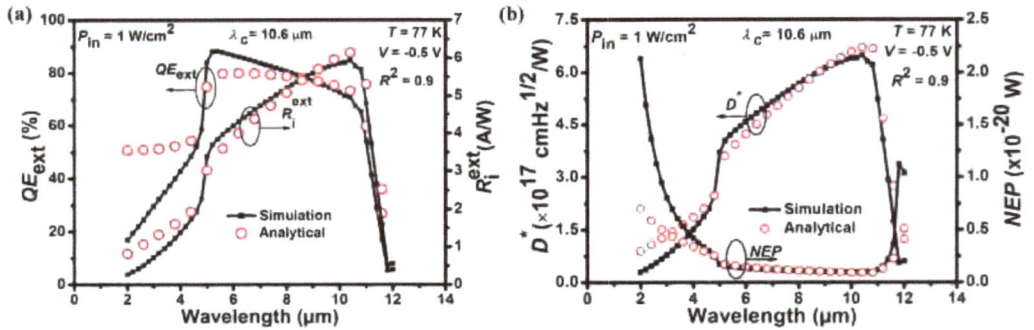

Fig. (14). The simulated and analytical results of **(a)** QE_{ext}, R_i^{ext}, **(b)** D^* and *NEP* of the p$^+$-Hg$_{0.69}$Cd$_{0.31}$Te/n$^-$-Hg$_{0.7783}$Cd$_{0.2217}$Te LWIR heterojunction detector with varying wavelength.

Fig. **(14a)** illustrates the variation in the QE_{ext} and R_i^{ext} with wavelength as found from both simulation and analytical models. The variation in D^* and *NEP* with wavelength is demonstrated in Fig. **(14b)**. The heterojunction detector demonstrates QE_{ext} of 67.6%, R_i^{ext} of 5.78 A/W, D^* of 6.31×10^{17} cmHz$^{1/2}$/W, and *NEP* of 1.0×10^{-21} W with an incident wavelength of 10.6 µm at -0.5 V. It is clear that the p$^+$-n$^-$ heterojunction detector outperforms the p$^+$-n$^-$ homojunction detector, and this superiority is attributed to the substantial electric field at the heterojunction.

n$^+$-n$^-$-p$^+$ Dual-Junction Detector

The dual-junction LWIR detector under consideration is depicted in Fig. **(15)**. The detector comprises a central, lightly doped n-type intrinsic/active layer, responsible for absorbing LWIR radiations, along with two heavily doped layers of Hg$_{1-x}$Cd$_x$Te material with distinct stoichiometric ratios of Hg and Cd. The doping concentrations of 10^{17}, 10^{12}, and 10^{17} cm^{-3} are taken for n$^+$-, n$^-$-, and p$^+$-regions, respectively. The LWIR radiations impinge on the 0.5 µm thick cladding layer of n$^+$-Hg$_{0.68}$Cd$_{0.32}$Te placed on 10.0 µm thick n$^-$-Hg$_{0.7783}$Cd$_{0.2217}$Te active layer. A majority of the incident radiations get absorbed by the n$^-$-Hg$_{0.7783}$Cd$_{0.2217}$Te due to its greater thickness compared to the cladding layer. Photons with energy larger than the energy bandgap generate photogenerated carriers in n$^-$, p$^+$, and depletion layers, resulting in the net photocurrent. The overall performance of the detector is notably affected by the J_{dark} originating from thermally generated carriers. The energy bandgap diagram of the n$^+$-n$^-$-p$^+$ dual-junction detector device without biasing under dark conditions is depicted in Fig. **(16a)**.

(a)

(b)

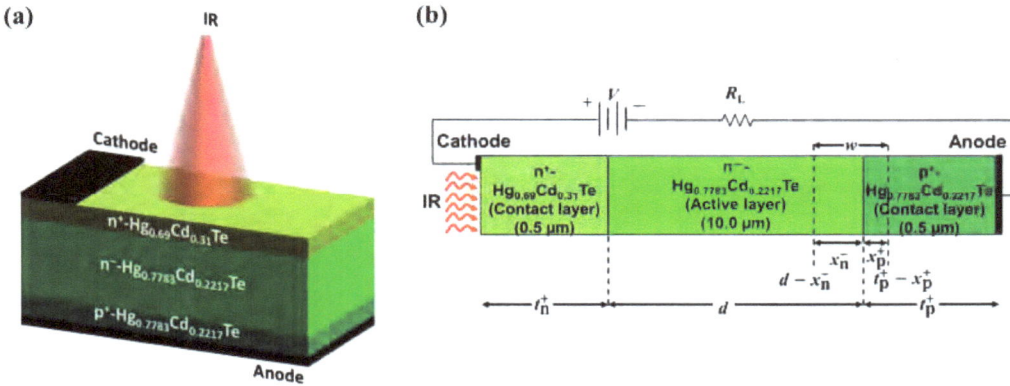

Fig. (15). The schematic representation of n^+-$Hg_{0.68}Cd_{0.32}Te$/n^--$Hg_{0.7783}Cd_{0.2217}Te$/p^+-$Hg_{0.7783}Cd_{0.2217}Te$ based dual-junction LWIR detector. **(a)** 3D view and **(b)** 2D view showing the depletion region in n^-- and p^+-regions.

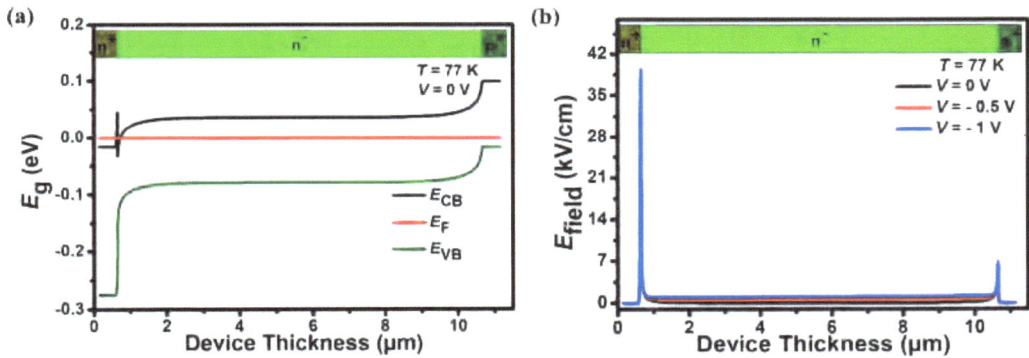

(a)

(b)

Fig. (16). (a) The energy bandgap depiction and **(b)** the electric field profile across the n^+-$Hg_{0.68}Cd_{0.32}Te$/n^--$Hg_{0.7783}Cd_{0.2217}Te$/p^+-$Hg_{0.7783}Cd_{0.2217}Te$ based LWIR dual-junction detector.

Results and Discussion

Fig. (**16b**) illustrates the triangular-shaped profile of the E_{field} within the detector under various external biasing. The maximum E_{field} of 39.32 and 6.78 kV/cm is found at the n^+-n^- and n^--p^+ junction of the detector, respectively, at –1.0 V.

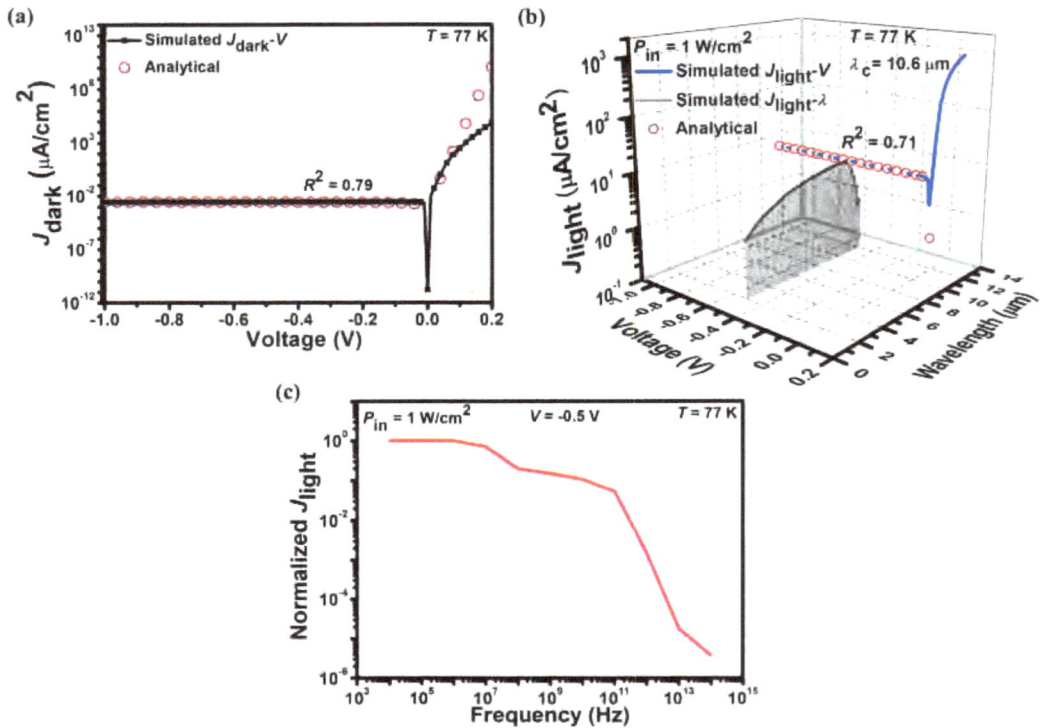

Fig. (17). (a) The J_{dark}-V characteristics of the n$^+$-Hg$_{0.68}$Cd$_{0.32}$Te/n$^-$-Hg$_{0.7783}$Cd$_{0.2217}$Te/p$^+$-Hg$_{0.7783}$Cd$_{0.2217}$Te LWIR dual-junction detector. **(b)** The J_{light}-V and J_{light}-λ characteristics of the n$^+$-Hg$_{0.68}$Cd$_{0.32}$Te/n$^-$-Hg$_{0.7783}$Cd$_{0.2217}$Te/p$^+$-Hg$_{0.7783}$Cd$_{0.2217}$Te LWIR dual-junction detector. **(c)** The normalized frequency response of the n$^+$-Hg$_{0.68}$Cd$_{0.32}$Te/n$^-$-Hg$_{0.7783}$Cd$_{0.2217}$Te/p$^+$-Hg$_{0.7783}$Cd$_{0.2217}$Te LWIR dual-junction detector.

The J_{dark}-V characteristic of the dual-junction detector is demonstrated in Fig. **17(a)**. Whereas the J_{light}-V and J_{light}-λ characteristics are depicted in Fig. **17(b)**. The results obtained from the simulations closely align with the results derived from analytical modeling having R^2 of 0.79 for the J_{dark}-V characteristic and 0.71 for the J_{light}-V characteristic. The estimated J_{dark}, J_{light}, J_{light}/J_{dark} ratio, and $(R_0A)_{NET}$ is of the order of $1.77{\times}10^{-5}$ pA/cm^2, 7.69 µA/cm^2, $4.34{\times}10^{11}$, and 1.8 MΩcm^2, respectively, at 0 V. The frequency response of the n$^+$-n$^-$-p$^+$ dual-junction detector is depicted in Fig. **17(c)**. The SRH carrier generation-recombination lifetime is found to be 0.57 µs. The estimated $f_{3\text{-}dB}$ is 1.28 THz with τ_r of 0.27 ps for the dual-junction detector.

Fig. **(18a)** demonstrates the variation in the QE_{ext} and R_i^{ext} with the wavelength of the dual-junction detector as found from both simulation and analytical models at –0.5 V. The variation in D^* and *NEP* with the detector's wavelength is depicted in Fig. (**18b**). The dual-junction detector shows QE_{ext} of 84.92%, R_i^{ext} of 7.26 A/W, D^* of 1.49×10^{14} cmHz$^{1/2}$/W, and *NEP* of 4.24×10^{-18} W and is better than the single-junction detectors. Such enhancement is attributed to the lowest thermal generation rate compared to single-junction-based detectors. The dual-junction configuration results in a high built-in E_{field}, facilitating the drift of photoinduced charge carriers to the external electrodes, thereby improving the detector's overall response [22].

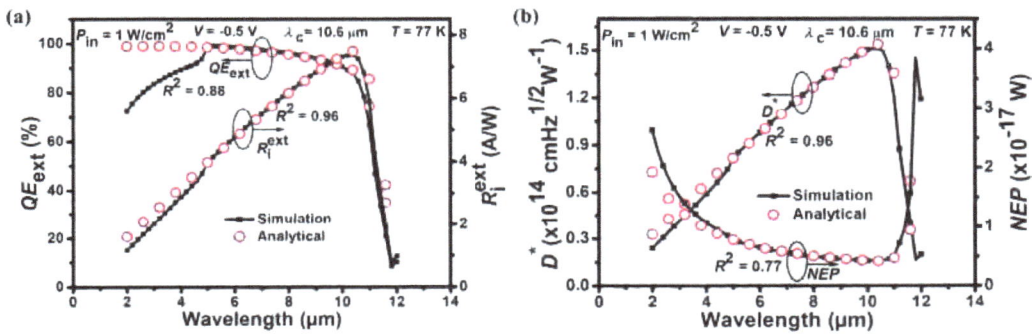

Fig. (18). The simulated and analytical results of (**a**) QE_{ext}, R_i^{ext}, (**b**) D^* and *NEP* of the n$^+$-Hg$_{0.68}$Cd$_{0.32}$Te/n$^-$-Hg$_{0.7783}$Cd$_{0.2217}$Te/p$^+$-Hg$_{0.7783}$Cd$_{0.2217}$Te based LWIR dual-junction detector with varying wavelength.

CONCLUSION

In conclusion, the two-dimensional architecture models of p$^+$-Hg$_{0.7783}$Cd$_{0.2217}$Te/n$^-$-Hg$_{0.7783}$Cd$_{0.2217}$Te (homojunction), p$^+$-Hg$_{0.69}$Cd$_{0.31}$Te/n$^-$-Hg$_{0.7783}$Cd$_{0.2217}$Te (heterojunction), and n$^+$-Hg$_{0.68}$Cd$_{0.32}$Te/n$^-$-Hg$_{0.7783}$Cd$_{0.2217}$Te/p$^+$-Hg$_{0.7783}$Cd$_{0.2217}$Te (dual-junction) are presented in LWIR spectral region at 77 K by using Silvaco Atlas TCAD software. The simulation results are compared with those attained from the analytical model, revealing alignment with the analytical expressions. The p$^+$-Hg$_{0.7783}$Cd$_{0.2217}$Te/n$^-$-Hg$_{0.7783}$Cd$_{0.2217}$Te detector shows J_{light}/J_{dark} of 1.9×107, QE_{ext} of 58.29%, R_i^{ext} of 4.98 A/W, D^* of 3.96×10^{11} cmHz$^{1/2}$/W, and a $f_{3\text{-}dB}$ of 104 GHz with a response time of 3.3 ps. On the other hand, the p$^+$-Hg$_{0.69}$Cd$_{0.31}$Te/n$^-$-Hg$_{0.7783}$Cd$_{0.2217}$Te detector exhibits a low dark current density with J_{light}/J_{dark} of 2.82×10^{14}, QE_{ext} of 67.6%, R_i^{ext} of 5.78 A/W, D^* of 6.31×10^{17} cmHz$^{1/2}$/W, a $f_{3\text{-}dB}$ of 265 GHz with a response time of 1.3 ps. Whereas the n$^+$-Hg$_{0.68}$Cd$_{0.32}$Te/n$^-$-Hg$_{0.7783}$Cd$_{0.2217}$Te/p$^+$-Hg$_{0.7783}$Cd$_{0.2217}$Te detector exhibits a J_{light}/J_{dark} of 4.34×10^{11},

QE_{ext} of 84.92%, R_i^{ext} of 7.26 A/W, D^* of 1.49×10^{14} cmHz$^{1/2}$/W, a $f_{3\text{-}dB}$ of 1.28 THz with a response time of 0.27 ps, further confirms the suitability of the dual-junction detector for low noise operations. The improvement in the dual-junction detector can be attributed to the high built-in electric field and the lowest thermal generation rate compared to single-junction-based detectors. The high electric field is advantageous for effectively drifting the photoinduced carriers toward the electrodes, thereby enhancing the performance of the detector.

Multiple Choice Questions (MCQs)

1. Which material is commonly used as the active layer in HgCdTe-based photodetectors?

a) Silicon

b) Gallium arsenide

c) HgCdTe

d) Aluminum nitride

Answer: c) HgCdTe

2. What is the main advantage of using HgCdTe-based photodetectors over silicon-based ones?

a) Higher operating temperature

b) Lower cost

c) Lower noise

d) Larger pixel size

Answer: a) Higher operating temperature

3. HgCdTe-based photodetectors are most commonly used in which spectral range?

a) Ultraviolet

b) Visible

c) Infrared

d) X-ray

Answer: c) Infrared

4. What does the abbreviation "HgCdTe" stand for?

a) High-grade Cadmium Telluride

b) Heterogeneous Cadmium Telluride

c) Mercury Cadmium Telluride

d) Hybrid Cadmium Telluride

Answer: c) Mercury Cadmium Telluride

5. Which of the following is a key parameter for characterizing the performance of HgCdTe-based photodetectors?

a) Bandwidth

b) Refractive index

c) Resistivity

d) Quantum efficiency

Answer: d) Quantum efficiency

6. The composition of Cd in HgCdTe can be varied to tune its:

a) Optical absorption

b) Mechanical strength

c) Electrical conductivity

d) Thermal conductivity

Answer: a) Optical absorption

7. Which of the following applications typically require HgCdTe-based photodetectors?

a) Digital cameras

b) X-ray machines

c) Night vision devices

d) Solar panels

Answer: c) Night vision devices

8. HgCdTe-based long-wavelength photodetectors are sensitive to which range of wavelengths?

a) 200 nm to 400 nm

b) 400 nm to 700 nm

c) 700 nm to 1200 nm

d) 1100 nm to 2000 nm

Answer: c) 700 nm to 1200 nm

9. In HgCdTe-based photodetectors, the bandgap can be adjusted by varying the:

a) Temperature

b) Voltage

c) Composition

d) Thickness

Answer: c) Composition

10. Which of the following factors contributes to the high sensitivity of HgCdTe-based photodetectors?

a) High dark current

b) Low quantum efficiency

c) Direct bandgap

d) High responsivity

Answer: c) Direct bandgap

11. What is the primary difference between homojunction and heterojunction photodetectors?

a) Bandgap energy

b) Composition of the active layer

c) Operating temperature

d) Quantum efficiency

Answer: b) Composition of the active layer

12. In a homojunction photodetector, the p-n junction is formed within

a) A single semiconductor material

b) Different semiconductor materials

c) A metal-semiconductor interface

d) An insulating layer

Answer: a) A single semiconductor material

13. Which of the following statements is true about heterojunction photodetectors?

a) They have a constant bandgap throughout the device.

b) They utilize two different semiconductor materials with different bandgaps.

c) They have a lower dark current compared to homojunction photodetectors.

d) They are less sensitive to temperature variations.

Answer: b) They utilize two different semiconductor materials with different bandgaps.

14. What does the "p" in p-n and p-i-n photodetectors represent?

a) Positive charge carriers

b) Negative charge carriers

c) Photo-generated charge carriers

d) Semiconductor material

Answer: a) Positive charge carriers

15. In a p-i-n photodetector, what does the "i" region typically represent?

a) Intrinsic semiconductor material

b) Extrinsic semiconductor material

c) Insulating material

d) Interfacial layer

Answer: a) Intrinsic semiconductor material

16. Which type of photodetector is more suitable for high-speed applications?

a) Homojunction

b) Heterojunction

c) Both are equally suitable

d) Not suitable for high-speed applications

Answer: b) Heterojunction

17. Which of the following statements is true about p-n photodetectors?

a) They have a lower quantum efficiency compared to p-i-n photodetectors.

b) They have a wider depletion region compared to p-i-n photodetectors.

c) They have a higher dark current compared to p-i-n photodetectors.

d) They have a lower responsivity compared to p-i-n photodetectors.

Answer: c) They have a higher dark current compared to p-i-n photodetectors.

18. In a heterojunction photodetector, which of the following factors affects the band alignment at the interface?

a) Operating voltage

b) Temperature

c) Composition of the materials

d) Light intensity

Answer: c) Composition of the materials

19. Which type of photodetector is more commonly used in applications requiring high sensitivity and low noise?

a) Homojunction

b) Heterojunction

c) P-N

d) P-I-N

Answer: d) P-I-N

20. What advantage do heterojunction photodetectors offer over homojunction photodetectors?

a) Higher operating temperature

b) Lower dark current

c) Greater versatility in bandgap engineering

d) Larger active area

Answer: c) Greater versatility in bandgap engineering

REFERENCES

[1] L.H. Zeng, M.Z. Wang, H. Hu, B. Nie, Y.Q. Yu, C.Y. Wu, L. Wang, J.G. Hu, C. Xie, F.X. Liang, and L.B. Luo, "Monolayer graphene/germanium Schottky junction as high-performance self-driven infrared light photodetector", *ACS Appl. Mater. Interfaces,* vol. 5, no. 19, pp. 9362-9366, 2013.

http://dx.doi.org/10.1021/am4026505 PMID: 24040753

[2] Shen, Z. Yang, Z. Dai, and B. Chen, "Multi-stage infrared detectors," *Semicond. Sci. Technol.,* vol. 40, no. 3, 2025.

[3] M. Amirmazlaghani, F. Raissi, O. Habibpour, J. Vukusic, and J. Stake, "Graphene-Si Schottky IR detector", *IEEE J. Quantum Electron.,* vol. 49, no. 7, pp. 589-594, 2013.

[4] S. Assefa, F. Xia, and Y.A. Vlasov, "Reinventing germanium avalanche photodetector for nanophotonic on-chip optical interconnects", *Nature,* vol. 464, no. 7285, pp. 80-84, 2010.

http://dx.doi.org/10.1038/nature08813 PMID: 20203606

[5] J. Yoon, S. Jo, I.S. Chun, I. Jung, H.S. Kim, M. Meitl, E. Menard, X. Li, J.J. Coleman, U. Paik, and J.A. Rogers, "GaAs photovoltaics and optoelectronics using releasable multilayer epitaxial assemblies", *Nature,* vol. 465, no. 7296, pp. 329-333, 2010.

http://dx.doi.org/10.1038/nature09054 PMID: 20485431

[6] J. Miao, W. Hu, N. Guo, Z. Lu, X. Liu, L. Liao, P. Chen, T. Jiang, S. Wu, J.C. Ho, L. Wang, X. Chen, and W. Lu, "High-responsivity graphene/InAs nanowire heterojunction near-infrared photodetectors with distinct photocurrent on/off ratios", *Small,* vol. 11, no. 8, pp. 936-942, 2015.

http://dx.doi.org/10.1002/smll.201402312 PMID: 25363206

[7] P.C. Eng, S. Song, and B. Ping, "State-of-the-art photodetectors for optoelectronic integration at telecommunication wavelength", *Nanophotonics,* vol. 4, no. 3, pp. 277-302, 2015.

http://dx.doi.org/10.1515/nanoph-2015-0012

[8]   A. Rogalski, "HgCdTe infrared detector material: history, status and outlook", *Rep Prog Phys.,* vol. 68, no. 10, pp. 2267-2336, 2005.

http://dx.doi.org/10.1088/0034-4885/68/10/R01

[9]   P.K. Saxena, and P. Chakrabarti, "Computer modeling of MWIR single heterojunction photodetector based on mercury cadmium telluride", *Infrared Phys. Technol,* vol. 52, no. 5, pp. 196-203, 2009.

http://dx.doi.org/10.1016/j.infrared.2009.07.009

[10] P.K. Saxena, "Numerical study of dual band (MW/LW) IR detector for performance improvement", *Def. Sci. J.,* vol. 67, no. 2, pp. 141-148, 2017.

http://dx.doi.org/10.14429/dsj.67.11177

[11] N. Dehdashti Akhavan, G.A. Umana-Membreno, R. Gu, M. Asadnia, J. Antoszewski, and L. Faraone, "Superlattice barrier HgCdTe nBn infrared photodetectors: validation of the effective mass approximation", *IEEE Trans Electron. Dev.,* vol. 63, no. 12, pp. 4811-4818, 2016.

http://dx.doi.org/10.1109/TED.2016.2614677

[12] M. Kopytko, A. Keblowski, W. Gawron, A. Kowalewski, and A. Rogalski, "MOCVD grown HgCdTe barrier structures for hot conditions (July 2014)", *IEEE Trans Electron. Dev.,* vol. 61, no. 11, pp. 3803-3807, 2014.

http://dx.doi.org/10.1109/TED.2014.2359224

[13] N. Dehdashti Akhavan, G.A. Umana-Membreno, R. Gu, J. Antoszewski, and L. Faraone, "Optimization of superlattice barrier HgCdTe nBn infrared photodetectors based on an NEGF approach", *IEEE Trans Electron. Dev.,* vol. 65, no. 2, pp. 591-598, 2018.

http://dx.doi.org/10.1109/TED.2017.2785827

[14] J. Wojtas, Z. Bielecki, T. Stacewicz, J. Mikołajczyk, and M. Nowakowski, "Ultrasensitive laser spectroscopy for breath analysis", *Opto-Electron. Rev.,* vol. 20, no. 1, pp. 26-39, 2012.

http://dx.doi.org/10.2478/s11772-012-0011-4

[15] E. Bellotti, and D. D'Orsogna, "Numerical Analysis of HgCdTe Simultaneous Two-Color Photovoltaic Infrared Detectors", *IEEE J. Quantum Electron.,* vol. 42, no. 4, pp. 418-426, 2006.

http://dx.doi.org/10.1109/JQE.2006.871555

[16] J. Hodgkinson, and R.P. Tatam, "Optical gas sensing: a review", *Meas Sci. Technol,* vol. 24, no. 1, pp. 012004-1, 012004-012059, 2013.

http://dx.doi.org/10.1088/0957-0233/24/1/012004

[17] "Atlas Simulation of a Long-Infrared P+-N Homojunction Photodiode," In: *Proc. 6th Edition of Int'l Conf. Wireless Networks & Embedded Systems (WECON 2018),* 2018, pp. 19–22.

http://dx.doi.org/10.1109/WECON.2018.8782077

[18] S. Bansal, M. Muthukumar, and S. Kumar, "Graphene/HgCdTe Heterojunction-based IR Detectors", In: *Handbook of II-VI Semiconductor-Based Sensors and Radiation Detectors.* Springer, Cham, G. Korotcenkov (Ed.), 2023, pp. 183-202.

http://dx.doi.org/10.1007/978-3-031-20510-1_8

[19] S. Bansal, K. Sharma, N. Gupta, and A.K. Singh, "Simulation and optimization of Hg1-xCdxTe based mid-wavelength IR photodetector", *2016 IEEE Uttar Pradesh Section International Conference on Electrical, Computer and Electronics Engineering (UPCON),* 2016pp. 422-425 Varanasi

[20] S. Bansal, P. Jain, N. Gupta, A.K. Singh, N. Kumar, S. Kumar, and N. Sardana, "A highly efficient bilayer graphene HgCdTe heterojunction based p+-n photodetector for long wavelength infrared (LWIR)", *in 2018 IEEE 13th Nanotechnology Materials and Devices Conference (NMDC), ,* 2018pp. 1-4

http://dx.doi.org/10.1109/NMDC.2018.8605848

[21] S. Bansal, K. Sharma, P. Jain, N. Sardana, S. Kumar, N. Gupta, and A.K. Singh, "Bilayer graphene/HgCdTe based very long infrared photodetector with superior external quantum efficiency, responsivity, and detectivity", *RSC Advances,* vol. 8, no. 69, pp. 39579-39592, 2018.

http://dx.doi.org/10.1039/C8RA07683A PMID: 35558011

[22] S. Bansal, A. Das, P. Jain, K. Prakash, K. Sharma, N. Kumar, N. Sardana, N. Gupta, S. Kumar, and A.K. Singh, "Enhanced optoelectronic properties of bilayer graphene/HgCdTe based single- and dual-junction photodetectors in long infrared regime", *IEEE Trans Nanotechnol.,* vol. 18, pp. 781-789, 2019.

[23] A.D.D. Dwivedi, "Analytical modeling and numerical simulation of P+-Hg0.69Cd0.31Te/n-Hg0.78Cd0.22Te/CdZnTe heterojunction photodetector for a long-wavelength infrared free space optical communication system", *J. Appl. Phys.,* vol. 110, no. 4, pp. 043101-1, 043101-043110, 2011.

[24] S. Bansal, K. Sharma, K. Soni, N. Gupta, K. Ghosh, and A.K. Singh, "Hg1−xCdxTe based p-i-n IR photodetector for free space optical communication", *in 2017 Progress In Electromagnetics Research Symposium-Spring (PIERS), ,* 2017

[25] A. Singh, A.K. Shukla, and R. Pal, "Performance of graded bandgap HgCdTe avalanche photodiode", *IEEE Trans Electron. Dev.,* vol. 64, no. 3, pp. 1146-1152, 2017.

[26] S. Bansal, K. Prakash, N. Sardana, S. Kumar, K. Sharma, P. Jain, N. Gupta, and A.K. Singh, "Bilayer graphene/HgCdTe based self-powered mid-wave IR nBn photodetector", *2019 IEEE 14th Nanotechnology Materials and Devices Conference (NMDC), ,* 2019pp. 1-4

[27] S. Bansal, A. Das, K. Prakash, K. Sharma, N. Sardana, G.M. Khanal, S. Kumar, N. Gupta, and A.K. Singh, *Bilayer Graphene/HgCdTe Heterojunction Based Novel GBn Infrared Detectors,* vol. Vol. 169. Micro and Nanostructures, 2022, pp. 207345-1, 207345-12.

[28] S. Bansal, "Long-wave Bilayer Graphene/HgCdTe based GBp Type-II Superlattice Unipolar Barrier Infrared Detector", In: *Results in Optics,* vol. 12. 2023, pp. 100425-1-100425-9.

[29] Y. Ren, G. Qin, S. Geng, J. Yang, S. J. Li, H. F. Li, ... and J. C. Kong, "Molecular beam epitaxial growth of HgCdTe mid-wave infrared dual-band detectors," *Infrared Physics & Technology*, vol. 145, p. 105641, 2025.

[30] A. Rogalski, *Infrared detectors,* 2nd ed CRC Press, 2010.

[31] A. Piotrowski, P. Madejczyk, W. Gawron, K. Kłos, J. Pawluczyk, J. Rutkowski, J. Piotrowski, and A. Rogalski, "Progress in MOCVD growth of HgCdTe heterostructures for uncooled infrared photodetectors", *Infrared Phys. Technol,* vol. 49, no. 3, pp. 173-182, 2007.

http://dx.doi.org/10.1016/j.infrared.2006.06.026

[32] M. Mohammadian, and H.R. Saghai, "Room temperature performance analysis of bilayer graphene terahertz photodetector", *Optik,* vol. 126, no. 11-12, pp. 1156-1160, 2015.

http://dx.doi.org/10.1016/j.ijleo.2015.03.021

[33] J. Brouckaert, G. Roelkens, D. Van Thourhout, and R. Baets, "Thin-film III-V photodetectors integrated on silicon-on-insulator photonic ICs", *J. Lightwave Technol,* vol. 25, no. 4, pp. 1053-1060, 2007.

http://dx.doi.org/10.1109/JLT.2007.891172

[34] L. Shi, and S. Nihtianov, "Comparative study of silicon-based ultraviolet photodetectors", *IEEE Sens. J.,* vol. 12, no. 7, pp. 2453-2459, 2012.

http://dx.doi.org/10.1109/JSEN.2012.2192103

[35] s

[36] A. Pospischil, M. Humer, M.M. Furchi, D. Bachmann, R. Guider, T. Fromherz, and T. Mueller, "CMOS-compatible graphene photodetector covering all optical communication bands", *Nat. Photonics,* vol. 7, no. 11, pp. 892-896, 2013.

http://dx.doi.org/10.1038/nphoton.2013.240

[37] M.H. Weiler, "Magnetooptical Properties of Hg1-xCdxTe Alloys", In: *Semiconductors and Semimetals,* vol. 16. 2008, no. 1981, pp. 119-191.

[38] Y. Özer, High performance HgCdTe photodetector designs via dark current suppression, 2018.

[39] H. Kocer, Numerical modeling and optimization of HgCdTe infrared photodetectors for thermal imaging, 2011.

[40] J. Chu, Z. Mi, and D. Tang, "Band-to-band optical absorption in narrow-gap Hg1-xCdxTe semiconductors", *J. Appl. Phys.,* vol. 71, no. 8, pp. 3955-3961, 1992.

[41] J. Chu, B. Li, K. Liu, and D. Tang, "Empirical rule of intrinsic absorption spectroscopy in Hg1-xCdxTe", *J. Appl. Phys.,* vol. 75, no. 2, pp. 1234-1235, 1994.

[42] K. Liu, J.H. Chu, and D.Y. Tang, "Composition and temperature dependence of the refractive index in Hg1−xCdxTe", *J. Appl. Phys.,* vol. 75, no. 8, pp. 4176-4179, 1994.

[43] *ATLAS user's manual version 5202R.* SILVACO International: Santa Clara, CA, USA, 2016.

[44] S. Bansal, K. Prakash, K. Sharma, N. Sardana, S. Kumar, N. Gupta, and A.K. Singh, "A highly efficient bilayer graphene/ZnO/silicon nanowire based heterojunction photodetector with

broadband spectral response", *Nanotechnology,* vol. 31, no. 40, pp. 405205-1, 405205-405210, 2020.

PMID: 32554900

[45] H. Selvi, N. Unsuree, E. Whittaker, M.P. Halsall, E.W. Hill, A. Thomas, P. Parkinson, and T.J. Echtermeyer, "Towards substrate engineering of graphene-silicon Schottky diode photodetectors", *Nanoscale,* vol. 10, no. 7, pp. 3399-3409, 2018.

PMID: 29388650

[46] A.D.D. Dwivedi, A. Mittal, A. Agrawal, and P. Chakrabarti, "Analytical modeling and ATLAS simulation of N+-InP/n0-In0.53Ga0.47As/p+-In0.53Ga0.47As p-i-n photodetector for optical fiber communication", *Infrared Phys. Technol,* vol. 53, no. 4, pp. 236-245, 2010.

http://dx.doi.org/10.1016/j.infrared.2010.03.003

[47] S. Bansal, "Simulation and optimization of Pt/ZnO Schottky UV photodetector with Al ohmic contact", *IEEE 3rd International Conference on Innovation in Technology (INOCON),* 2024pp. 1-6 India

http://dx.doi.org/10.1109/INOCON60754.2024.10512163

[48] M. Casalino, U. Sassi, I. Goykhman, A. Eiden, E. Lidorikis, S. Milana, D. De Fazio, F. Tomarchio, M. Iodice, G. Coppola, and A.C. Ferrari, "Vertically illuminated, resonant cavity enhanced, graphene-Silicon Schottky photodetectors", *ACS Nano,* vol. 11, no. 11, pp. 10955-10963, 2017.

http://dx.doi.org/10.1021/acsnano.7b04792 PMID: 29072904

[49] A. Rogalski, "Infrared detectors: status and trends", *Prog Quantum Electron.,* vol. 27, no. 2-3, pp. 59-210, 2003.

http://dx.doi.org/10.1016/S0079-6727(02)00024-1

A Review of Nanostructure Field Effect Transistor Devices in Healthcare Applications

N. Suthanthira Vanitha[1*], K. Radhika[2], D. Anbuselvi[3], C. Kathiravan[4], S. Grace Infantiya[3] and A. Kalaiyarasan[5]

[1]Department of Electrical and Electronics Engineering, Muthayammal Engineering College, Rasipuram, Namakkal, India

[2]Department of ECE, Muthayammal Engineering College, Rasipuram, Tamil Nadu, India

[3]Department of Physics, Muthayammal Engineering College, Rasipuram, Namakkal, India

[4]Department of Chemistry, Muthayammal Engineering College, Engineering College, Rasipuram, Tamil Nadu, India

[5]Department of Mechanical, Muthayammal Engineering, Rasipuram, Tamil Nadu, India

Abstract: The evolution of the Nanostructure Field Effect Transistor (Nano FET) has provided significant progress in healthcare applications. Inherent properties such as easy integration, high sensitivity, and better selectivity increased the role of Nano FET devices in wearable electronic devices. Nano FET biosensors have placed great attention in the biomedical field, which performs label-free biomolecule sensing to screen out various diseases. The detection includes cancer biomarkers, cardiovascular diseases, diabetes, HIV/AIDS, DNA and RNA, and viral and bacterial infections. This chapter discusses the overview of diverse applications in healthcare, challenges, and future technologies of NanoFET devices.

Keywords: Nano FET devices, biosensors, biomarkers, nanomaterial, clinical applications.

INTRODUCTION

Currently, sensors based on field-effect transistors have shown superiority because of their remarkable qualities like precision, low operational power requirements, label-free character, economical, and easy surface functionalization. Bio-FET has

* **Corresponding author N. Suthanthira Vanitha:** Department of Electrical and Electronics Engineering, Muthayammal Engineering College, Rasipuram, Namakkal, India; Tel: 9965531515; E-mails: anbuselvivpm@gmail.com and varmans03@gmail.com

Dharmendra Singh Yadav & Prabhat Singh (Eds.)

two layers, the bio-recognition layer and the transducer layer, which act as exemplary applicants for Point-Of-Care Testing (POCT) because of its progress in fabrication techniques. Conventionally, Bio-FET sensors are based on novel functional materials such as metal–oxide–semiconductors, organic semiconductors, one-dimensional nanostructured materials (or) silicon nanowires, carbon nanotubes, and two-dimensional nanostructured materials (or) graphene [1-6].

Nanostructured materials are used as channel materials in FET sensors owing to their electrical characteristics, high sensitivity, chemical stability, biocompatibility, and high surface-to-volume ratio. The nanomaterials like nanoparticles, nanowires, and carbon nanotubes have a new approach to biosensing compared to analytical targets. This characteristic facilitates the detection of biomolecules, including proteins, nucleic acids, neurotransmitters, and cancer biomarkers. The scaling of MOSFET offers many field effect transistors on a single chip package facility. One-dimensional nanowire structures are used for the development of a variety of nanoscale semiconductor devices such as FETs, photodetectors, LEDs, biosensors, etc. Biosensors play a significant role in the agriculture and marine industry, surveillance, environmental monitoring, and clinical diagnosis. Biosensors are functional in monitoring health, such as glucose and oxygen sensors, enzyme-based sensors,immunosensors, and pH biosensors [6-12].

A substitute for minimizing short-channel effects in the structure of MOSFET is the nanowire FET (NW-FET). Due to the tunneling effect, drain current and subthreshold swing at extremely small VG increase when nanoHUB simulations reduce the channel length from 12 nm to 4 nm. Silicon nanowires (SiNWs) have superior application prospects in the field of biomedical sensing with outstanding electronic properties for improving the sensitivity detection of biosensors [2]. The combination of silicon nanowires and field effect transistors shapes real-time biosensors with great sensitivity and selectivity [13-17].

Examples:

1. Zinc Oxide (ZnO), 3, 4 Ethylene Dioxythiophen, Silicon Nanowires (SiNWs), Graphene

2. Indium Gallium (PEDOT), Carbon Nanotubes (CNTs), BlackPhosphorous

3. Zinc Oxide (IGZO)

In recent years, SiNW-FETs have been more focused on the biomedical field for detection. Research groups have been effective in highlighting the primary properties and applications of nanowires to fabricate semiconductor nanowires with controlled diameters. Nanowire detects diseases in the early stage through sensors with high sensitivity Some classification shown in Fig. (1). The material size decreases in the order of a nanometer. The physical and chemical material properties are determined by the large surface area to volume ratio and quantum size. The main use of nanowires in biomedicine is to identify chemical and biological substances, diagnose illnesses, and aid in drug discovery [5]. Semiconducting nanowires facilitate easier interaction with various electrical signals [18, 19].

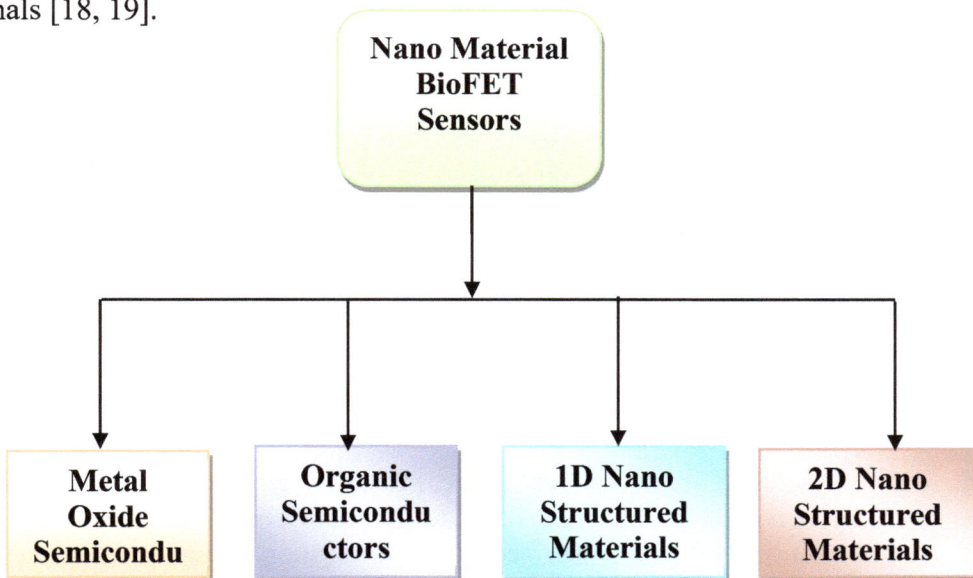

Fig. (1). Classification of BioFET Sensors.

NANO-FET DEVICES

Nanoscale electrical devices are called nanostructure field effect transistors   or nano-FET biosensors. Nano-FET detects and measures the presence and concentration of particular biological molecules [20, 21].

MOSFET and Nanowire FET

The four terminals of a MOSFET are the source, drain, gate, and body. Conduction does not occur unless a positive voltage is provided between the gate and the source.

A channel is formed in the uppermost layer of the material between p-type silicon and oxide by the positive gate voltage. The channel's thickness increases as the gate voltage rises. Channel, therefore, grows as the gate voltage increases, increasing the flow of current from source to drain. The characteristics of a MOSFET vary when the channel length decreases because of short channel effects. Short gate lengths cause MOSFETs to exhibit increased output conductance, higher leakage current, and threshold voltage shift. Quantum capacitance is a property of the channel element that has a significant impact on nanoscale devices [22, 35].

Silicon/ Germanium /GaAs NanowireFET

The gate terminal is responsible for the depletion of the semiconductor in the long channel length devices, whereas short channel devices utilize source and drain bias to achieve depletion. Maximum scaling is driven by reducing the channel length. Leakage power rises with respect to the threshold voltage, and devices are scaled down in nanometers [6-7]. The possible uses of semiconductor nanowire-FETs (NW-FETs) are as logic and memory circuits. NW-FET are classified into Si-NW-FET, Ge-NW-FET, and GaAs-NW-FET. The fabrication of NW-FET is done by placing a nanowire on top of an oxidized silicon wafer, and metal contacts as the source and drain electrodes are attached to the silicon substrate and then serve as a large-area back gate [8].

Silicon Nanowire Field Effect Transistor (SiNW-FET)

The sensitivity, selectivity, label-free, and real-time detection capabilities of Silicon Nanowire-FET made them an exceptional tool for biosensor design. SiNW-FETs are created using a reversible surface functionalization process to make them reusable devices. Based on biological research, SiNW-FETs are employed in the study on early cancer detection, biomarker detection, viral infection monitoring, protein-protein interaction, and DNA hybridization. They synthesize a 3D-localized bioprobe to record intracellular potential and examine electrical and transmitter signals from living cells [36, 37].

Fig. (**2**) illustrates the three-electrode system of SiNW-FET biosensors, which serve as the source, drain, and gate. The SiNW semiconductor channel is bridged by the source and drain electrodes, and the channel conductance modulation is carried out via the gate electrode. SiNW-FETs have two major fabrication techniques: The primary is the "top-down" approach carried out through lithographic processes combined with an electron-beam technique by physically etching a single-crystalline silicon wafer [9,52]. SiNWs are formed in the secondary "bottom-up" phase, which usually begins with a Chemical Vapor Deposition (CVD) reaction [38, 39].

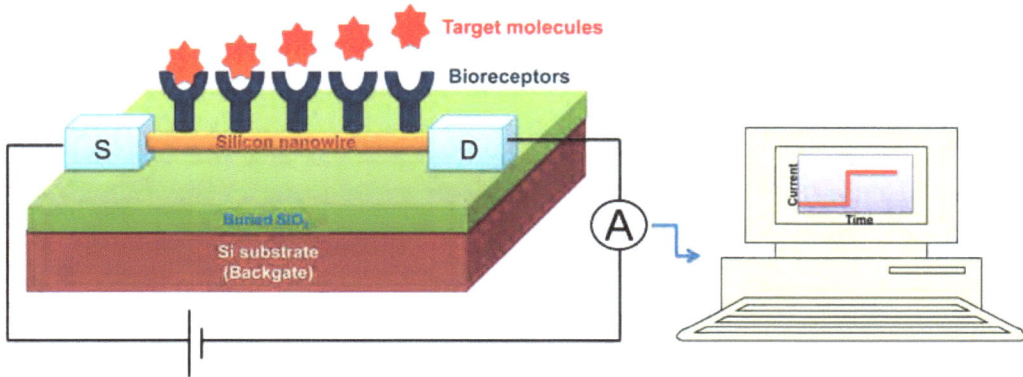

Fig. (2). Schematic Diagram of SiNW-FET [3].

Ge-Nanowire FET

Generally, Ge is used as a semiconductor in electronic devices. The semiconductor electronics industry entirely uses germanium, However, today's production of semiconductor electronics uses about 2% of silicon. Si-Ge alloys show promise as a significant semiconductor material for high-speed integrated circuits. Faster speed can be achieved by Si-Si Ge junctions in a circuit [14]. It is suggested that the Ge Nanowire Field-Effect Transistor (Ge-NW-FET) provides the foundation for cutting-edge electronic devices due to its potential electrical properties, low-temperature fabrication procedures, and device characteristics [40-45].

GaAs-Nanowire FET

GaAs, or semiconductor gallium arsenide, is a mixture of gallium and arsenic components utilized in solar cells and laser diodes, among other semiconductor devices. At frequencies of up to 250 GHz, arsenide transistors can operate because of their higher saturated electron velocity, electron mobility, and direct bandgap in comparison to silicon. A viable structure for charge-qubit operations at ambient temperature involves the double gate nanowire FET [46-51].

Calculations are made on the electrical properties and parameters of silicon and gallium arsenide nanowires with multi-gate structural arrangements using the Non-Equilibrium Green Function (NEGF) technique. For [001] orientations, in order to calculate the transmission coefficient of silicon and gallium arsenide nanowires, a semi-empirical tight-binding technique (sp3d5s*) is used. GaAs NW produces a visually appealing simulation for a few parameter findings, including electron density and transmission. Compared to gallium arsenide NWs, silicon NWs are better suited for leakage current reduction issues. Compounds containing gallium arsenide have potential uses in medical devices, portable electronics, and energy storage. The creation of unique compositions and enhanced characteristics of nanowires might arise from the advancement of new synthesis processes in the future [10].

Indium arsenide Nanowire FET (InAsFET)

A semiconductor substance is indium arsenide composed of indium and arsenic, having high electron mobility, narrow energy bandgap, and direct bandgap material, and is similar to gallium arsenide(GaAs). To simulate InAs-NWFET, a drain bias of 0.6 V is fixed for the conduction behavior of the drain currents with dissimilar gate voltages. The dielectric constant of the channel is 15.15, and the similarity of channel material is considered to be 4.9 eV. InAs is a high-speed metal-oxide-semiconductor channel material with very high electron mobility and saturation velocity [52].

Ion Sensitive Field-Effect Transistors

One type of MOSFET that is comparable is the Ion Sensitive Field-Effect Transistor (ISFET) used to detect pH values as shown in Fig. (**3a**). Biosensing can be used with ISFETs that have complementary bio-probes attached to their surface. With a common design platform, new ISFET technologies can achieve notable advancements in design based on CMOS technology.

An insulating layer serves as the sensing membrane in the conventional ISFET, replacing the silicon dioxide MOSFET. The dielectric layer capacitance determines the overall sensitivity, which directly influences the amount of generated charges per potential change. The CMOS ISFET configuration is the extended-gate architecture; here, the gate dielectric is isolated from the solution, but it uses gate metal extending from the transistor to the top passivation layer of the device that reduces the steps of post-processing [12]. The Extended-Gate Design involves adding an extra passivation serial capacitor (Silicon Nitride, for example) to reduce the overall sensitivity of the sensor.

A back gate is held by a dual-gate ISFET (DG-ISFET) in relation to the active channel, such as the silicon substrate as shown in Fig. (**3b**). Because of the capacitance imbalance between the top and bottom gate oxide, it is highly sensitive to pH. A SOI substrate containing a high-k sensing dielectric substance is used to create the DG-ISFET device. Signal-to-noise ratio (SNR), drift rate, and sensitivity are significantly improved.

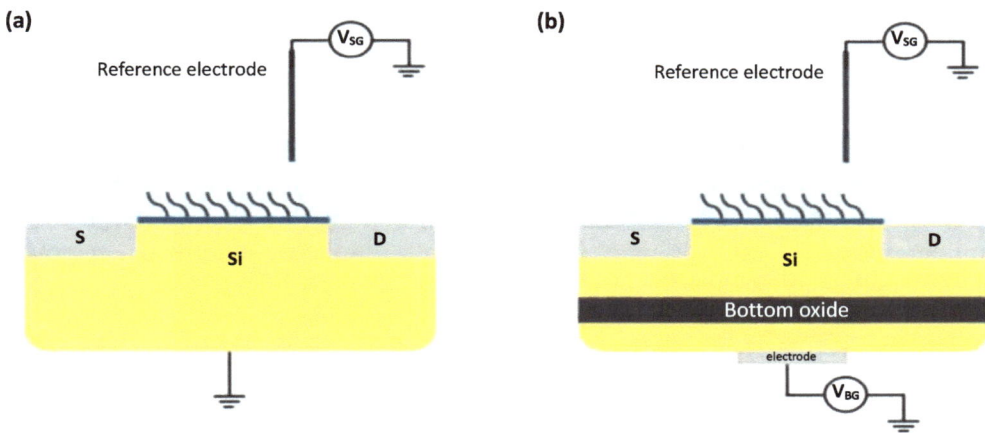

Fig. (3): (a) Conventional ISFET [48] **(b)** Double-Gate ISFET [48].

ISFET biosensors are used for detecting protein concentrations, DNA identification, genetic diagnostics, DNA amplification assays, and research on cell physiology.. The development of analog front-end design techniques has made it simple to implement ISFET arrays for multiplex and multi-functional sensing. Recently, the applications of ISFET have been transformed from molecular detection to other emerging applications [48].

Organic FETBiosensor (OFET)

An organic field-effect transistor is a thin-film transistor using organic material semiconductors. It can be either P-type or N-type, depending on material selection. OFET operates under accumulation mode. Typically, OFET mobility is $10^{-1} \sim 10^{-2}$ cm^{2}Vs^{-1}, which is much less than the mobility of crystalline silicon. The primary carrier transition pathways in OSC are the π bonds and non-covalent bonds, which lead to reduced mobility. This discrepancy results from the material composition of OSC. Despite its drawbacks, OSC is very compatible with flexible substrates, offering the development of wearable and flexible electronics. A reference electrode is added as the solution gate electrode to control the device, and the EDL structure appears as soon as the OFET is submerged in an electrolyte environment. In addition, EGOFET makes wearable technology more suitable for nano-electronic biosensing devices.

Graphene FET Biosensor

Graphene is considered a next-generation semiconductor device. It has various material properties like a high electron, hole mobility, transparency, mechanical strength, etc. Graphene is a zero bandgap material from a band structure standpoint that generates electrons and holes with positive and negative electric fields. Due to its unique physicochemical properties, graphene possesses a large surface to volume ratio, excellent thermal and electrical conductivity, biocompatibility, and attention for biosensing applications. With the advances in electrochemical biosensors, graphene nanomaterial has applications in healthcare in the form of graphene oxide, CVD graphene, nanostructures graphene, nanomesh, nanowalls, *etc.*

The main areas of interest include the synthesis of materials, the creation of devices, and the biofunctionalization of graphene electrodes in biosensing through the use of conductometry, potentiometry, amperometry/voltammetry, and electrochemical impedance. Moreover, there is a plethora of creative biosensing techniques that use graphene biosensors in clinical diagnosis to detect proteins as disease biomarkers, nucleic acids to analyze mutations in genetic diseases, small molecules to detect disease metabolites like glucose and lactic acid, and pathogens to detect biological and viral infections.

In GFET, the channel's activity and control are determined by the back-side potential, which is modulated through the gate dielectric as depicted in Fig. (**4**). The conductance in an electrolyte environment is 30 times higher than in a vacuum, as observed in an electrolyte-gated GFET biosensor used for protein adsorption detection and pH sensing. When the pH value increases, the conductance exhibits a linear relationship. However, the chemical vapor deposition-grown graphene is employed in GFET biosensor fabrication, while compared to the exfoliation method, the CVD-growth method raises the graphene film quality and undergoes a long process, leading to real-time practical applications [8].

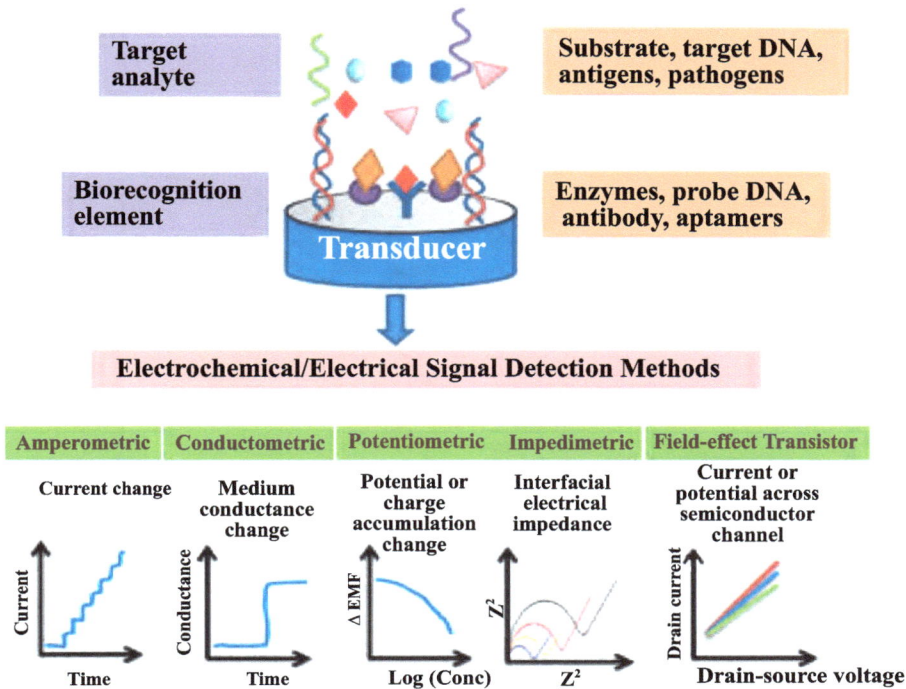

Fig. (4). Graphene-based Biosensors for Healthcare[8].

Nano-FET biosensors such as SiNWs, OFET, and GFET play a vital role in healthcare. In ultrasensitive applications, SiNWs and GFETs provide a Femtomolar Limit of Detection (LoD) and a greater dynamic range, but OFETs are limited in terms of LoD and dynamic range due to the material's properties. The possibilities of flexible electronics integration are the biggest strength and value of OFET[25]. A comparison of OFET and graphene biosensors is shown in Fig. (**5a** and **5b**).

Novel Field-Effect Transistors Biosensor

Novel FET-based biosensors have emerged to be potential candidates for next-generation biosensors. A new SiNW tunneling field-effect transistor biosensor that can identify the bipolar behavior of protein CYFRA21-1 is produced by doping the two sides of the nanowire with opposing amounts of p+ and n+. Consequently, this unique design improves the detection limit while lowering sensor noise. By using a multi-gate field-effect design, better sensitivity, structural enhancements, high mobility compound, and dynamic range are achieved. For example, AlGaN/GaN biosensors detect protein-peptide binding affinity and cardiac biomarkers.

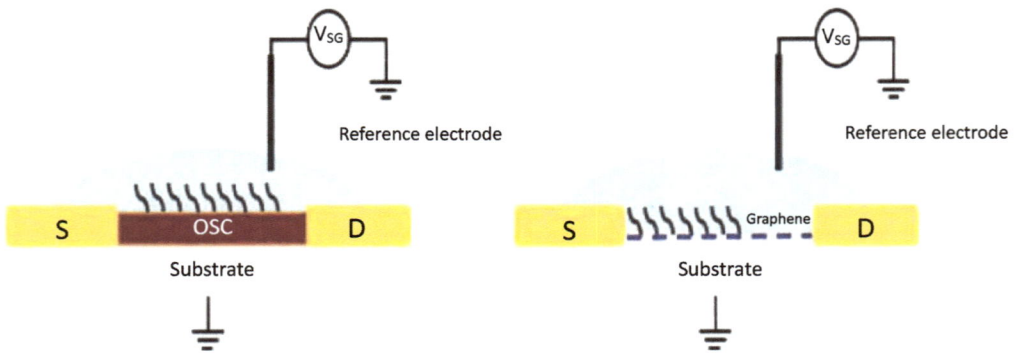

Fig. (5). (**a**).Organic FET(OFET) Biosensor[25] (**b**). Graphene FET Biosensor [8].

Carbon Nanotube Field-Effect Transistor (CNTFET)

 CNTFET is a transistor suitable for the environment where more carbon nanotubes are used. CNTFETs have power consumption and latency and are hence considered viable options to switch CMOS circuits. Due to unique electrical structures, they have a high potential to drive several technological advances in the industry. They hold certain merits, high sensitivity, high selectivity, easy operation, low operating temperature, fast response speed, short recovery time, and good stability[13].

CNTFET is a three-terminal device, namely, gate, drain, and source, similar to MOSFET, as shown in Fig. (**6**). It consists of a silicon dioxide substrate layer where carbon nanotubes are present between the semiconducting channels; the channels are turned ON/OFF electrostatically through the gate. CNTFETs are categorized as Schottky barrier CNTFETs (SB-CNTFETs) when they function with metal electrodes to generate Schottky contacts and as Ohmic contacts when they function with doped CNT electrodes. The current transmission method and CNTFET output properties are determined by this form of contact. The n-doped CNT contact is the Ohmic contact CNTFET type.

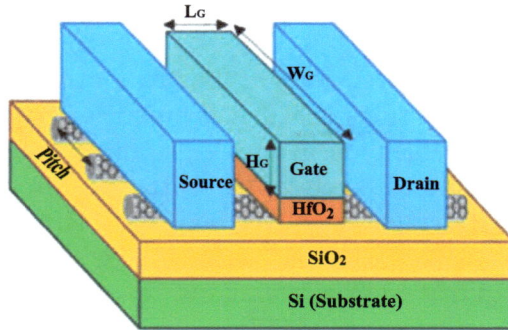

Fig. (6). Structure of Carbon Nanotube Field-Effect Transistor[13].

NANO FET APPLICATIONS IN HEALTHCARE

With the rapid growth of the human population, a lot of civilized diseases are rising. Early detection of diseases has become important, mainly in achieving personalized precision medicine and providing better medical assistance in remote areas. NanoFET-based biosensing devices are widely used in clinical applications, such as cardiovascular diseases, cancers, diabetes, AIDS/HIV, viruses and bacteria, COVID-19, DNA sequence, and Wearable Healthcare. Field effect transistor-based nanoelectronic biosensors such as glucose sensors, DNA sensors, dopamine sensors, and cancer biomarker sensors, shown in Fig. (**7**), afford a practical means of detecting a sensitive and targeted manner of diseases.

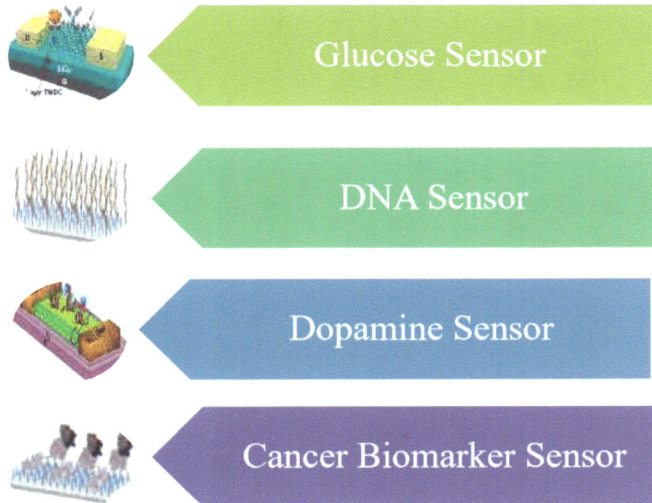

Fig. (7). NanoFET Sensors in Healthcare Applications.

NanoFET-Based Wearable Biosensors for Healthcare

Wearable biosensors have transformed the healthcare industry due to their rapid advancements. Among various sensing techniques, Nano Field-Effect Transistor (NanoFET)-based wearable biosensors are gaining increasing attention because of their advantageous features, such as label-free detection, quick response times, user-friendliness, and integration capabilities. The popularity of wearable biosensors has surged in recent years, coinciding with the rise of smartphones and other mobile devices that enable continuous, real-time collection of physiological data. This provides valuable insights into an individual's health, as illustrated in Fig. (**8**).

The design of NanoFET biosensors includes ion-sensitive membrane sensors, aptamers, antibodies, and nanobodies. These components work together to monitor biomarkers present in physiological fluids such as saliva, tears, sweat, and interstitial fluid from the skin. NanoFET-based wearable biosensors have demonstrated remarkable results in both diagnostics and health monitoring applications.

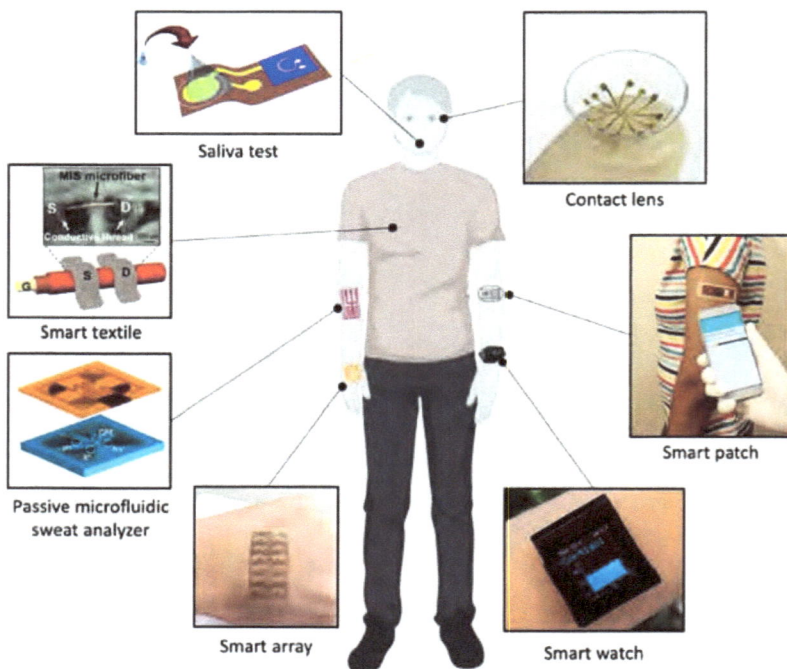

Fig. (8). Nano- FET-based Wearable Biosensors for Healthcare Applications [47].

NanoFET-based Biosensor for Detection of Cardiovascular Diseases

In today's world, cardiovascular diseases have become the major killers, among which Coronary Artery Diseases (CADs) contribute to morbidity and mortality. Certain limitations exist in the cardiovascular system with the current CAD diagnosis and therapy approaches as shown in Fig. (**9**). Nanotechnology, a new transdisciplinary technique, has demonstrated promise for application in medicine. The use of nanomaterials in biosensors for molecular imaging strategies, magnetic resonance imaging, optical imaging, nuclear scintigraphy, and multimodal imaging has made significant progress in diagnosis. These advancements in nanoFET are evident in their sensitivity and specificity.

Nanomaterials like liposomes, polymers, natural particles, and biomimetics are used as nucleic acids, chemicals, proteins, and peptides. Hence, the carriers target pathological sites based on their optimal physicochemical properties and surface modification potential. Owing to their inherent photoelectric and antioxidative properties in a complicated plaque environment, some of these nanomaterials serve as medications for the treatment of atherosclerosis.

Globally, the primary cause of morbidity and death is cardiovascular disorders. 17.9 million fatalities worldwide are predicted to occur from CVDs each year, accounting for 32% of all deaths. Heart attacks and strokes account for 85% of these deaths. The most common cause of CVDs is atherosclerosis. Atherosclerotic lesions in the coronary arteries affect lumen obstruction and stenos that result in myocardial ischemia, hypoxia, and even necrosis, known as coronary atherosclerotic heart diseases [40].

Nanomaterials creatively function as medicinal medications, and specially created nanomaterials facilitate the targeted administration of various treatments to thrombi and atherosclerotic plaques. The latest study has pointed out that nanomaterials with intrinsic antioxidative and anti-inflammatory activities [41] are promising future-generation therapies for the treatment of atherosclerosis in pathogenesis and other inflammatory diseases. Nanotechnology provides feasible attention for finding and handling cardiovascular diseases. The diagnosis's sensitivity and specificity are increased with the use of molecular imaging and aided biosensing. Furthermore, because of their surface change and physicochemical characteristics, nanomaterials allow for the direct exertion of therapeutic activity on the cardiovascular system.

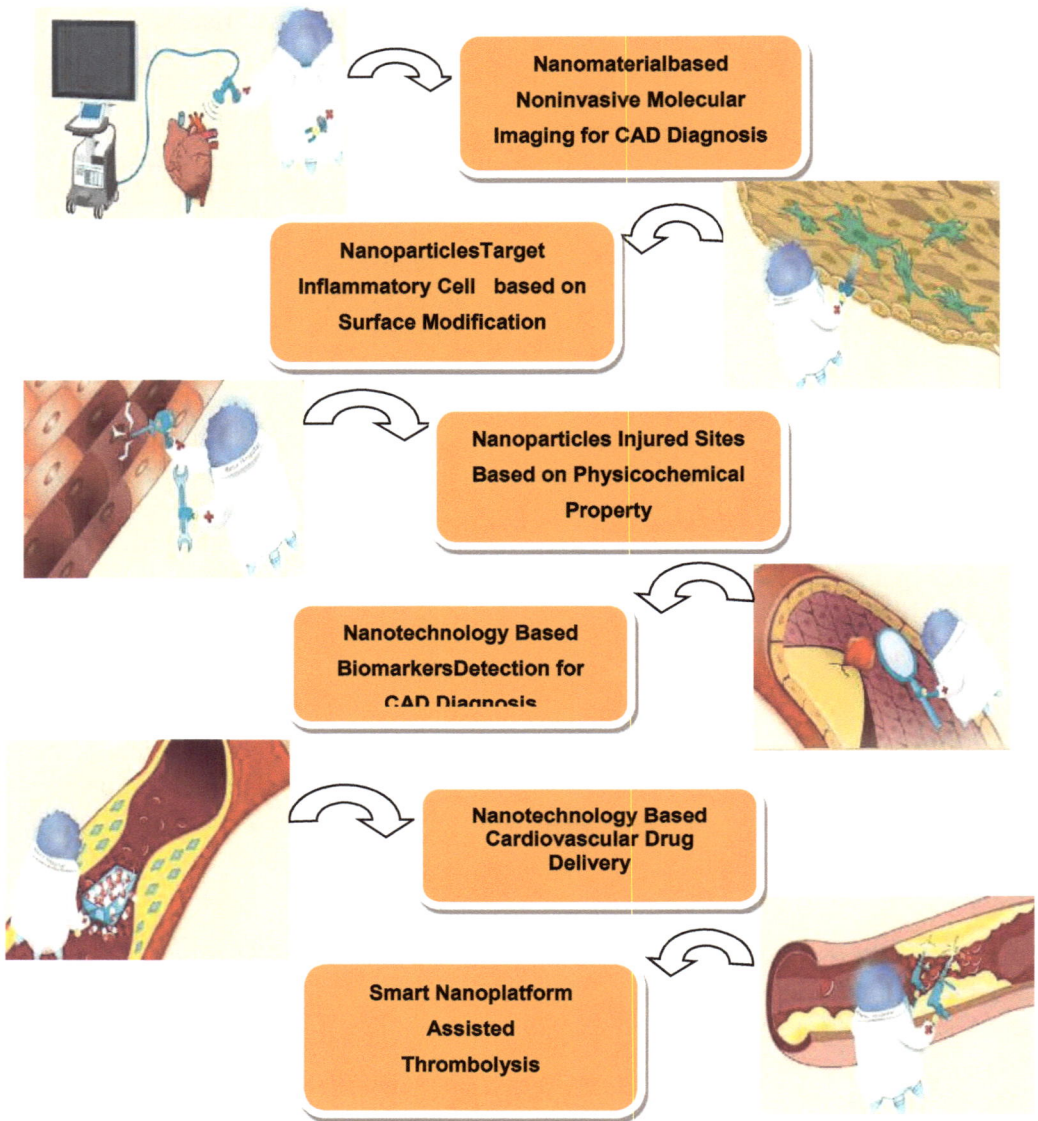

Fig. (9). Different Nanotechnology Approaches for Cardiovascular Diseases [40].

NanoFET-Based Biosensors for Cancer Detection

Worldwide, cancer is considered one of the primary causes of death that embodies a serious healthcare crisis. Cancer is expected to increase due to the aging of the world population. Based on a distinctive pattern of transmutation gathered in every patient, cancer is a complex and variable disease; hence, it requires individual care and precision medicine. Recent developments in the field of Micro/Nanoelectromechanical Systems (MEMs/NEMs) offer promise for addressing these drawbacks and creating a compact device capable of very sensitive and selective therapeutic detecting tasks. FET devices are used for point-of-care diagnostics and monitoring of environmental pollution, food quality, and pharmaceuticals that directly interpret the analyte-receptor interaction into electrical signals [18].

In the medical field, biomarkers are used to evaluate the conditions of a patient in multistage medical inquiry, estimating the risk of exact kinds of malignancy diseases. Different areas provide stratification of the malignancy type. As a special case, biomarkers are used for noninvasive screening, early detection of cancer, and prognosis of disease following the treatment. Implementation of NanoFET in cancer diagnostics addresses issues of operational conditions, reproducibility, large-scale production relevant sensitivity, selectivity, anti-interference, reusability, disposability, and economic viability.

NANO FET-BASED BIOSENSORS FOR DETECTION OF AIDS/HIV

According to WHO statistics, 1.0 million people died from HIV-related problems worldwide. The virus known as the Human Immunodeficiency Virus (HIV) causes AIDS by slowly destroying the immune system, which in turn makes many infectious illnesses easier to spread. It provides support for numerous clinical symptoms. RNA aptamers are used as a solid-surface sensing element in Diamond Field-Effect Transistors. In recent years, the application of nanotechnology for HIV/AIDS treatment and prevention has gained consideration. Emerging nanotechnology approaches enhance treatment through advanced therapeutic strategies such as antiretroviral therapy, gene therapy, immunotherapy, vaccinology, and microbicides [16]. The numerous therapy modalities concentrate on treating HIV/AIDS by specifically addressing HIV at the living cell stage.

Immunotherapy for HIV/AIDS

Immunotherapy is a medical intervention that uses immune modulatory drugs to alter the immune system in addition to curing a disease. It is predicated, like vaccinations, on an individual's immunization against a range of immunologic compositions. On the other hand, the goal is to safeguard healthy individuals and treat HIV-positive people. Ex vivo-generated autologous dendritic cell therapy and viral agent-based immunotherapy for HIV/AIDS have drawbacks because of the virus and the intricate processes involved in the production and use of ex vivo dendritic cells.

Nanotechnology-Based Drug Delivery for HIV/AIDS Treatment

Several antiretroviral medications are delivered both in vitro and in vivo for HIV/AIDS treatment using nanotechnology-based drug delivery systems. Nanotechnology-based targeted delivery of antiretroviral drugs to CD4+ T cells and macrophages, as well as delivery to the brain and lymphocytes, ensures that drugs reach the latent reservoir [16].

Nanomaterials as Therapeutic Agents

Many nanomaterials have demonstrated anti-HIV properties in vitro, including fullerenes, dendrimers, and silver and gold nanoparticles. HIV is inhibited by attaching to CD4+ T cells via size-dependent interaction between silver and gold nanoparticles, while the latter is coupled to the virus SDC -1721.

Gene Therapy for HIV/AIDS Treatment

Methods for delivering siRNA to HIV-specific cells include single-walled nanotubes,peptide-antibody conjugates, dendrimers, and fusion proteins. siRNA in gene therapy has shown promise for HIV/AIDS treatment[16].

Nanotechnology-Based Preventive HIV/AIDS Vaccine

Nanoparticles improve the delivery of antigens to enhance the immune response, which aids in encapsulating antigens in their core from antigen-presenting cells that can process, present, and cross-present antigen to CD4+ and CD8+ T cells and absorb the antigens on their surfaces, allowing B cells to generate humoral responses[16]. Nanoparticle vaccines optimize for different means of direction. Delivering DNA, protein, and peptide-based antigens in vivo and stimulating robust humoral, cellular, and mucosal immune responses are made possible by various polymeric and lipid-based nanoparticles.

Nanotechnology-Based Intravaginal Microbicides

Nanotechnology-based approaches are being developed to use dendrimers, siRNA, and nanoparticles for microbicidal functions [17]. To stop the spread of HIV/AIDS and other STDs, topical applications of intravaginal microbicides are used as a preventative measure.

Nano FET-based Biosensors for the Detection of Diabetes

As per the WHO statistics, 1.6 million deaths occur due to diabetes. Furthermore, WHO says that diabetes will be the seventh leading reason of death in 2030. Glycated hemoglobin, or HbA1c, is a reliable indicator for diabetes diagnosis that is unaffected by sporadic fluctuations and allows for long-term surveillance. In clinical diagnosis, an HbA1c result that falls between 4% and 6% is regarded as normal. Diabetes mellitus has been considered one of the major causes of human death because of poor quality of life and disability Mainly, it is caused by high blood glucose levels and insulin shortage [46]. Diabetes cases are of two types. Insulin-dependent diabetes, often known as type 1 diabetes, is an autoimmune disease in which the islets of Langerhans' beta cells fail to produce enough insulin, which raises blood sugar levels. Type 2 diabetes, noninsulin-dependent diabetes, is a metabolic disorder involving insulin resistance that is characterized by hyperglycemia [36].

Within ten years of delivery, high blood sugar levels during pregnancy cause gestational diabetes mellitus (GDM), which raises the risk of type 2 diabetes in both the mother and the child. Therefore, the diagnosis, self-monitoring, and treatment of the condition must have continuous and accurate glucose sensing in the blood and serum. The ability to develop portable devices with the necessary sensitivity for non-invasive diabetes detection via saliva, urine, or tears is made possible by advancements in nanomaterial sensors. Between a variety of nanomaterials, carbon nanotubes are considered promising materials for the expansion of future devices to assist in the premature sensing of hyperglycemia in diabetes [35]

Diabetes Mellitus is a metabolic disease characterized by long-term blood glucose levels. The pancreas' inability to make enough insulin is one of the two basic causes of diabetes. An important tool for such at-home chronic illness monitoring is the nanoFET-based biosensor. The FET-based immune sensor has a standard CMOS process and a disposable extended-gate electrode chip by the Micro-Electro-Mechanical-Systems technique [38].

BioFET for Dopamine Detection

In the human body, dopamine (DA) is an important neurotransmitter that is essential for the operation of the cardiovascular, hormone-regulatory, and central neurological systems. The human body typically has between 10^{-6} and 10^{-8} M of dopamine. Many neurological conditions, including schizophrenia, Parkinson's disease, and Alzheimer's disease, can be brought on by abnormally high amounts of dopamine (DA). For diagnosis and illness prevention, it is crucial to have accurate, quick, portable, affordable, and reliable measurement of DA levels in the human body[21-22].

There are several reports of high selectivity and sensitivity DA detection using Bio-FET-based devices. The accumulation or depletion of charge carriers inside the SiNW-FET indicates the presence of local charge transmission at the wire's surface, which is made possible by an electric field effect and a SiNW's large area-to-volume ratio. The conductance of the nanowire is altered when analyte molecules bind selectively to the bio-recognition elements on its surface. Both a top-down and bottom-up method are used to create SiNWs, applicable for metal-catalytic growth and optical and e-beam lithography, which is quite expensive. For example, a poly-SiNW Bio-FET for DA recognition supports the poly-Si sidewall spacer method to create a poly-SiNW channel. The use of 3-Amino Propyl Tri-Ethoxy-Silane (APTES) is to functionalize the SiNW channel by introducing an amino group subsequently cross-linked to carboxyphenyl boronic acid (CPBA). The selective binding of CPBA to dopamine generates a dopamine-boronate ester complex, which results in a change in the drain current and an overall negative charge at the poly-SiNW-FET interface. The sensitivity range of this method is 10-15 M.

Deoxyribonucleic acid (DNA)-aptamer-modified multiple parallel-connected (MPC) SiNW-FET is used for selective detection [23]. Hundreds of p-type single crystals make up the MPC SiNW-FET device, whereas standard SiNW-FETs have few SiNWs. In comparison with conventional SiNW-FETs, the numerous connections of SiNW allow for higher detection sensitivity, higher transconductance, and larger signal-to-noise ratio. Being smaller than other bio-receptors like enzymes or antibodies, the aptamer binds to the transducer channel closer and has a higher immobilization density producing a stronger electrical signal. Consequently, a relatively higher sensitivity is demonstrated by the ability to detect DA down to 10-11 M. In the detection of DA, SiNW-FET has shown excellent sensitivity owing to its size and high surface-to-volume ratio; however, it still experiences selectivity issues. Apart from SiNWs, single-walled (SWCNT) and multi-walled CNTs have been investigated for their potential role in building the Bio-FET channel [25]. Electrolyte Gate FET (EG-FET) detects DA level in contact with a gate electrode through an electrolyte. The conductance of the semiconducting channel varies using the electrolyte/channel interface via the capacitive field-effect method.

CNTFET Biosensor for the Detection of COVID-19

To enable proper containment and treatment protocols, early detection of severe Acute Respiratory Syndrome Coronavirus 2 (SARS-CoV-2) is crucial. Superior quality semiconducting single-walled carbon nanotube field-effect transistors possess specific chemical binding capabilities that enable the identification of SARS-CoV-2 antigens in nasopharyngeal clinical specimens. SAb's efficient FET sensors confirmed sensory efficacy in discriminating positive and negative clinical samples, indicating evidence of a system that could be used as a diagnostic tool for COVID-19 antigen because of its high sensitivity for analysis and low cost [14]. SWCNT FET sensor rate is lowered since it is more readily available and significantly less expensive than chemical vapor deposition graphene sheets. Comparing the nucleic acid amplification test with the SARS-CoV-2 Ag FET machine has shown effective differentiation between positive and negative samples that enhances the potential for diagnosing COVID-19[34].

NanoFET Sensors for DNA Detection

One popular technique for detecting DNA without labeling is the nanoFET approach. The two most popular designs of bio-FETs, TMDC- and G-FET-based, are widely used because of their high sensitivity, outstanding visibility, ease of manufacture, and ability to be integrated into the fabrication of on-chip devices. CVD-grown single-layer graphene is the most secure for detecting DNA down to 100 fM since it has a significant challenge in preparing and transferring into the device architecture of high-quality graphene without surface contamination[28]. In addition, the limited on/off current ratio and absence of a bandgap in graphene limit their practical application and reduce the device's sensitivity because there is a greater chance of interfering species triggering the sensing response. In contrast to mono-layer graphene creation, TMDC materials guarantee simplicity and maturity of fabrication procedures, and their sensing capability demonstrates great potential for the development of DNA sensors. With a high sensitivity of 17 mV/dec, a MoS2-based Bio-FET device senses down to 10 fM DNA in comparison to GFET-based DNA sensors. Increased TMDC device performance, particularly in terms of enhanced sensing characteristics like repeatability and reaction time, makes TMDCs useful for DNA detection applications [32-33].

CNTFET for Detection of Cholesterol

High cholesterol levels in the human body can cause hypertension and hypotension, which can lead to heart problems. Therefore, there is a need for new measurements that provide accurate, fast, and straightforward cholesterol numbers. By using CNTFET, blood cholesterol levels may be measured, which helps in the diagnosis and management of several conditions, including diabetes, liver, and heart disease. Generally, the normal cholesterol level of a person's serum is between 130-260 mg/dL[29].CNT FET-based biosensors have proved reliable, fast, and inexpensive methods of diagnostic techniques when compared with existing methods. Innovation in technology has led to CNTFETs as a better alternative to conventional CMOS[19].

CNTFET Diagnosis of Prostate Cancer

Generally, cancer is detected by using biosensors based on carbon nanotube field effect transistors [9]. CNTFET-based biosensors are used to diagnose cervical cancer where the sensitivity of the response can be raised. The cancer-causing agent's intensity is determined by contrasting the response from a reference analyte with a cancer analyte; this allows for the analysis of current flow deviation, which is compatible with the analyte's cancer cell severity. Prostate Specific Antigen (PSA) is specially produced by prostatic tissue and is currently available for the detection and treatment of prostate cancer[30]. Two molecules, namely PSA-ACT and f-PSA (free form), are used to treat prostate cancer with a range of 4.0 to 20 ng/ml, which is considered a "gray dose". Medical examinations are performed before the detection of diseases. Hence, physicians strongly advise PSA dose tests for the detection or tracking of prostate cancer. The charged proteins are approached easily where CNT-FETs are modified based on various molar concentrations of linkers to spacers [31, 37].

CHALLENGES AND FUTURE OF NANOFET DEVICES

To enhance sensing performance, novel materials with continually changing morphologies, chemistries, and architectures have been used in nanomaterial-based Bio-FET sensing over the past ten years. The major challenges in nanomaterial Bio-FET remain unsolved. A promising new sensing platform for biological detection is provided by two-dimensional nanomaterials like graphene and MoS2, which have variable bandgaps, high surface areas, and unique electrical characteristics. In the laboratory, MoS2-based FET biosensors demonstrate superior sensitivity, flexibility, transparency, scalability, ease of fabrication, and large-scale integrability when juxtaposed with 1D nanomaterials, including carbon nanotubes (CNTs), SiNW, and G-FET biosensors. High-throughput device fabrication is a major difficulty for 2D nanomaterial Bio-FETs. Because 2D materials enable effective molecular recognition layer regeneration at the lab scale, the majority of

2D Bio-FETs are only possible at a higher cost. Another challenge is to achieve better commercialization to understand the specifics of the interaction mechanisms between biological molecules and the sensing materials [42].

Compared to conventional detection techniques, biomarkers with smaller sizes and lower concentrations are now suitable for FET-based biosensors due to the advancement of nanotechnology methodologies. The sensitivity of FET-based biosensors is a crucial characteristic for future improvement. Ion entrapment in dielectric substrates results in a drifting feature that impairs detection accuracy. The design and multistep fabrication of the FET device, surface activation with a target-specific bioreceptor/recognition element, and the identification of biomarkers for particular diseases such as cancer, COVID-19, hepatitis B virus, influenza, etc., are among the parameters taken into account during the fabrication and assessment of the sensing performances of FET biosensors. Owing to enhanced sensibility, specificity, and related characteristics, sophisticated techniques and methods may enhance their multidisciplinary applications in various domains, including clinical healthcare testing, biomedical disciplines, and nutritional and pharmaceutical industries.

Despite this, nanotechnology is marching towards its role from translational medicine to scientific application. Novel biomarkers are evaluated for molecular imaging techniques and nanomaterial-based biosensors to figure out the development of atherosclerosis in the context of clinical diagnosis. In the case of treatment strategy, approval for clinical trials was obtained by evaluating the safety, pharmacokinetics, and biocompatibility of nanomaterials in nonhuman primates. Additional research is needed to understand the relationship between different components and outcomes of nanomedicines in diverse plaque microenvironments. In the case of the large-scale manufacturing of nanomedicines with stable physicochemical properties, the entire industrial operation is due to the position of nanomedicines in clinical application [45].

Nanotechnology has an innovative impact on HIV/AIDS treatment and prevention in a variety of ways. Treatment options are enhanced with nanoplatforms for antiretroviral drug delivery [43]. Drugs that are released gradually and under control may help patients stick to their prescribed dosages, which would improve the efficacy of treatment. Governmental and nonprofit organizations must assist in the effective dissemination of nanotechnology-enabled microbicides to developing nations. Shortly, as clinical transformation continues, new developments in nanotechnology-based research will likely broaden the scope of nanoscale technologies for use in diagnosis and treatment.

Future-generation FET biosensors that focus on wearable technology with AI-based Internet of Things integration, fast and simultaneous detection of multiple biomarkers, self-sufficient, adaptable and portable with ultrahigh sensitivity and selectivity, recyclable, repeatable, easily disposable, affordable, etc., are the primary concern of the field due to several research efforts in this area. Complete information and continuous human health monitoring are made possible by the captivating properties of FET devices. The integration of FET devices into IoT allows for the early identification of patient ailments through the use of high-quality data obtained from individuals. With the right pre-treatment techniques and the appropriate drugs, it creates a path for improving their health problems. Hence, these various demanding specifications and features open up novel opportunities for FET-based nanoelectronic devices in multidisciplinary applications [44,49-50].

CONCLUSION

Field-effect transistors, or Bio-FETs based on nanomaterials, are used in many different domains due to their many qualities and advantages, which include high sensitivity, mobility, easy integration, and excellent sensitivity. The development of FETs in biomedical applications has advanced dramatically, with nanomaterial-based Bio-FETs demonstrating remarkable performance in biosensing. The majority of the time, label-free biomolecule sensing using nanomaterial FET biosensors is used in the healthcare industry to screen out a variety of disorders. A FET-based biosensor can be used to detect infectious diseases like bacterial and viral infections, as well as glucose and diabetes levels. An exciting new direction for healthcare monitoring is the use of wearable field-effect transistor biosensors. Recent advances in FET sensor technology have investigated the field's possible uses in diagnosis. These devices present non-invasive monitoring of biomarkers in interstitial fluids such as saliva, tears, and perspiration, thereby offering affordable, dependable, and readily accessible real-time health observations. This chapter discussed the basics of NanoFET devices, the application of NanoFET in healthcare, challenges, and future perspectives of NanoFET..

Learning Objectives/Key Points

(1) Nanoparticles: A particle of matter 1 to 100 nanometers (nm) in diameter.

(2) Nanowires: A nanostructure in the form of a wire with a diameter of the order of a nanometer.

(3) NanoFET: A semiconducting quasi-one-dimensional nanostructure bridging the source and drain electrodes.

(4) Carbon Nanotube FET (CNTFET): Field-effect transistor that utilizes a single carbon nanotube or an array of carbon nanotubes as the channel material instead of bulk silicon, as in the traditional MOSFET structure.

(5) Organic FET(OFET): Field-effect transistor using an organic semiconductor in its channel.

(6) Nanosensor: Nanoscale devices that measure physical quantities and convert these to signals that can be detected and analyzed.

(7) Wearable Sensor: Sensors that are directly attached to human skin or clothes for gathering biological signals for the purpose of personal healthcare.

(8) Point-of-Care Testing (POCT): Medical diagnostic testing at or near the point of care.

(9) MEMS: Micro-electromechanical system, which is a miniature machine that contains mechanical and electronic components.

(10) NEMS: Nanoelectromechanical systems are a class of devices integrating electrical and mechanical functionality on the nanoscale.

Multiple Choice Questions (MCQs)

1. Which one of the following is an example of human origin nanomaterial?

a)Metals **b)**Viruses **c)**Forest fires   **d)**Both b and c

Answer: d) Both b and c

2. Give one example of the electrical properties of nanostructure

a)Melting temperature **b)**Tunneling current **c)**Both a and b **d)**None of the above

Answer: b) Tunneling current

3. The carbon nanotubes, graphene, and fullerenes are the _____-based nanoparticles?

a)Organic **b)**Inorganic **c)**Carbon **d)**None of the above

Answer: c) Carbon

4........ is the main drawback of silicon nanowires used in FET devices for sensing

a) Control over surface receptor density

b) Exponential decay of electrostatic potential

c) Fragmentation of antibody-capturing units

d) Screening of dissolved molecules by counterions

Answer: d) Screening of dissolved molecules by counterions

5. Which one of the following is used in cancer therapeutics?

a) Carbon nanotubes **b)**Nanorods **c)**Nanobots **d)** All of the above

Answer: a) Carbon nanotubes

6. The _____ to the ceramics are superior coatings.

a) Nanoparticles **b)** Nano powder) Nanocrystal coating **d)** Nanogel

Answer: c) Nanocrystal coating

7. DNA detection through the _____ by using the oligonucleotide functionalized gold nanocrystals is developed.

a) Colorimetric **b)** Diathermy**c)** Electrotherapy **d)** Treatment tables

Answer: a) Colorimetric

8. Which one of the following is an example of a one-dimensional nanomaterial?

a) Colloids   **b)** Nanowires   **c)** Thin films **d)** All of the above

Answer: a) Colloids

9. The size of polymeric nanoparticle nanosystem is around _____ ?

a)1-300 cm   **b)**1-500 mm **c)**10-1000 nm **d)**None of the above

Answer: c) 10-1000 nm

10. Which one of the following nanosensors measures the wavelength?

a) Mechanical **b)** Thermal **c)** Magnetic **d)** Optical

Answer: d) Optical

11. Which of the following is an example of a bottom-up approach?

a) Attrition **b)** Colloidal dispersion **c)** Milling **d)** Etching

Answer: b) Colloidal dispersion

12. Nanoparticles target the rare _____-causing cells and remove them from blood.

a) Tumor   **b)** Fever   **c)** Infection   **d)** Cold

Answer: a) Tumor

13. The size of a human red blood cell is _____?

a) 40 microns **b)** 4.6 microns **c)** 200 – 300 nanometers **d)** None of the above

Answer: b) 4.6 microns

14. What are the approaches used in making nanosystems?

a) Top-down **b)** Bottom-up **c)** Both a and b **d)** Neither a nor b

Answer: c) Both a and b

15. Which of the following is the biomedical application of quantum dots?

a) LEDs **b)** Solar cells **c)** Qubits **d)** Medical imaging

Answer: d) Medical imaging

REFERENCES

[1]   L. Manjakkal, D. Szwagierczak, and R. Dahiya, "Metal oxides based electrochemical pH sensors: Current progress and future perspectives," *Prog. Mater. Sci.*, vol. 109, 2020.

http://dx.doi.org/10.1016/j.pmatsci.2019.100635

[2]   R. Ahmad, T. Mahmoudi, M.S. Ahn, and Y.B. Hahn, "Recent advances in nanowires-based field-effect transistors for biological sensor applications", *Biosens Bioelectron,* vol. 100, pp. 312-325, 2018.

http://dx.doi.org/10.1016/j.bios.2017.09.024 PMID: 28942344

[3]   T. Manimekala, R. Sivasubramanian, and G. Dharmalingam, "Nanomaterial-Based Biosensors using Field-Effect Transistors: A Review", *J. Electron. Mater.,* vol. 51, no. 5, pp. 1950-1973, 2022.

http://dx.doi.org/10.1007/s11664-022-09492-z PMID: 35250154

[4]   S. Su, J. Chao, D. Pan, L. Wang, and C. Fan, "Electrochemical Sensors Using Two-Dimensional Layered Nanomaterials", *Electroanalysis,* vol. 27, no. 5, pp. 1062-1072, 2015.

http://dx.doi.org/10.1002/elan.201400655

[5]   Sanjeet K. Sinha, 2K. Shyam and 3L. Nishanthini Nanowire-FET Devices for future Nanotechnology Conference Paper, 2016 KPRIT Hyderabad, India

[6]   H. Li, D. Li, H. Chen, X. Yue, K. Fan, L. Dong, and G. Wang, "Application of Silicon Nanowire Field Effect Transistor (SiNW-FET) Biosensor with High Sensitivity", *Sensors (Basel),* vol. 23, no. 15, p. 6808, 2023.

http://dx.doi.org/10.3390/s23156808 PMID: 37571591

[7]   P. Paramasivam, N. Gowthaman, and V.M. Srivastava, "Design and Analysis of Gallium Arsenide-Based Nanowire Using Coupled Non-Equilibrium Green Function for RF Hybrid Application", *Nanomaterials,* vol. 13, pp. 959-959, 2023.

http://dx.doi.org/10.3390/nano13060959,2023

[8]   Trupti Terse-Thakoor, Sushmee Badhulika, and Ashok Mulchandani, "Graphene Based Biosensors for Healthcare", *Journal of Materials Research,* vol. 32, pp. 2905-2905, 2017.

http://dx.doi.org/10.1557/jmr.2017.175

[9]   D. Wang, V. Noël, and B. Piro, "Electrolytic Gated Organic Field-Effect Transistors for Application in Biosensors—A Review", *Electronics,* vol. 5, no. 1, p. 9, 2016.

http://dx.doi.org/10.3390/electronics5010009

[10]   S.K. Arya, S. Saha, J.E. Ramirez-Vick, V. Gupta, S. Bhansali, and S.P. Singh, "Recent advances in ZnO nanostructures and thin films for biosensor applications: Review", *Anal. Chim Acta.,* vol. 737, pp. 1-21, 2012.

http://dx.doi.org/10.1016/j.aca.2012.05.048 PMID: 22769031

[11]   K. Ogata, K. Koike, T. Tanite, T. Komuro, S. Sasa, M. Inoue, and M. Yano, "High resistive layers toward ZnO-based enzyme modified field effect transistor", *Sens. Actuators B Chem.,* vol. 100, no. 1-2, pp. 209-211, 2004.

http://dx.doi.org/10.1016/j.snb.2003.12.022

[12]   C.D. Chen, B-R. Yang, C. Liu, X-Y. Zhou, Y-J. Hsu, Y-C. Wu, J-S. lm, P-Y. Lu, M. Wong, H-S. Kwok, and H-P.D. Shieh, "Integrating Poly-Silicon and InGaZnO Thin-Film Transistors for CMOS Inverters", *IEEE Trans Electron. Dev.,* vol. 64, no. 9, pp. 3668-3671, 2017.

http://dx.doi.org/10.1109/TED.2017.2731205

[13] K.M. Puneeth, R. Shruthi, and A. Veditha, "B Poorvi Baliga," Review on CNTFET: Recent Developments in Biosensors for Health Care Applications", *Int. J. Res. Appl. Sci. Eng. Technol,* vol. 10, p, 2022.

[14] W. Shao, M.R. Shurin, S.E. Wheeler, X. He, and A. Star, "Rapid Detection of SARS-CoV-2 Antigens Using High-Purity Semiconducting Single-Walled Carbon Nanotube-Based Field-Effect Transistors", *ACS Appl. Mater. Interfaces,* vol. 13, no. 8, pp. 10321-10327, 2021.

      http://dx.doi.org/10.1021/acsami.0c22589 PMID: 33596036

[15] G. Keshwani, H.R. Thakur, and J.C. Dutta, "Fabrication and electrical characterization of carbon nanotube based enzyme field effect transistor for cholesterol detection", In: *IEEE Region 10 Conference.* TENCON, 2019, pp. 1821-1824.

      http://dx.doi.org/10.1109/TENCON.2019.8929330

[16] T. Mamo, E.A. Moseman, N. Kolishetti, C. Salvador-Morales, J. Shi, D.R. Kuritzkes, R. Langer, U. Andrian, and O.C. Farokhzad, "Emerging Nanotechnology Approaches for HIV/AIDS Treatment and Prevention", *Nanomedicine (Lond),* vol. 5, no. 2, pp. 269-285, 2010.

      http://dx.doi.org/10.2217/nnm.10.1

[17] N. Anzar, R. Hasan, M. Tyagi, N. Yadav, and J. Narang, "Carbon nanotube – A review on synthesis, properties and plethora of applications in the field of biomedical science," *Sensors Int.*, vol. 1, p. 100003, 2020.

      http://dx.doi.org/10.1016/j.sintl.2020.100003

[18] K. Mohana Sundaram, P. Prakash, D. Karthikeyan, and W.D. Mammo,, "Improved Carbon Nanotube Field Effect Transistor for Designing a Hearing Aid Filtering Application", *Journal of Nanomaterials,* p. 12, 2021.

      http://dx.doi.org/10.1155/2021/7024032

[19] A.S. Vidhyadharan, and S. Vidhyadharan, "An ultra-low-power CNFET -based improved Schmitt trigger design for VLSI sensor applications", *Int. J. Numer Model,* vol. 34, no. 4, p. e2874, 2021.

      http://dx.doi.org/10.1002/jnm.2874

[20] X. Yao, Y. Zhang, W. Jin, Y. Hu, and Y. Cui, "Carbon Nanotube Field-Effect Transistor-Based Chemical and Biological Sensors", *Sensors,* vol. 21, no. 3, p. 995, 2021.

      http://dx.doi.org/10.3390/s21030995 PMID: 33540641

[21] S.J. Park, H. Yang, S.H. Lee, H.S. Song, C.S. Park, J. Bae, O.S. Kwon, T.H. Park, and J. Jang, "Dopamine Receptor D1 Agonism and Antagonism Using a Field-Effect Transistor Assay", *ACS Nano,* vol. 11, no. 6, pp. 5950-5959, 2017.

      http://dx.doi.org/10.1021/acsnano.7b01722 PMID: 28558184

[22] S.M. Siddeeg, "Electrochemical detection of neurotransmitter dopamine: a review", *Int. J. Electrochem Sci.,* vol. 15, no. 1, pp. 599-612, 2020.

http://dx.doi.org/10.20964/2020.01.61

[23] T.K. Aparna, R. Sivasubramanian, and M.A. Dar, "One-pot synthesis of Au-Cu2O/rGO nanocomposite based electrochemical sensor for selective and simultaneous detection of dopamine and uric acid", *J. Alloys Compd,* vol. 741, pp. 1130-1141, 2018.

http://dx.doi.org/10.1016/j.jallcom.2018.01.205

[24] D.C. Li, P.H. Yang, and M.S.C. Lu, "CMOS Open-Gate Ion-Sensitive Field-Effect Transistors for Ultrasensitive Dopamine Detection", *IEEE Trans Electron. Dev.,* vol. 57, no. 10, pp. 2761-2767, 2010.

http://dx.doi.org/10.1109/TED.2010.2063330

[25] S. Casalini, F. Leonardi, T. Cramer, and F. Biscarini, "Organic field-effect transistor for label-free dopamine sensing", *Org Electron.,* vol. 14, no. 1, pp. 156-163, 2013.

http://dx.doi.org/10.1016/j.orgel.2012.10.027

[26] T. Vijayaraghavan, R. Sivasubramanian, S. Hussain, and A. Ashok, "A Facile Synthesis of LaFeO 3 -Based Perovskites and Their Application towards Sensing of Neurotransmitters", *ChemistrySelect,* vol. 2, no. 20, pp. 5570-5577, 2017.

http://dx.doi.org/10.1002/slct.201700723

[27] B.R. Li, Y.J. Hsieh, Y.X. Chen, Y.T. Chung, C.Y. Pan, and Y.T. Chen, "An ultrasensitive nanowire-transistor biosensor for detecting dopamine release from living PC12 cells under hypoxic stimulation", *J. Am Chem. Soc.,* vol. 135, no. 43, pp. 16034-16037, 2013.

http://dx.doi.org/10.1021/ja408485m PMID: 24125072

[28] T.T. Tran, and A. Mulchandani, "Carbon nanotubes and graphene nano field-effect transistor-based biosensors", *Trends Analyt Chem.,* vol. 79, pp. 222-232, 2016.

http://dx.doi.org/10.1016/j.trac.2015.12.002

[29] M.A. Barik, and J.C. Dutta, "Fabrication and characterization of junctionless carbon nanotube field effect transistor for cholesterol detection", *Appl. Phys. Lett,* vol. 105, no. 5, pp. 053509-053509, 2014.

http://dx.doi.org/10.1063/1.4892469

[30] A. Sarkar, S. Maity, P. Chakraborty, and S. Chakraborty, "Characterization of carbon nanotubes and its application in biomedical sensor for prostate cancer detection," *Sensor Letters*, vol. 17, pp. 17–24, 2019.

[31] K. Teker, R. Sirdeshmukh, and K. Sivakumar, "Applications of Carbon Nanotubes for Cancer Research", *Nanobiotechnol,* vol. 1, pp. 171-182, 2005.

http://dx.doi.org/10.1385/NBT:1:2:171,2005

[32] PV. R. Anitha Gopinath, P. G. Gopinath, and S. Aruna Mastani, "Modeling and simulation of carbon nanotubes based FET for cervical cancer detection," *Int. J. Innov. Res. Sci. Eng. Technol.*, vol. 5, no. 7, pp. 13817–13823, 2016.

[33] J.P. Kim, B.Y. Lee, J. Lee, S. Hong, and S.J. Sim, "Enhancement of sensitivity and specificity by surface modification of carbon nanotubes in diagnosis of prostate cancer

based on carbon nanotube field effect transistors", *Biosens Bioelectron,* vol. 24, no. 11, pp. 3372-3378, 2009.

http://dx.doi.org/10.1016/j.bios.2009.04.048 PMID: 19481922

[34]   M. Thanihaichelvan, S.N. Surendran, T. Kumanan, U. Sutharsini, P. Ravirajan, R. Valluvan, and T. Tharsika, "Selective and electronic detection of COVID-19 (Coronavirus) using carbon nanotube field effect transistor-based biosensor: A proof-of-concept study", *Mater. Today Proc,* vol. 49, pp. 2546-2549, 2022.

http://dx.doi.org/10.1016/j.matpr.2021.05.011

[35]   H. Ehtesabi, "Application of carbon nanomaterials in human virus detection", *J. Sci. Adv. Mater. Devices,* vol. 5, no. 4, pp. 436-450, 2020.

http://dx.doi.org/10.1016/j.jsamd.2020.09.005

[36]   B.L. Allen, P.D. Kichambare, and A. Star, "Carbon nanotube field-effect transistor based biosensors", *Adv. Mater.,* vol. 19, no. 11, pp. 1439-1451, 2007.

http://dx.doi.org/10.1002/adma.200602043

[37]   R. Varghese, S. Salvi, P. Sood, J. Karsiya, and D. Kumar, "Carbon nanotubes in COVID-19: A critical review and prospects", *Colloid Interface Sci. Commun,* vol. 46, p. 100544, 2022.

http://dx.doi.org/10.1016/j.colcom.2021.100544 PMID: 34778007

[38]   K. Radhika, "N SuthanthiraVanitha," Applications of Nanofluids in Healthcare: Future Perspectives", In: *Sustainable Utilization of Nanoparticles and Nanofluids in Engineering Applications.* IGI Global Publishers, 2023.

http://dx.doi.org/10.4018/978-1-6684-9135-5.ch011

[39]   T. Mamo, E.A. Moseman, N. Kolishetti, C. Salvador-Morales, J. Shi, D.R. Kuritzkes, R. Langer, U. Andrian, and O.C. Farokhzad, "Emerging Nanotechnology Approaches for HIV/AIDS Treatment and Prevention", *Nanomedicine,* vol. 5, no. 2, pp. 269-285, 2010.

http://dx.doi.org/10.2217/nnm.10.1

[40]   Q. Hu, Z. Fang, J. Ge, and H. Li, "Nanotechnology for cardiovascular diseases", *Innovation,* vol. 3, no. 2, p. 100214, 2022.

http://dx.doi.org/10.1016/j.xinn.2022.100214 PMID: 35243468

[41]   N. Csaba, M. Garcia-Fuentes, and M.J. Alonso, "Nanoparticles for nasal vaccination", *Adv. Drug Deliv Rev.,* vol. 61, no. 2, pp. 140-157, 2009.

http://dx.doi.org/10.1016/j.addr.2008.09.005 PMID: 19121350

[42]   Jing Guo, Jing Wang, E. Polizzi, S. Datta, and M. Lundstrom, "Electrostatics of nanowire transistors", *IEEE Trans Nanotechnol.,* vol. 2, no. 4, pp. 329-334, 2003.

http://dx.doi.org/10.1109/TNANO.2003.820518

[43]   D.L. John, L.C. Castro, and D.L. Pulfrey, "Quantum capacitance in nanoscale device modeling", *J. Appl. Phys.,* vol. 96, no. 9, pp. 5180-5184, 2004.

http://dx.doi.org/10.1063/1.1803614

[44] N. S. Vanitha, K. Radhika, G. Sudarmozhi, G. Kavitha, and M. Shenbagapriya, "Nanotechnology in cosmetics and cosmeceuticals: Future direction and safety," in *Sustainable Utilization of Nanoparticles and Nanofluids in Engineering Applications*, IGI Global. 2023.

http://dx.doi.org/10.4018/978-1-6684-9135-5.ch006,2023

[45] K. Gunasekaran, N. S. Vanitha, K. Radhika, and P. Suresh, "Nanomedicine in healthcare: Impact and challenges for future generation," In: *Networking Technologies in Smart Healthcare*, 1st ed., P. Singh, O. Kaiwartya, N. Sindhwani, V. Jain, and R. Anand, Eds. Boca Raton, FL: CRC Press, 2022, pp. 17.

http://dx.doi.org/10.1201/9781003239888-12

[46] S. Same, and G. Samee, "Carbon Nanotube Biosensor for Diabetes Disease", *Crescent J. Med. Biol. Sci.,* vol. 5, pp. 1-6, 2018.

[47] T.T.H. Nguyen, C.M. Nguyen, M.A. Huynh, H.H. Vu, T.K. Nguyen, and N.T. Nguyen, "Field effect transistor based wearable biosensors for healthcare monitoring", *J. Nanobiotechnology,* vol. 21, no. 1, p. 411, 2023.

http://dx.doi.org/10.1186/s12951-023-02153-1 PMID: 37936115

[48] Y.C. Syu, W.E. Hsu, and C.T. Lin, "Review—Field-Effect Transistor Biosensing: Devices and Clinical Applications", *ECS J. Solid State Sci. Technol,* vol. 7, no. 7, pp. Q3196-Q3207, 2018.

http://dx.doi.org/10.1149/2.0291807jss

[49] P. Singh, and D.S. Yadav, "Impact of work function variation for enhanced electrostatic control with suppressed ambipolar behavior for dual gate L-TFET", *Curr. Appl. Phys.,* vol. 44, pp. 90-101, 2022.

http://dx.doi.org/10.1016/j.cap.2022.09.014

[50] A. Raman, K.J. Kumar, D. Kakkar, R. Ranjan, and N. Kumar, "Performance investigation of source delta-doped vertical nanowire TFET", *J. Electron. Mater.,* vol. 51, no. 10, pp. 5655-5663, 2022.

http://dx.doi.org/10.1007/s11664-022-09840-z

[51] P. Singh, and D.S. Yadav, "Assessing the Impact of Drain Underlap Perspective Approach to Investigate DC/RF to Linearity Behavior of L-Shaped TFET", *Silicon,* vol. 14, no. 17, pp. 11471-11481, 2022.

http://dx.doi.org/10.1007/s12633-022-01814-4

[52] N. Kumar, and A. Raman, "Design and analog performance analysis of charge-plasma based cylindrical GAA silicon nanowire tunnel field effect transistor", *Silicon,* vol. 12, no. 11, pp. 2627-2634, 2020.

http://dx.doi.org/10.1007/s12633-019-00355-7

SUBJECT INDEX

A

Adsorption 175, 176, 185, 186, 195
　metal ion 176
　system 175
AFM techniques 186
Alumina devices 84
Alzheimer's disease 336
Anthracycline combinations 178
Anti-inflammatory activities 332
Applications 121, 191
　immunolabeling 121
　of sensors 191
Arc discharge 45, 182
　electric 182
Architectures, electrical biosensing 181
Atherosclerotic 332
　lesions 332
　plaques 332
Atomic layer deposition (ALD) 47, 93, 268

B

Behavior, electrostatic 23
Binding 80, 132, 135, 146, 147, 149, 150, 173, 175
　antibody-antigen 173
　macromolecule 175
Bio-FET(s) 169, 194, 195, 320, 321, 337, 338, 339, 340, 341
　based devices 337
　-FET sensors 321
　nanomaterial 194, 339
　nanomaterial-based 194, 195, 339, 341
Biochemical 145, 173
　changes 145
　communications 173
Biocompatible nanomaterials 110, 111
　carbon-based 110
Bioelectronic protein nanowire sensors 191
BioFET 322, 336
　for dopamine detection 336

sensors 322
BioNano electro-mechanical systems (BioNEMS) 125
Biosensing 78, 98, 110, 111, 115, 120, 121, 123, 124, 125, 144, 146, 147, 148, 149, 150, 151, 153, 154, 155, 169, 190, 321, 325, 327
　amplified 148
　applications 78, 98, 121, 147, 149, 150, 153, 154, 155, 169, 190
　technologies 98, 110, 115, 120, 125, 144, 150, 169
Biosensing devices 110, 147, 327
　nano-electronic 327
Biosensor(s) 78, 83, 84, 97, 98, 110, 111, 112, 117, 120, 122, 123, 124, 125, 133, 134, 136, 137, 138, 139, 141, 144, 145, 148, 149, 150, 154, 321, 330
　applications 110, 124, 139
　development 98, 112, 123, 124, 125, 148
　glucose-based 97
　photoelectrochemical 134
　technology 112, 120, 125, 138, 148, 150
　transistor-based nanoelectronic 330
Bore gauging displacement sensors (BGDS) 115

C

Cancer 178, 330, 333, 334, 339, 340
　breast 178
　cervical 339
Cancer biomarker 149, 330
　breast 147
　sensor 330
Capacitive displacement sensors (CDS) 115
Carbon nanomaterials in diagnosis and therapy 177
Carbon nanotube 40, 41, 43, 44, 45, 46, 47, 52, 53, 63, 64, 72, 80, 92, 96, 97, 98, 175, 179, 183, 194, 219, 222, 329, 339
　applications 175, 194

www.ingramcontent.com/pod-product-compliance
Lightning Source LLC
Chambersburg PA
CBHW050803220326
41598CB00006B/107